Methods in Molecular Biology

For further volumes:
http://www.springer.com/series/7651

For over 35 years, biological scientists have come to rely on the research protocols and methodologies in the critically acclaimed *Methods in Molecular Biology* series. The series was the first to introduce the step-by-step protocols approach that has become the standard in all biomedical protocol publishing. Each protocol is provided in readily-reproducible step-by-step fashion, opening with an introductory overview, a list of the materials and reagents needed to complete the experiment, and followed by a detailed procedure that is supported with a helpful notes section offering tips and tricks of the trade as well as troubleshooting advice. These hallmark features were introduced by series editor Dr. John Walker and constitute the key ingredient in each and every volume of the *Methods in Molecular Biology* series. Tested and trusted, comprehensive and reliable, all protocols from the series are indexed in PubMed.

Prediction of Protein Secondary Structure

Second Edition

Edited by

Andrzej Kloczkowski

The Steve and Cindy Rasmussen Institute for Genomic Medicine, Nationwide Children's Hospital, Columbus, OH, USA

Lukasz Kurgan

Department of Computer Science, Virginia Commonwealth University, Richmond, VA, USA

Eshel Faraggi

Department of Physics, Indiana University, Indianapolis, IN, USA

Humana Press

Editors
Andrzej Kloczkowski
The Steve and Cindy Rasmussen
Institute for Genomic Medicine
Nationwide Children's Hospital
Columbus, OH, USA

Lukasz Kurgan
Department of Computer Science
Virginia Commonwealth University
Richmond, VA, USA

Eshel Faraggi
Department of Physics
Indiana University
Indianapolis, IN, USA

ISSN 1064-3745 ISSN 1940-6029 (electronic)
Methods in Molecular Biology
ISBN 978-1-0716-4195-8 ISBN 978-1-0716-4196-5 (eBook)
https://doi.org/10.1007/978-1-0716-4196-5

This Humana imprint is published by the registered company Springer Science+Business Media, LLC, part of Springer Nature.
The registered company address is: 1 New York Plaza, New York, NY 10004, U.S.A.

Preface

The first edition of our book *Prediction of Protein Secondary Structure* was published in October 2016. Subsequently, Springer-Nature asked us to prepare the second edition of this book. From the original Editorial team (Yaoqi Zhou, Andrzej Kloczkowski (A.K.), Eshel Faraggi (E.F.), and Yeudong Yang), two (A.K. and E.F.) have accepted this invitation and invited Lukasz Kurgan as a new Co-Editor. Although this is a second edition of the book, it differs substantially from second editions of other published titles that typically cover the same scope as the original edition; this book focuses on adding new developments in the field. While the title of this volume remains *Prediction of Protein Secondary Structure*, it adds completely new materials written chiefly by different authors when compared to the coverage of its first edition.

Since 2016, there has been enormous amount of progress in protein structure prediction due to the development, release, and applications of deep learning-based methods, such as AlphaFold and Protein Language Models that use Natural Language Processing (NLP) techniques.

In the post-AlphaFold age, one of the most critical challenges in molecular biology is to understand how boundary conditions and the one-dimensional (1D) sequence of amino acid determines three-dimensional (3D) structure and function of a given protein and how these aspects change as the sequence changes. After more than 50 years of efforts, reliable computational protein models that reach quality of experimental structure are now available for most proteins. But even when static computational 3D models of proteins are available, and more so when they do not (e.g., because these proteins are intrinsically disordered), connecting sequence and structure to dynamic and function remains challenging. Secondary structure (SS) and other one-dimensional (1D) structural properties, such as the accessible surface area and intrinsic disorder, help predict 3D models. Many methods that predict 1D structure from sequence have been developed as an intermediate step or a substitute for the 3D structure prediction. These 1D characteristics can be structural and functional properties that are expressed by a one-dimensional vector along the protein sequence. They provide ways to represent information that can be more readily processed, for example, by machine learners.

In the secondary structure, the protein backbone structure is annotated by a few states, such as helices, sheets, or coils. Torsion angles can also characterize the protein's local structure. In addition to the structural properties of the backbone, protein structures can be described by global structural properties. These properties depend on interactions between multiple residues far apart in the sequence. Examples include the solvent-accessible surface area and the number of residues in a given spatial region, which are relevant to proteins' tertiary packing and function. More recently, predicting intrinsic disorder (lack of well-defined structure under physiological conditions) and one-dimensional functional properties (e.g., residues that bind ligands and posttranslational modification sites) and the effects of sequence variation on function have received increasing attention. Correspondingly, this book expands the first edition to cover a broad spectrum of related topics ranging from those related directly to protein secondary structure, to methods for the prediction of 3D structure, the deep learning, and large-scale protein language models for the prediction of other 1D properties of proteins that include protein-nucleic acid and protein-protein interactions, intrinsic disorder, and noncoding RNA fragments and their secondary structure and applications as medical biomarkers and therapeutic targets.

The following chapters are included in this volume of *Methods in Molecular Biology*:

The first chapter is written by George D. Rose of the Johns Hopkins University, USA, one of the pioneers in protein structure prediction. This chapter gives an excellent overview of proteins' most important secondary structure element, the α-helix. It starts from the historically first model of α-helix, developed by Pauling, and goes to the most recent developments in understanding the α-helix by deep learning and AlfaFold2 predictions.

Adam Liwo from the University of Gdansk, Poland, and his collaborators discuss the role of secondary structure as a scaffold for building the tertiary structure of protein. Both template-based (assisted) modeling and free modeling for proteins with novel folds are overviewed. Particular emphasis is given to the physics-based UNRES force field developed by Liwo and Scheraga and the energy terms of UNRES that govern secondary structure formation.

In Chap. 3, Jianlin Cheng from the University of Missouri-Columbia, USA, with collaborators, describes how protein secondary structure prediction can be improved by applying deep language models and transformer networks. They developed a novel TransPross method based on the transformer network and attention mechanism commonly used in natural language processing that significantly improves protein secondary structure prediction. The TransPross algorithm is freely available on GitHub.

Aleksandra E. Badaczewska from the Iowa State University, USA, and Andrzej Kolinski from the University of Warsaw, Poland, discuss the significant role of secondary structure in large-scale protein modeling with coarse-grained methods. The CABS model developed in the past by the Kolinski group is too slow for modeling large-scale transitions in large protein systems. Therefore, they created a lower-resolution, faster, coarse-grained modeling tool called SURPASS. The application and performance of the SURPASS methodology are discussed in the chapter.

Chapter 6, authored by Michal Brylinski and his colleagues from the Louisiana State University, USA, surveys contributions of machine learning to the development of computational methods for the prediction of protein structure and function. Motivated by recent developments in this area, they focus on the deep neural networks and natural language processing inspired algorithms and cover a broad range of predictive targets, such as secondary and 3D structures, residue-residue contacts, subcellular localization, and protein function annotations.

Daisuke Kihara from the Purdue University, USA, with collaborators, describe how the identification of protein secondary structure and the detection of DNA/RNA location in cryo-electron microscopy (cryo-EM) and cryo-electron tomography (cryo-ET) maps at intermediate or low-resolution enables significantly better interpretations of such maps. Kihara's lab developed Emap2sec software to identify the secondary protein structures and Emapsec+ to identify additional locations of DNA or RNA molecules in the cryo-EM maps. Both methods are publicly available as web servers.

In Chap. 7, Liam J. McGuffin and Ahmet Gurkan Genc from the University of Reading, UK, discuss impact of AlphaFold2 and the subsequent and related tools on the protein structure prediction field. This chapter provides easy to read history of the protein structure prediction and structure refinement areas and covers key methods and advancements, which include generative models, protein language models, co-evolutionary approaches, and graph neural network. The authors examine impact of the Critical Assessment of Structure Prediction (CASP) experiments.

Eshel Faraggi with colleagues from across the USA analyze targets from the latest CASP15 experiment for the blind prediction of 3D protein structures from amino acid sequences. They focus on cases where the AlphaFold2 method, the top predictor for CASP15, failed. They identify six such targets and compare these results with predictions of SPINE-X and Seder; two methods developed by the authors of this chapter and participated in CASP15. They find that the poor quality of the AlphaFold2 predictions results from incorrectly assigned secondary structures. They also observe that SPINE-X and Seder sometimes give better predictions by improving detection of the secondary structure elements.

Robert L. Jernigan and collaborators at the Iowa State University, USA, focus their chapter on an analysis of a minimal organism capable of self-replication, JCVI-Syn3. This synthetic minimal organism contains only 473 genes that code 434 proteins, while most of these genes still have unknown functions. They used their novel and highly successful PROST method that relies on the large protein language model, ESM1b, for functional annotation of the protein-coding genes. They use PROST to annotate 93% of the JCVI-Syn3 genes with high accuracy and found new functions for other genes.

In Chap. 10, Lukasz Kurgan and his colleagues from the Virginia Commonwealth University, USA, describe the DescribePROT database, a large online resource that provides easy access to 13 diverse 1D protein structure and function descriptors for 1.4 million proteins from 83 complete proteomes. They summarize the scope of this comprehensive database and different ways to access and query its contents, and comment on future plans to expand it.

M. Michael Gromiha and coworkers from the Indian Institute of Technology Madras and University of the Witwatersrand, South Africa, focus on the HIV protease, a 99-residue protein protease that is recognized as one of the potent targets for treating HIV infections. They overview various HIV protease subgroups, the structure of HIV protease, the catalytic mechanism of HIV-1 protease and its specificity, and the movement of the flaps, which is crucial for HIV-1 PR enzymatic activity. They discuss drug resistance in the HIV protease and the application of 3D-QSAR methods to study HIV protease–inhibitor interactions.

The next two chapters focus on the prediction of intrinsic disorder and its arguably most extreme form, which are disordered linkers. In Chap. 12, Lukasz Kurgan and his collaborators from the Nankai University, China, describe the flDPnn method that targets prediction of the intrinsic disorder. This is one of the most accurate disorder predictors based on the results from the last two Critical Assessment of Intrinsic Disorder (CAID) experiments, which is additionally capable of predicting several major functions of disorder. In Chap. 13, Zhenling Peng from the Shandong University, China, and her colleagues summarize the disordered linker predictor, APOD. The disordered linkers connect protein domains and structural elements within domains and they perform these functions without ever folding into a well-defined structure. This is in contrast to other disordered regions that undergo folding under certain circumstances, such as binding. APOD provides accurate predictions and is available as a web server. In both chapters, the authors explain the underlying predictive models and use case studies to showcase how to read and understand their predictions.

Jörg Gsponer and Nawar Malhis from the University of British Columbia, Canada, provide insights into the area of the analysis and prediction of linear interacting peptides (LIPs), a generic class of disordered regions that are involved in inter-molecular interactions. They perform empirical comparison of several relevant predictors using benchmark data from the CAID1 and CAID2 experiments. They find significant differences between predictors and recommend using multiple methods and performing manual analysis of their results to minimize false positive rate.

In Chap. 15, Jian Zhang and his collaborators from the Xinyang Normal University, China, and the Virginia Commonwealth University, USA, describe the SCRIBER tool that accurately predicts protein binding residues in protein sequences. The key feature of this method is the fact that it minimizes cross-predictions, which are defined as predictions of residues that interact with other ligand types (ions, nucleic acids, lipids, etc.) as protein binding. The authors include a case study that illustrates how to read and interpret predictions generated by the web server of SCRIBER.

Dukka B. KC and his research group at the Rochester Institute of Technology, USA, introduce their LMPTMSite (Language Model-based Post-Translational Modification Site) prediction platform. This platform includes accurate predictors of two types of the post-translational modification sites, S-nitrosylation and succinylation. The authors explain the underlying deep learning architectures, empirically compare their predictors with related tools, and present a case study to demonstrate how to use their tools.

Chapter 17, authored by Mubashir Hassan and collaborators, overviews protein aggregation phenomena and discusses various computational methods, databases, algorithms, and tools that predict the aggregation of proteins and peptides. Protein aggregation is directly related to misfolding and secondary structure. Proteins aggregate, forming complex amyloid structures that contain alpha helices and beta sheets. The authors discuss secondary structure propensity in the protein aggregation and diseases related to the protein aggregation.

M. Michael Gromiha and his coworkers discuss the recognition mechanism in protein-nucleic acid interactions, particularly focusing on the binding affinities of protein-DNA and protein-RNA complexes. They describe various databases of protein-nucleic acid complexes and interactions between proteins and nucleic acids. They also overview multiple algorithms and tools for extracting features from protein-nucleic acid complexes and nucleic acid structures, analyzing protein-DNA and protein-RNA complexes, and predicting protein-nucleic acid complex structures. The authors also discuss databases on protein-nucleic acid binding affinities and methods that analyze and predict these binding affinities.

The last chapter, written by Andrzej Kloczkowski and collaborators, differs from the other chapters since it covers noncoding RNAs (ncRNAs) rather than proteins. The ncRNAs are nowadays one of the hottest topics, being linked to various diseases including cancers, neurodegenerative disorders, and psychiatric disorders. This chapter overviews multiple types of ncRNAs, secondary structures of RNAs linked to intramolecular base matching, databases, tools used to analyze ncRNAs, and ncRNA target prediction. The authors discuss applications of ncRNAs in treating different diseases, including their use as diagnostic and prognosis biomarkers and therapeutic targets. Finally, they also summarize clinical implementations of ncRNAs in cancer, neurodegenerative, cardiovascular, and infectious diseases.

This volume covers a comprehensive collection of methods, resources, and studies that target analysis and prediction of various structural and functional aspects of proteins and ncRNAs. We hope it will help readers to gain a better understanding of this dynamic and growing field of research and guide them in the quest to identify and use the best computational methods and resources.

Columbus, OH, USA *Andrzej Kloczkowski*
Richmond, VA, USA *Lukasz Kurgan*
Indianapolis, IN, USA *Eshel Faraggi*

Contents

Preface .... *v*
Contributors .... *xi*

1 The Iconic α-Helix: From Pauling to the Present 1
George D. Rose

2 Secondary Structure in Free and Assisted Modeling of Proteins with the Coarse-Grained UNRES Force Field 19
Emilia A. Lubecka, Cezary Czaplewski, Adam K. Sieradzan, Agnieszka G. Lipska, Łukasz Dziadek, and Adam Liwo

3 Improving Protein Secondary Structure Prediction by Deep Language Models and Transformer Networks 43
Tianqi Wu, Weihang Cheng, and Jianlin Cheng

4 Importance of Secondary Structure Data in Large Scale Protein Modeling Using Low-Resolution SURPASS Method 55
Aleksandra E. Badaczewska-Dawid and Andrzej Kolinski

5 Machine Learning Techniques to Infer Protein Structure and Function from Sequences: A Comprehensive Review 79
Gopal Srivastava, Mengmeng Liu, Xialong Ni, Limeng Pu, and Michal Brylinski

6 Protein Secondary Structure and DNA/RNA Detection for Cryo-EM and Cryo-ET Using Emap2sec and Emap2sec$^+$ 105
Javad Baghirov, Han Zhu, Xiao Wang, and Daisuke Kihara

7 Beyond AlphaFold2: The Impact of AI for the Further Improvement of Protein Structure Prediction 121
Ahmet Gurkan Genc and Liam J. McGuffin

8 AlphaFold2, SPINE-X, and Seder on Four Hard CASP Targets 141
Eshel Faraggi, Robert L. Jernigan, and Andrzej Kloczkowski

9 Improving the Annotations of JCVI-Syn3a Proteins 153
Mesih Kilinc, Kejue Jia, and Robert L. Jernigan

10 DescribePROT Database of Residue-Level Protein Structure and Function Annotations 169
Bi Zhao, Sushmita Basu, and Lukasz Kurgan

11 Structural and Functional Studies on HIV Protease: Mechanism of Action, Subtypes, Inhibitors, and Drug Resistance 185
Sankaran Venkatachalam, Sowmya Ramaswamy Krishnan, Yasien Sayed, and M. Michael Gromiha

12 Accurate and Fast Prediction of Intrinsic Disorder Using flDPnn 201
Kui Wang, Gang Hu, Zhonghua Wu, and Lukasz Kurgan

13 Prediction of Disordered Linkers Using APOD 219
Zhenling Peng, Haiyan Wu, Yuxian Luo, and Lukasz Kurgan

14 Computational Prediction of Linear Interacting Peptides 233
Nawar Malhis and Jörg Gsponer

15 Accurate Prediction of Protein-Binding Residues in Protein Sequences Using SCRIBER 247
Jian Zhang, Feng Zhou, Xingchen Liang, and Lukasz Kurgan

16 LMPTMSite: A Platform for PTM Site Prediction in Proteins Leveraging Transformer-Based Protein Language Models 261
Pawel Pratyush, Suresh Pokharel, Hamid D. Ismail, Soufia Bahmani, and Dukka B. KC

17 Prediction and Evaluation of Protein Aggregation with Computational Methods 299
Mubashir Hassan, Saba Shahzadi, Mai Suan Li, and Andrzej Kloczkowski

18 Bioinformatics Approaches for Understanding the Binding Affinity of Protein–Nucleic Acid Complexes 315
K. Harini, Masakazu Sekijima, and M. Michael Gromiha

19 A Glimpse of Noncoding RNAs: Secondary Structure, Emerging Trends, and Potential Applications in Human Diseases 331
Mubashir Hassan, Amal Malik, Zainab Yaseen, Saba Shahzadi, Muhammad Yasir, and Andrzej Kloczkowski

Index *345*

Contributors

ALEKSANDRA E. BADACZEWSKA-DAWID • *Office of Biotechnology, Iowa State University, Ames, IA, USA*
JAVAD BAGHIROV • *Department of Computer Science, Purdue University, West Lafayette, IN, USA*
SOUFIA BAHMANI • *Computer Science Department, Rochester Institute of Technology, Rochester, NY, USA; Michigan Technological University, Comptuer Science Department, Houghton, MI, USA*
SUSHMITA BASU • *Department of Computer Science, Virginia Commonwealth University, Richmond, VA, USA*
MICHAL BRYLINSKI • *Department of Biological Sciences, Louisiana State University, Baton Rouge, LA, USA; Center for Computation and Technology, Louisiana State University, Baton Rouge, LA, USA*
JIANLIN CHENG • *Electrical Engineering and Computer Science Department, University of Missouri, Columbia, MO, USA*
WEIHANG CHENG • *Department of Chemistry, Hubei University, Wuhan, Hubei, China*
CEZARY CZAPLEWSKI • *Faculty of Chemistry, University of Gdańsk, Fahrenheit Union of Universities, Gdańsk, Poland*
ŁUKASZ DZIADEK • *Faculty of Chemistry, University of Gdańsk, Fahrenheit Union of Universities, Gdańsk, Poland*
ESHEL FARAGGI • *Research and Information Systems, LLC, Indianapolis, IN, USA; Department of Physics, Indiana University, Indianapolis, IN, USA*
AHMET GURKAN GENC • *School of Biological Sciences, University of Reading, Reading, UK*
M. MICHAEL GROMIHA • *Department of Biotechnology, Bhupat and Jyoti Mehta School of Biosciences, Indian Institute of Technology Madras, Chennai, Tamil Nadu, India; International Research Frontiers Initiative, School of Computing, Tokyo Institute of Technology, Yokohama, Japan*
JÖRG GSPONER • *Michael Smith Laboratories, the University of British Columbia, Vancouver, BC, Canada*
K. HARINI • *Department of Biotechnology, Bhupat and Jyoti Mehta School of Biosciences, Indian Institute of Technology Madras, Chennai, Tamil Nadu, India*
MUBASHIR HASSAN • *The Steve and Cindy Rasmussen Institute for Genomic Medicine, Nationwide Children Hospital, Columbus, OH, USA*
GANG HU • *NITFID, School of Statistics and Data Science, LPMC and KLMDASR, Nankai University, Tianjin, China*
HAMID D. ISMAIL • *Computer Science Department, Rochester Institute of Technology, Rochester, NY, USA; North Carolina A&T State University, Computational Data Science and Engineering, Greensboro, NC, USA*
ROBERT L. JERNIGAN • *Roy J. Carver Department of Biochemistry, Biophysics and Molecular Biology, Iowa State University, Ames, IA, USA; Bioinformatics and Computational Biology Program Iowa State University, Ames, IA, USA*
KEJUE JIA • *Molecular, Cellular and Developmental Biology, Yale University, New Haven, CT, USA*

DUKKA B. KC • *Computer Science Department, Rochester Institute of Technology, Rochester, NY, USA*

DAISUKE KIHARA • *Department of Computer Science, Purdue University, West Lafayette, IN, USA; Department of Biological Sciences, Purdue University, West Lafayette, IN, USA*

MESIH KILINC • *Bioinformatics and Computational Biology Program Iowa State University, Ames, IA, USA*

ANDRZEJ KLOCZKOWSKI • *The Steve and Cindy Rasmussen Institute for Genomic Medicine, Nationwide Children's Hospital, Columbus, OH, USA; Department of Pediatrics, The Ohio State University, Columbus, OH, USA; Department of Biomedical Informatics, The Ohio State University, Columbus, OH, USA*

ANDRZEJ KOLINSKI • *Faculty of Chemistry, University of Warsaw, Warsaw, Poland*

SOWMYA RAMASWAMY KRISHNAN • *Department of Biotechnology, Bhupat and Jyoti Mehta School of Biosciences, Indian Institute of Technology Madras, Chennai, Tamil Nadu, India*

LUKASZ KURGAN • *Department of Computer Science, Virginia Commonwealth University, Richmond, VA, USA*

MAI SUAN LI • *Institute of Physics, Polish Academy of Sciences, Warsaw, Poland*

XINGCHEN LIANG • *School of Computer and Information Technology, Xinyang Normal University, Xinyang, China*

AGNIESZKA G. LIPSKA • *Centre of Informatics Tri-city Academic Supercomputer and Network (CI TASK), Gdańsk University of Technology, Fahrenheit Union of Universities in Gdańsk, Gdańsk, Poland*

MENGMENG LIU • *Division of Electrical and Computer Engineering, Louisiana State University, Baton Rouge, LA, USA*

ADAM LIWO • *Faculty of Chemistry, University of Gdańsk, Fahrenheit Union of Universities, Gdańsk, Poland*

EMILIA A. LUBECKA • *Faculty of Electronics, Telecommunications and Informatics, Gdańsk University of Technology, Fahrenheit Union of Universities in Gdańsk, Gdańsk, Poland*

YUXIAN LUO • *Research Center for Mathematics and Interdisciplinary Sciences, Shandong University, Qingdao, China*

NAWAR MALHIS • *Michael Smith Laboratories, the University of British Columbia, Vancouver, BC, Canada*

AMAL MALIK • *Institute of Molecular Biology and Biotechnology, The University of Lahore, Lahore, Pakistan*

LIAM J. MCGUFFIN • *School of Biological Sciences, University of Reading, Reading, UK*

XIALONG NI • *Department of Biological Sciences, Louisiana State University, Baton Rouge, LA, USA*

ZHENLING PENG • *Research Center for Mathematics and Interdisciplinary Sciences, Shandong University, Qingdao, China; Frontier Science Center for Nonlinear Expectations, Ministry of Education, Shandong University, Qingdao, China*

SURESH POKHAREL • *Computer Science Department, Rochester Institute of Technology, Rochester, NY, USA*

PAWEL PRATYUSH • *Computer Science Department, Rochester Institute of Technology, Rochester, NY, USA*

LIMENG PU • *Center for Computation and Technology, Louisiana State University, Baton Rouge, LA, USA*

GEORGE D. ROSE • *Johns Hopkins University, Baltimore, MD, USA*

YASIEN SAYED • *Protein Structure-Function Research Unit, School of Molecular and Cell Biology, University of the Witwatersrand, Johannesburg, South Africa*

MASAKAZU SEKIJIMA • *Department of Computer Science, Tokyo Institute of Technology, Yokohama, Japan*
SABA SHAHZADI • *The Steve and Cindy Rasmussen Institute for Genomic Medicine, Nationwide Children Hospital, Columbus, OH, USA*
ADAM K. SIERADZAN • *Faculty of Chemistry, University of Gdańsk, Fahrenheit Union of Universities, Gdańsk, Poland*
GOPAL SRIVASTAVA • *Department of Biological Sciences, Louisiana State University, Baton Rouge, LA, USA*
SANKARAN VENKATACHALAM • *Department of Biotechnology, Bhupat and Jyoti Mehta School of Biosciences, Indian Institute of Technology Madras, Chennai, Tamil Nadu, India*
KUI WANG • *NITFID, School of Statistics and Data Science, LPMC and KLMDASR, Nankai University, Tianjin, China*
XIAO WANG • *Department of Computer Science, Purdue University, West Lafayette, IN, USA*
HAIYAN WU • *Research Center for Mathematics and Interdisciplinary Sciences, Shandong University, Qingdao, China*
TIANQI WU • *Electrical Engineering and Computer Science Department, University of Missouri, Columbia, MO, USA*
ZHONGHUA WU • *School of Mathematical Sciences and LPMC, Nankai University, Tianjin, China*
ZAINAB YASEEN • *Faculty of Science and Technology, University of Central Punjab, Lahore, Pakistan*
MUHAMMAD YASIR • *Department of Pharmacology, College of Medicine, Kangwon National University, Chuncheon, Republic of Korea*
JIAN ZHANG • *School of Computer and Information Technology, Xinyang Normal University, Xinyang, China*
BI ZHAO • *Genomics program, College of Public Health, University of South Florida, Tampa, FL, USA*
FENG ZHOU • *School of Computer and Information Technology, Xinyang Normal University, Xinyang, China*
HAN ZHU • *Department of Computer Science, Purdue University, West Lafayette, IN, USA*

Chapter 1

The Iconic α-Helix: From Pauling to the Present

George D. Rose

Abstract

The protein folding problem dates back to Pauling's insights almost a century ago, but the first venture into actual protein structure was the Pauling–Corey–Brandson α-helix in 1951, a proposed model that was confirmed almost immediately using X-ray crystallography. Many subsequent efforts to predict protein helices from the amino acid sequence met with only partial success, as discussed here. Surprisingly, in 2021, these efforts were superseded by deep-learning artificial intelligence, especially AlphaFold2, a machine learning program based on neural nets. This approach can predict most protein structures successfully at or near atomic resolution. Deservedly, deep-learning artificial intelligence was named Science magazine's 2021 "breakthrough of the year." Today, ~200 million predicted protein structures can be downloaded from the AlphaFold2 Protein Structure Database. *Deep learning* represents a *deep conundrum* because these successfully predicted macromolecular structures are based on methods that are completely devoid of a hypothesis or of any physical chemistry. Perhaps we are now poised to transcend five centuries of reductive science.

Key words Helix capping, Helix propensities, Hydrogen bonding, Protein scaffolds, Chirality, Neural nets, Reductionist science

1 Introduction

The protein α-helix was initially proposed as a model structure by Pauling, Corey, and Branson (PCB) [1] in 1951. Pauling had launched an exhaustive search for all possible repetitive, single-chain configurations of a polypeptide chain that satisfied two primary restraints: (1) the backbone geometry was specified using values that Pauling and Corey had previously elucidated by X-ray crystallography of amino acids and peptides [2] and (2) a requirement that all N-H•••O=C hydrogen bonds were satisfied.

A repeating structure of amino acid residues—one in which all component residues are equivalent—automatically results in a helix, where every residue is related to its successor in the chain by a rotation about an axis and a translation. As such, the search for

Andrzej Kloczkowski et al. (eds.), *Prediction of Protein Secondary Structure*, Methods in Molecular Biology, vol. 2867, https://doi.org/10.1007/978-1-0716-4196-5_1,

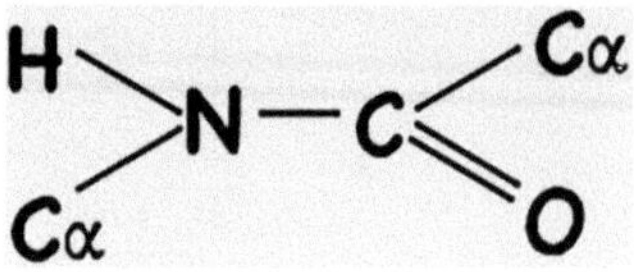

Fig. 1 Planarity of the peptide bond

viable models boils down to finding rotations and translations that are compatible with intra-chain hydrogen-bond satisfaction.

Earlier unsuccessful attempts to model the structure of fibrous proteins using related ideas [3–5] had already conditioned expectations that viable candidates would be helical. Why then was Pauling successful when others were not?

Pauling, together with Corey, had previously determined residue interatomic distances using X-ray crystallography [6], and he recognized that the peptide bond, C-N, had partial double-bond character, thereby inhibiting free, single-bond-type rotation. Consequently, the six backbone atoms (below) are essentially coplanar, a simplifying realization for model-building because this 6-atom unit eliminated a rotatable angle. This is illustrated in Fig. 1.

In addition, earlier proposals had assumed an integer number of residues per helical turn, an arbitrary assumption that Pauling rejected. The PCB model predicted a non-integer value of 3.7 residues per helical turn, with a spacing of 1.5 Å from each residue to the next and repeating values of the backbone dihedral angles $\phi,\psi \cong (-57°, -47°)$.

In practice, helices in the protein data bank (PDB) [7] are found to have backbone dihedral angles with values closer to $\phi,\psi = (-60°, -40°)$ [8] on average, significantly different from the PCB model. Marshall et al. rationalized this difference by noting that "*the carbonyl oxygen of the amide bond in protein helices is not restrained to linear hydrogen bonds with the amide hydrogens in a protein helix, but prefers, instead, to form bifurcated hydrogen bonds simultaneously to adjacent amides (i to i+3 and i to i+4) across the helical axis*" [9].

Motivated by Pauling's landmark publication, Perutz, using X-ray crystallography [11], searched for and found a reflection corresponding to the 1.5 Å spacing in three different structures: a polypeptide, a fibrous protein, and a globular protein, specifically poly-γ-benzyl-L-glutamate, α-keratin, and hemoglobin. The finding was consistent with the PCB model and inconsistent with other proposed models, including his own [11]. Pauling's proposed structure of the α-helix and its immediate confirmation by experiment was a triumph for the PCB model and for molecular model-making in general. See Eisenberg for an engaging account of this historical discovery [12].

2 Helix Capping

α-helices in globular proteins range in size from a single helical turn to more than 50 residues. However, the average length of an α-helix is approximately 12 residues [13], as illustrated in Fig. 2. α-helices are characterized by hydrogen bond donors between successive amide hydrogens and corresponding carbonyl oxygen acceptors in the previous helical turn, N_i-H•••O=C_{i-4}, as sketched above. Consequently, for the helix of average length, only 50% of the backbone polar groups can be satisfied by intra-helical partners because the initial four amide hydrogens and final four carbonyl oxygens necessarily lack intra-helical partners [8], as seen in Fig. 2b. Hydrogen-bond satisfaction of the backbone polar groups in these flanking residues [8] is known as *helix-capping* [14]. An earlier analysis had also noted distinct residue preferences in protein helices [15].

A useful terminology for the helical and flanking residues is

. . . -N″-N′-Ncap-N1-N2-N3- . . . C3-C2-C1-Ccap-C′-C″- . . .

where N1 through C1 belong to the helix proper, and the primed residues belong to turns that bracket the helix at either end. Ncap and Ccap are boundary residues that establish a bridge between the helix and the adjacent turn that terminates the helix.

Charlotte Schellman was the first to document a recurring glycine-based termination pattern at the C-terminus of α-helices [16], an illustrative example of a capping motif. The Schellman motif is a doubly hydrogen-bonded pattern between backbone partners, specifically hydrogen bonds between (i) a glycine N-H donor at C′ with the O=C acceptor at C2 and (ii) the N-H donor at C″ with the O=C acceptor at C3 (Fig. 3a). In this motif, the glycine

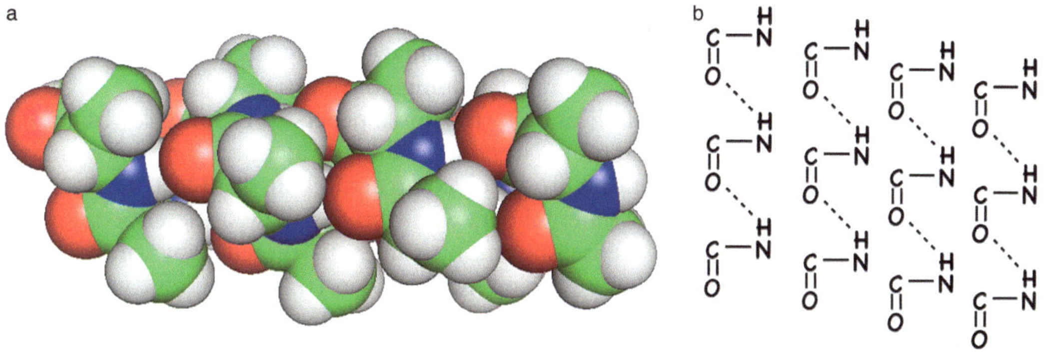

Fig. 2 (**a**) A poly-alanine α-helix shown as a space-filling atomic model [10]. Atom types are rendered in color: oxygen (red), nitrogen (blue), hydrogen (white), and carbon (green). The structure is close-packed, with no voids the size of an atom. Note the hydrogen bonds between N-H and C=O groups. (**b**) Schematic diagram of hydrogen bonding in a 12-residue α helix

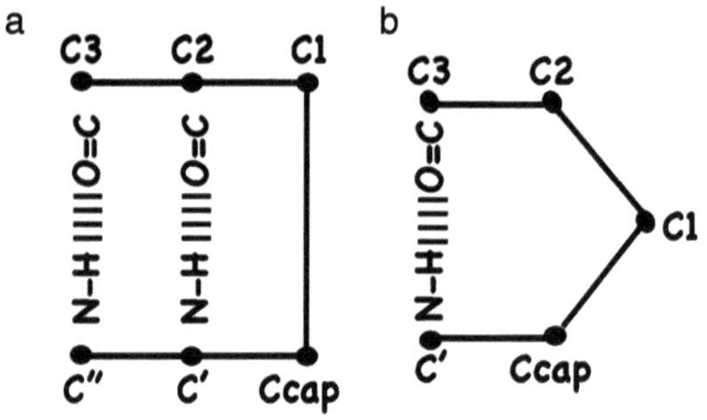

Fig. 3 (**a**) Schellman motif. (**b**) α_L motif

at C′ adopts a left-handed conformation (i.e., $\phi > 0$), effectively terminating the series of repeating right-handed helical turns.

The other frequently-observed C-capping motif, similar to the Schellman motif, is the α_L motif [17, 18]. This motif is defined by a hydrogen bond between the N-H at C′ and the C=O at C3 (Fig. 3b). Like the Schellman motif, the C′ residue is typically glycine, which adopts a left-handed conformation (i.e., $\phi > 0$), effectively terminating the series of repeating right-handed helical turns. Many investigators identified glycine-based helix termination motifs [17–27].

Additionally, side chains interactions were also found to function as capping motifs [28], for example, in the "capping box" [24, 29–32], an N-capping interaction involving two of the initial four backbone amide hydrogen donors of the helix. In this motif, the side chain of Ncap forms a hydrogen bond with the backbone of N3 and, reciprocally, the side chain of N3 forms a hydrogen bond with the backbone of Ncap. The normalized sequence preferences in a capping box are threonine > serine > asparagine at Ncap and glutamic acid > glutamine at N3 [29].

The ongoing publication of newly recognized capping motifs [33] together with an associated glossary of their proper names (e.g., Schellman motif, capping box, ...) prompted a review that includes a "wall-chart" summary of the then extant motifs and introduces a systematic naming convention [34].

After publication of the initial papers on capping [8, 14], many new capping motifs were discovered by identifying motifs observed in helices taken from solved structures in the PDB, as discussed above. Experimental data was soon to follow, as described next.

Early skepticism among some was dispelled by a report that tested the helix-capping hypothesis in 10 single-residue variants of a 12-residue alanine-based peptide [35]. The authors chose an alanine-based peptide because alanine was known to be a helix-promoting residue [36]. The variants included six peptides that differed only by a single residue at the N-terminus (viz., alanine, asparagine, aspartate, glutamine, glutamate, and serine) and four other peptides with single residue substitutions at other positions.

Helicity was measured by circular dichroism, and it ranged between 25.1% and 2.4% for the six N-terminal variants. In detail, helicity was >20% for aspartate, asparagine, and serine, whereas the alanine peptide was essentially random coil; glutamine and glutamic acid had values of 4.8% and 8.6%, respectively. Quoting the authors, "*these results are completely consistent with the hypothesis that side-chain-main-chain hydrogen bonds at the amino terminus are helix stabilizing.*" Results for the remaining four peptides were also consistent with capping.

Not everyone agreed with this conclusion. For example, Serrano and Fersht [37] introduced eight Ncap mutations in the two helices of barnase, positions threonine 6 and threonine 26, and they compared the measured stabilization energy of each mutation (relative to threonine) with the observed frequency of occurrence [14] of that residue at the Ncap position in proteins. Most mutations were as expected, with the notable exception of asparagine, which was destabilizing in both Ncaps despite being strongly preferred at this position in proteins.

In retrospect, side chain capping interactions may involve more than one residue. In barnase, the Oγ oxygen in threonine 6 is hydrogen-bonded to the backbone N-H of glycine 9, like half a capping box. Asparagine may not be able to form this hydrogen bond. Similarly, the backbone N-H in threonine 26 is hydrogen-bonded to the side chain Oε1 of glutamate 29, again like half a capping box. See also [38].

Many investigators tested capping in succeeding years [for example, see, [24–26, 28, 30, 31, 38–45].

An unanticipated result from Lecomte and Moore reported the remarkable discovery that histidine (80) at the N-cap position of a helix in a water-soluble fragment of apocytochrome b_5 can function as a pH-dependent switch. In particular, the neutral imidazole group forms a hydrogen bond with the backbone amide at N2, and it "*directs the folding at physiological pH by introducing a ca. 2 kcal mol^{-1} preference for the native helix and survives in the final structure*" [46]. Later, this initial finding was shown to be general: according to Preimesberger et al. [47], "*the side chain of histidine often serves as a hydrogen bond (H-bond) acceptor. These H-bonds provide a convenient pH-dependent switch for local structure and functional motifs. In hundreds of instances, a histidine caps the N-terminus of α- and 3_{10}-helices by forming a backbone NH···Nδ1 H-bond.*"

Helices terminating in a glycine can be classified by their hydrogen bonding patterns into either a Schellman or an α_L motif (Fig. 3b). These two primary motifs differ in their respective patterns of stabilizing apolar residues, as described in [23]. Specifically, the primary (but not only) distinguishing criterion is the presence or absence of a hydrophobic interaction between C″ and C3, which selects for either a Schellman or an α_L motif, respectively. Despite minor exceptions, these reference patterns are clearly evident in solved structures, akin to rules for a C-capping grammar. How ironclad are these "rules"? Are they reliable enough to distinguish between the two motifs from sequence alone [23]?

This question prompted an experimental test using T4 lysozyme [48]. Matthews and coworkers introduced two mutations that were intended to change the α_L capping motif in helix J into a Schellman motif, in accord with the glycine-based C-capping rules [23]. However, it did not work. According to the authors, although the mutations did "*result in some large-scale changes in the structure of the protein, the backbone conformation in the immediate vicinity of the mutation, and including the helix-capping region, changes very little.*" At least in this case, the test indicates that mutations in a protein helix are constrained by the global environment, in contrast to corresponding mutations in a peptide helix, where all mutations are local.

3 Amino Acid Preferences in Helices

Soon after a few solved structures became available, it was evident that specific residues are selected preferentially in α-helices. This observation led to the concept of *residue propensities.* The idea, as formalized by Chou and Fasman [49], was to calculate the normalized frequency of occurrence for each of the 20 residues in α-helices, β-sheet, and β-turns, by using residue census data from solved structures, an empirical approach. For example:

$$f_\alpha(\text{ala}) = \text{fraction of alanines in } \alpha\text{-helices}/\text{fraction of alanines in proteins},$$

where f_α (ala) is the residue propensity for alanine. Normalized in this way, an f_α (ala) = 1 when alanines are found in α-helices and in entire proteins with equal likelihood. When f_α (ala) > 1, alanines are found in α-helices more frequently than in entire proteins, that is, alanines are favored in α-helices, and conversely, when f_α (ala) < 1, alanines are disfavored in α-helices. Many propensity scales were developed in ensuing years, and they varied depending upon ever-accumulating additions to the PDB and different approaches to normalization: see Koehl and Levitt [50] and references therein.

A major step forward in secondary structure prediction was introduced by Rost and Sander in a program amusingly dubbed PHD [51, 52] because it was the topic of Rost's PhD thesis. They extended the earlier single-residue approach by using evolutionary information, namely patterns of residue substitutions in families with a known 3-dimensional structure. These data were then taken as input to a neural net, resulting in secondary predictions with greater than 70% accuracy. Later improvements using the same basic approach [53] achieved even greater predictive accuracy, around 80–90%.

An alternative approach to determining helix propensities was that of Scheraga and coworkers [54]. Applying helix-coil theory, they extracted Zimm-Bragg [55] nucleation and propagation parameters, σ and s, respectively, by introducing each amino acid, in turn, as a "guest" residue into a poly-hydroxyalkylglutamine "host." These propagation parameters measure helix propensities.

Helix-coil theory stems from the plausible model that conformational opportunities for a polypeptide chain in a disorganized coil far exceed those when the chain adopts a helical conformation. Consequently, nucleating the first turn of helix is entropically disfavored because five consecutive residues must adopt ϕ,ψ-angles that are compatible with an $i - (i + 4)$ hydrogen bond, which is presumed to be enthalpically favorable. However, once a turn is nucleated, the enthalpic gain from adding additional hydrogen bonds in successive residues more than compensates for the added entropic cost. Were this not the case, the chain would remain coil. In helix-coil theory, σ and s are the nucleation and propagation parameters, respectively. An excellent treatment of helix-coil theory can be found in almost any textbook on proteins, such as [56, 57]. Undeniably, helix-coil theory [55, 58] has played an influential role in modeling α-helix formation, but it is important to realize that results from experiment depict a less idealized mechanism [59].

Helix stability is expected to increase as a function of helix length, owing to the fact that enthalpy overcomes competing entropy once the helix has been nucleated. Of course, the length needed for a stable helix depends on σ and s for a given amino acid in a specific host. The helix-forming propensities in empirical studies [49] were known to be moderately correlated with propagation parameters from host-guest experiments. Both lines of investigation concur in the finding that the weak helical propensities of the strongest helix formers in either peptides or proteins would require chains in excess of 100 residues to achieve stability at detectable levels. Consequently, the weak helix-forming tendencies of natural amino acids led to the unwelcome conviction that helices lack sufficient stability to serve as autonomous folding units during protein self-assembly, where the length of an average protein helix is approximately 12 residues [8].

Thus, it came as a revelation when Baldwin and coworkers [60] demonstrated the existence of stable, short (13-residues), isolated helices in water. Their results prompted re-evaluation of earlier work and changed the existing paradigm: helices can function as independent folding units after all.

Baldwin's intuition was guided by an earlier landmark discovery: limited proteolytic cleavage of bovine pancreatic ribonuclease (RNase-A) by subtilisin resulted in two fragments, residues 1–20 and 21–124, S-peptide and S-protein, respectively [61]. Although the individual fragments lack enzymic activity, they can re-associate non-covalently into RNase-S, with full enzymic activity and a structure that is essentially superimposable with RNase-A [62]. At least in the example of RNase, the two isolated segments, S-peptide and S-protein, can self-assemble with high fidelity and exquisite complementarity. Many subsequent studies using complementing fragments were able to reproduce this result in other protein systems, for example, Oas and Kim [63].

Of particular interest, S-peptide, which appeared to lack a persisting structure in isolation, includes a prominent α-helix (residues 2–12) in the context of the full protein. Later, Brown and Klee [64] isolated an even smaller peptide, C-peptide, using cyanogen bromide to cut the polypeptide chain at methonine-13. C-peptide is almost exactly coincident with the α-helix between residues 2–12 in the native protein, and using circular dichroism, the investigators detected a minor population of α-helix in aqueous solution [64]. These telltale clues were well known to Baldwin, and they conditioned his intuition about the stability of short α-helices.

A pressing question about S-peptide:S-protein association is whether the order of events is folding, then binding or binding, then folding. It is tempting to speculate that the S-peptide helix (residues 2–12) forms initially and facilitates complementary docking between the two moieties: folding, then binding. A definitive answer to this question had to wait more than half a century for a crucial experiment by Bachmann and Kiefhaber [65]. Contrary to expectations, they found that "*the S-peptide from ribonuclease S forms its α-helical structure only upon binding to the folded S-protein*"—that is, binding, then folding.

Short peptides afford an ideal background for conducting host-guest studies. In 1990, four groups produced experimental scales of helix propensities using stable host peptides. Baldwin and coworkers measured the helix-forming tendency of five natural, apolar residues [66] and three non-natural residues [67] by circular dichroism (CD) spectroscopy of a soluble, alanine-based peptide. O'Neil and DeGrado obtained the free energy contribution to helix formation-relative to Gly-for all naturally occurring amino acids by measuring changes in the helix ⇌ coil equilibrium constant upon residue substitution at a solvent-exposed site in a coiled-coil system [68]. Kallenbach and co-workers derived the free energy contribution to helix formation relative to Gly for 10 natural [69] and 4 nonnatural residues [70] based on Zimm–Bragg [55] theory. Finally, Stellwagen and coworkers measured the helix-forming tendency using CD for 19 different guests residues in a stable peptide host designed by Marqusee and Baldwin [71].

Comparing results, the rank order of helix preferences in these four scales is similar, although not identical—this despite differing host systems and different methods for assessing helicity (i.e., CD, coiled-coil equilibrium, helix-coil theory). However, as noted by Merutka [72], "*the relative stability of the helical forms of these peptides does not correlate with the preference of the variable amino acid to occupy a middle position in a protein helix*" based on the preference scale of Argos and Palau [15]. In sharp contrast to this finding, Pace and coworkers compared the propensities between a helix in RNase T_1 and a peptide model of identical sequence [73, 74], and they "*found that for most amino acids the propensities are very similar in each system.*"

Creamer and Rose hypothesized that the helix-forming propensities in these four scales are largely due to constraints imposed upon side chain rotations by steric interference arising from bulky helix backbone [75]. To test the hypothesis, they calculated difference in configuration entropy, ΔS_{Xaa}, between the α-helical state and a model of the unfolded state, for Xaa selected from a set of eight different hydrophobic residues. The ΔS_{Xaa} values were compared to experimentally determined helix preference values from three of the four scales and found to be in close agreement. Surprisingly, among these eight residues, none favor helix formation: alanine is neutral, and all remaining residues are unfavorable to varying degrees because they lose rotational freedom in the helix compared to the unfolded state. Apparently, the drive to form helices originates in the backbone.

4 Did Nature Have a Choice in Using α-Helices for Protein-Based Life?

Of thermodynamic necessity, globular proteins are built on backbone scaffolds of α-helices and β-sheet [76] (with the minor exception of small, metal-binding polypeptides). This conclusion is corroborated by analysis or visualization of structures in the PDB. These are the only two repetitive secondary structure elements where the number of hydrogen bond donors and acceptors is equal, at least in ideal cases. Consequently, both scaffold elements can be completely hydrogen-bond satisfied, as necessitated [77], and both can be extended indefinitely without encountering a steric clash.

For approximately two decades, Banavar, Maritan and their colleagues have been seeking to develop a theory that can account for these fundamental protein scaffolds from a purely geometric foundation [78–81]. Their ultimate goal is to show that simple geometric primitives alone are sufficient to generate α-helices and both parallel and anti-parallel strands, which, of course, are also helical. Remarkably, their effort has now culminated in the demonstration that these secondary structure building blocks can be produced without invoking sequence information, chemical particulars, or energy terms such as hydrogen bonding. The only adjustable parameter is the average $C\alpha_i - C\alpha_{i+1}$ distance, a scaling parameter taken from proteins of known structure [81]. In their approach, a helix is modeled as a discretized tube with a set of equally spaced points, analogous to the Cα atoms in a protein. Coincidentally, their coordinate system is similar to one described by Haas [82], which can capture a protein backbone quite accurately, albeit not exactly, with three parameters—a dihedral angle, a scalar bend angle, and an inter-residue distance—for each residue.

In the Škrbić et al. theory [81], there are two possible solutions for any helix, left-handed and right-handed, mirror images of each other. The handedness of a helix can be determined by orienting one's thumb along the helix axis and observing whether the curl of the helix matches the fingers of the left hand or the right hand. Handedness can also be determined mathematically, by comparing the sign of the determinant of a set of basis vectors to a reference set of known handedness.

Chirality, that is, handedness, is of fundamental significance in biology: for example, α-helices are right-handed, as is B-DNA. When Pauling proposed his model, the handedness of the α-helix was unknown, nor was it known whether protein helices even had a preferred handedness. For specificity, Pauling illustrated his model with an arbitrary guess [1], an incorrect one as it turned out, not that it mattered for his purposes.

The issue of handedness is not confined to biochemistry. For example, in the seventeenth century, Gottfried Leibniz, co-inventor of the calculus, proposed that the existing world is the *best* of all possible worlds that God could have created. In the eighteenth century, Immanuel Kant disproved Leibniz's thesis, arguing that left-handed and right-handed coordinate systems—Kant used left and right gloves—are equivalent with respect to the relation of their parts to one another. Regardless of the underlying handedness of our existing world, it cannot be uniquely the best possible world; there must be another that is equivalent.

Similarly, both left- and right-handed helices are commonplace in art and architecture. They are often used together in symmetrical arrangements, as in the Taj Mahal, a world-famous example of a building designed to have bilateral symmetry along a central axis, as illustrated in Fig. 4.

Does the prevalence of right-handed α-helices imply an evolutionary advantage compared to their left-handed counterpart? A far more likely explanation originates in nature's choice of L-amino acids (Fig. 5), which, in turn, seems arbitrary. In theory, D-amino acids would have been an equivalent choice, and the resultant preference for left-handed α-helices would then have been equally viable and equally possible [81]. Turning from 17th–eighteenth century philosophical theology to contemporary molecular biology, the arguments of Leibniz and Kant, suitably reframed, still resonate today.

Fig. 4 Entrance to the Taj Mahal, a seventeenth-century white marble mausoleum in Agra, India. Top: the bilaterally symmetric entrance, framed on either side by helices. Bottom left: close-up of the left-handed helices on the left side of the entrance. Bottom right: close up of the right-handed helices on the right side of the entrance. (Picture taken by the author in 2004)

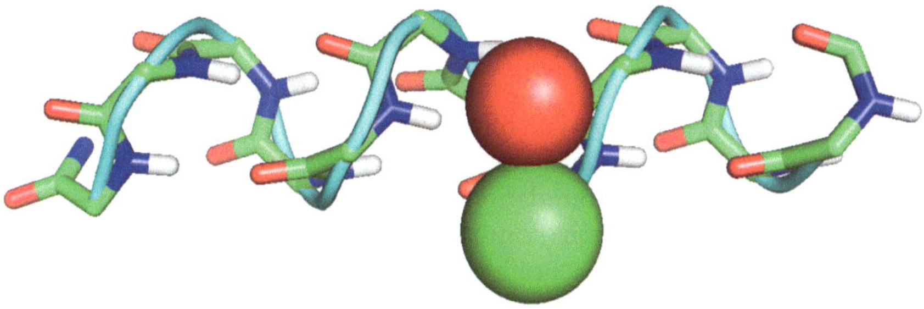

Fig. 5 A 12-residue left-handed poly-alanine α-helix. The helix backbone is rendered in CPK colors, as in Fig. 2b. Unavoidably, the Cβ carbon at residue i (green sphere) clashes with the backbone oxygen at residue *i*-1 (red sphere) in a left-handed helix of L-amino acids. The cyan ribbon traces helix curvature

Paul Dirac was responsible for another fundamental derivation with two possible solutions: the electron and, as he predicted, the positron. Ultimately, his equation led to the discovery of antimatter. In this era of the James Webb space telescope, it is tempting to speculate that D-amino acids may prevail on an exo-planet somewhere in the universe, giving rise to "anti-life."

5 Artificial Intelligence

In 2021, protein fold prediction was transformed by using artificial intelligence (AI) [83–85]. Remarkably—and unexpectedly—these machine-learning algorithms can now predict protein structures successfully at or near atomic resolution.

Of course, if a protein's fold can be predicted, then its secondary structure is also predicted, including the α-helices. Most, but not all, proteins are successfully predicted by AlphaFold2 [83, 84]. Notable exceptions include intrinsically disordered proteins [86], fold-switching proteins [87], and protein with too few available homologs. Nevertheless, these computational methods have solved more than 200 million protein structures, and they can be downloaded from the AlphaFold Protein Structure Database [88].

AlphaFold2 marks a radical departure from past efforts, which emphasized the predominant role of physical chemistry in folding studies. This machine-learning approach *is not hypothesis driven, nor does it rely on any underlying physical chemistry.* Instead, AlphaFold is based on neural nets, multiple sequence alignment (MSA) from protein homologs, and data from the PDB and its approximately 200,000 protein structures.

Physics, chemistry, and biology were all of central importance during WWII, overwhelmingly so. The somber ashes of war spawned a fresh vision of the role of science in society, captured memorably in Vannevar Bush's forward-looking report to President Roosevelt [89]. With support from government agencies, like the National Institutes of Health and the newly formed National Science Foundation, intertwined research in neuroscience and computation flourished, as exemplified by the McCulloch–Pitts formal model of a neuron [90]. These early beginnings did much to set the stage for the present success [91].

The concept of a neural net is based on (and named for) our understanding of the brain. As described in Wolfram's lucid book:

> In human brains there are about 100 billion neurons (nerve cells), each capable of producing an electrical pulse up to perhaps a thousand times a second. The neurons are connected in a complicated net, with each neuron having tree-like branches allowing it to pass electrical signals to perhaps

> thousands of other neurons. And in a rough approximation, whether any given neuron produces an electrical pulse at a given moment depends on what pulses it's received from other neurons—with different connections contributing with different 'weights'. [92]

In mammals, the brain's neocortex (Hercule Poirot's "little gray cells") is a folded sheet of neurons in six, highly interconnected layers, a real-life neural net. In corresponding deep-learning networks—like AlphaFold or large language models like ChatGPT—such an organization gives rise to a high-dimensional space with many billions of parameters.

Initially, a neural net is trained on a data set, one with sufficiently many examples. During incremental rounds, its parameters are successively updated so as to minimize a loss function that measures how well the training set can be reproduced. AlphaFold2 incorporates many special-purpose neural nets as it trains on data from the PDB [83].

MSA is another critical component in AlphaFold2. MSA alone is sufficient to reconstruct a protein's three-dimensional topology using only the sequences of homologs, provided that an adequate number of homologs is available [93]. In detail, when pairs of residues are observed to coevolve, it is often, but not always, the case that the residues interact chemically and must therefore be proximate in space. The problem is to distinguish valid evolutionary couplings from adventitious correlations. Using maximum entropy methods, this problem was solved by Marks et al. [93]. Once established, a set of true couplings is similar to a set of nuclear Overhauser effects in nuclear magnetic resonance spectroscopy.

Machine learning using neural nets can be likened to the way newborn children learn a language. Through practice, imitation, and empirical input, they acquire language proficiency incrementally, notably without any explicit reference to a finite set of rules, that is, a grammar.

This contrast between machine learning using neural nets and traditional physical science is dramatic. The traditional goal is to explain a broad range of phenomena from a few "first principles" and a correspondingly small number of parameters. Newton's second law, $F = ma$, is a familiar example. On reflection, our cognitive ability to describe complex phenomena in this way seems utterly *unreasonable* [94], but we've been doing so for centuries and have come to expect it. John von Neumann made fun of overfitting data by using too many parameters, saying, "*with four parameters I can fit an elephant, and with five I can make him wiggle his trunk.*" No longer. AlphaFold2 can usually deduce a protein's fold from its sequence but it takes billions of adjustable parameters.

6 Have We Moved Beyond Reductionist Science?

Reductionism his been the cornerstone of scientific thinking since Galileo. The great scientific revolutions, relativity theory, and quantum mechanics, ushered in profound paradigm shifts, but the conviction that an underlying theory could account for complex phenomena remained intact until now.

Machine-learning using neural nets is not grounded in a unifying theory. No hypothesis is under scrutiny. Instead, the goal is to find parameters—often billions of parameters—that can capture the phenomenon under consideration and to then utilize those parameters predictively. It's pattern recognition applied to large scale phenomena, akin to a child learning a language. Undoubtedly, the approach has met with stunning success in multiple venues, but it is not science as we have come to know it.

Is this the end of reductionist science? A suitable neural net could surely reproduce planetary motions from Tycho Brahe's data; Kepler did that. But could the method ever be harnessed to produce the concept of gravitation or, for that matter, a theory of gravitation? And, if not, do we care? Some might even question whether large-scale biological phenomena, both self-modifying and contingent, are compatible with any meaningful theory beyond evolution.

In celebrating the success of AlphaFold2, it is important to realize that it solves the *fold problem*, but not the *folding problem*, that is, the physical process by which an unfolded protein attains its native structure [95]. This is not a failure; indeed, it is not a goal. Here, the process of determining the fold is completely unphysical, although at the very end a force field is invoked to touch-up the final structure.

Pauling deduced the structure of the α-helix before the structure of any protein was known, and here we are today, with ~200 million predicted folds in the AlphaFold Protein Structure Database [88]. Although further work is in the offing, for most intents and purposes the *fold problem* has been solved. Shall we move on, or will the unsolved *folding problem* remain a pressing challenge [95]? Confronted with this conundrum, the very nature of scientific inquiry appears to be poised at a cusp as we now embark upon a liminal era for science.

References

1. Pauling L, Corey RB, Branson HR (1951) The structures of proteins: two hydrogen-bonded helical configurations of the polypeptide chain. Proc Natl Acad Sci USA 37:205–210
2. Corey RB (1948) X-ray studies of amino acids and peptides. Adv Protein Chem 4:385–406
3. Astbury WT, Bell FO (1941) Nature of the intramolecular fold in alpha-keratin and alpha-myosin. Nature 147:696–699
4. Huggins ML (1943) The structure of fibrous proteins. Chem Rev 32:195–218

5. Bragg WL, Kendrew JC, Perutz MF (1950) Polypeptide chain configurations in crystalline proteins. Proc R Soc Lond, Ser A 203:321–357
6. Corey RB, Pauling L (1953) Fundamental dimensions of polypeptide chains. Proc R Soc Lond B Biol Sci 141(902):10–20
7. Kouranov A et al (2006) The RCSB PDB information portal for structural genomics. Nucleic Acids Res 34(Database issue):D302–D305
8. Presta LG, Rose GD (1988) Helix signals in proteins. Science 240(4859):1632–1641
9. Marshall GR, Fang J, Kuster DJ (2010) Bifurcated H-Bonds—The Molten Helix, manuscript
10. Koltun WL (1965) Precision space-filling atomic models. Biopolymers 3(6):665–679
11. Perutz MF (1951) New x-ray evidence on the configuration of polypeptide chains. Nature 167(4261):1053–1054
12. Eisenberg D (2003) The discovery of the alpha-helix and beta-sheet, the principal structural features of proteins. Proc Natl Acad Sci USA 100(20):11207–11210
13. Schulz GE, Schirmer RH (1979) Principles of protein structure. Springer, New York
14. Richardson JS, Richardson DC (1988) Amino acid preferences for specific locations at the ends of alpha-helices. Science 240:1648–1652
15. Argos P, Palau J (1982) Amino acid distribution in protein secondary structures. Int J Pept Protein Res 19(4):380–393
16. Schellman C (1980) The alpha-L conformation at the ends of helices. In: Jaenicke R (ed) Protein folding. Elsevier/North Holland, pp 53–61
17. Baker EN, Hubbard RE (1984) Hydrogen-bonding in globular proteins. Prog Biophys Mol Biol 44:97–179
18. Milner-White EJ (1988) Recurring loop motif in proteins that occurs in right-handed and left-handed forms. Its relationship with alpha-helices and beta-bulge loops. J Mol Biol 199(3):503–511
19. Bork P, Preissner R (1991) On alpha-helices terminated by glycine. 2. Recognition by sequence patterns. Biochem Biophys Res Commun 180(2):666–672
20. Preissner R, Bork P (1991) On alpha-helices terminated by glycine. 1. Identification of common structural features. Biochem Biophys Res Commun 180(2):660–665
21. Dasgupta S, Bell JA (1993) Design of helix ends. Amino acid preferences, hydrogen bonding and electrostatic interactions. Int J Pept Protein Res 41(5):499–511
22. Nagarajaram HA et al (1993) Termination of right handed helices in proteins by residues in left handed helical conformations. FEBS Lett 321(1):79–83
23. Aurora R, Srinivasan R, Rose GD (1994) Rules for alpha-helix termination by glycine. Science 264(5162):1126–1130
24. Munoz V, Blanco FJ, Serrano L (1995) The hydrophobic-staple motif and a role for loop-residues in alpha-helix stability and protein folding. Nat Struct Biol 2(5):380–385
25. Sukumar M, Gierasch LM (1997) Local interactions in a Schellman motif dictate interhelical arrangement in a protein fragment. Fold Des 2(4):211–222
26. Kallenbach NR, Gong Y (1999) C-terminal capping motifs in model helical peptides. Bioorg Med Chem 7(1):143–151
27. Leader DP, Milner-White EJ (2011) The structure of the ends of alpha-helices in globular proteins: effect of additional hydrogen bonds and implications for helix formation. Proteins 79(3):1010–1019
28. Bruch MD, Dhingra MM, Gierasch LM (1991) Side chain-backbone hydrogen bonding contributes to helix stability in peptides derived from an alpha-helical region of carboxypeptidase A. Proteins 10(2):130–139
29. Harper ET, Rose GD (1993) Helix stop signals in proteins and peptides: the capping box. Biochemistry 32(30):7605–7609
30. Lyu PC et al (1993) Capping interactions in isolated alpha helices: position-dependent substitution effects and structure of a serine-capped peptide helix. Biochemistry 32(2): 421–425
31. Zhou HX et al (1994) Alpha helix capping in synthetic model peptides by reciprocal side chain-main chain interactions: evidence for an N terminal "capping box". Proteins 18(1):1–7
32. Seale JW, Srinivasan R, Rose GD (1994) Sequence determinants of the capping box, a stabilizing motif at the N-termini of alpha-helices. Protein Sci 3(10):1741–1745
33. Doig AJ et al (1997) Structures of N-termini of helices in proteins. Protein Sci 6(1):147–155
34. Aurora R, Rose GD (1998) Helix capping. Protein Sci 7:21–38
35. Forood B, Feliciano EJ, Nambiar KP (1993) Stabilization of alpha-helical structures in short peptides via end capping. Proc Natl Acad Sci USA 90(3):838–842
36. Marqusee S, Robbins VH, Baldwin RL (1989) Unusually stable helix formation in short alanine-based peptides. Proc Natl Acad Sci USA 86(14):5286–5290

37. Serrano L, Fersht AR (1989) Capping and alpha-helix stability. Nature 342(6247): 296–299
38. Bell JA et al (1992) Dissection of helix capping in T4 lysozyme by structural and thermodynamic analysis of six amino acid substitutions at Thr 59. Biochemistry 31(14):3590–3596
39. Lyu PC et al (1992) Position-dependent stabilizing effects in alpha-helices - N-terminal capping in synthetic model peptides. J Am Chem Soc 114:6560–6562
40. Serrano L et al (1992) Alpha-helix stability in proteins. I. Empirical correlations concerning substitution of side-chains at the N and C-caps and the replacement of alanine by glycine or serine at solvent-exposed surfaces. J Mol Biol 227(2):544–559
41. Chakrabartty A, Doig AJ, Baldwin RL (1993) Helix capping propensities in peptides parallel those in proteins. Proc Natl Acad Sci USA 90(23):11332–11336
42. Yumoto N et al (1993) Stabilization of alpha-helix in C-terminal fragments of neuropeptide Y. Biochem Biophys Res Commun 196(3): 1490–1495
43. Wang Y, Shortle D (1997) Residual helical and turn structure in the denatured state of staphylococcal nuclease: analysis of peptide fragments. Fold Des 2(2):93–100
44. Das C, Shankaramma SC, Balaram P (2001) Molecular carpentry: piecing together helices and hairpins in designed peptides. Chemistry 7(4):840–847
45. Bang D et al (2006) Dissecting the energetics of protein alpha-helix C-cap termination through chemical protein synthesis. Nat Chem Biol 2(3):139–143
46. Lecomte JTJ, Moore CD (1991) Helix formation in apocytochrome b5: the role of a neutral histidine at the N-cap position. J Am Chem Soc 113:9663–9665
47. Preimesberger MR et al (2015) Helix-capping Histidines: diversity of N-H...N hydrogen bond strength revealed by (2h)JNN scalar couplings. Biochemistry 54(46):6896–6908
48. Sagermann M et al (2002) A test of proposed rules for helix capping: implications for protein design. Protein Sci 11(3):516–521
49. Chou PY, Fasman GD (1974) Conformational parameters for amino acids in helical, beta-sheet, and random coil regions calculated from proteins. Biochemistry 13(2):211–222
50. Koehl P, Levitt M (1999) Structure-based conformational preferences of amino acids. Proc Natl Acad Sci USA 96(22):12524–12529
51. Rost B, Sander C (1993) Prediction of protein secondary structure at better than 70% accuracy. J Mol Biol 232(2):584–599
52. Rost B, Sander C (1993) Improved prediction of protein secondary structure by use of sequence profiles and neural networks. Proc Natl Acad Sci USA 90(16):7558–7562
53. Drozdetskiy A et al (2015) JPred4: a protein secondary structure prediction server. Nucleic Acids Res 43(W1):W389–W394
54. Go N, Go M, Scheraga HA (1968) Molecular theory of the helix-coil transition in polyamino acids, I. *Formulation*. Proc Natl Acad Sci U S A 59(4):1030–1037
55. Zimm BH, Bragg JK (1959) Theory of the phase transition between helix and random coil in polypeptide chains. J Chem Phys 31: 526–535
56. Almeida PF (2015) Proteins: concepts in biochemistry
57. Barrick DE (2018) Biomoleculear thermodynamics. CRC Press
58. Lifson S, Roig A (1961) On the theory of helix-coil transition in polypeptides. J Chem Phys 34:1963–1974
59. Neumaier S et al (2013) Testing the diffusing boundary model for the helix-coil transition in peptides. Proc Natl Acad Sci USA 110(32): 12905–12910
60. Bierzynski A, Kim PS, Baldwin RL (1982) A salt bridge stabilizes the helix formed by isolated C-peptide of RNase a. Proc Natl Acad Sci USA 79(8):2470–2474
61. Richards FM (1958) On the enzymic activity of Subtilisin-modified ribonuclease. Proc Natl Acad Sci USA 44(2):162–166
62. Kim EE et al (1992) Refinement of the crystal structure of ribonuclease S. Comparison with and between the various ribonuclease A structures. Biochemistry 31(49):12304–12314
63. Oas TG, Kim PS (1988) A peptide model of a protein folding intermediate. Nature 336(6194):42–48
64. Brown JE, Klee WA (1971) Helix-coil transition of the isolated amino terminus of ribonuclease. Biochemistry 10(3):470–476
65. Bachmann A et al (2011) Mapping backbone and side-chain interactions in the transition state of a coupled protein folding and binding reaction. Proc Natl Acad Sci USA 108(10): 3952–3957
66. Padmanabhan S et al (1990) Relative helix-forming tendencies of nonpolar amino acids. Nature 344(6263):268–270

67. Padmanabhan S, Baldwin RL (1991) Straight-chain non-polar amino acids are good helix-formers in water. J Mol Biol 219(2):135–137
68. O'Neil KT, DeGrado WF (1990) A thermodynamic scale for the helix-forming tendencies of the commonly occurring amino acids. Science 250:646–651
69. Lyu PC et al (1990) Side chain contributions to the stability of alpha-helical structure in peptides. Science 250(4981):669–673
70. Lyu PC et al (1991) Alpha-helix stabilization by natural and unnatural amino acids with alkyl side chains. Proc Natl Acad Sci USA 88(12): 5317–5320
71. Marqusee S, Baldwin RL (1987) Helix stabilization by Glu-...Lys+ salt bridges in short peptides of de novo design. Proc Natl Acad Sci USA 84(24):8898–8902
72. Merutka G et al (1990) Effect of central-residue replacements on the helical stability of a monomeric peptide. Biochemistry 29(32): 7511–7515
73. Myers JK, Pace CN, Scholtz JM (1997) Helix propensities are identical in proteins and peptides. Biochemistry 36(36):10923–10929
74. Myers JK, Pace CN, Scholtz JM (1997) A direct comparison of helix propensity in proteins and peptides. Proc Natl Acad Sci USA 94(7):2833–2837
75. Creamer TP, Rose GD (1992) Side-chain entropy opposes alpha-helix formation but rationalizes experimentally determined helix-forming propensities. Proc Natl Acad Sci USA 89(13):5937–5941
76. Rose GD et al (2006) A backbone-based theory of protein folding. Proc Natl Acad Sci USA 103(45):16623–16633
77. Fleming PJ, Rose GD (2005) Do all backbone polar groups in proteins form hydrogen bonds? Protein Sci 14(7):1911–1917
78. Banavar JR et al (2004) Unified perspective on proteins: a physics approach. Phys Rev E Stat Nonlinear Soft Matter Phys 70(4 Pt 1):041905
79. Hoang TX et al (2004) Geometry and symmetry presculpt the free-energy landscape of proteins. Proc Natl Acad Sci USA 101(21): 7960–7964
80. Banavar JR, Maritan A (2007) Physics of proteins. Annu Rev Biophys Biomol Struct 36: 261–280
81. Škrbić T et al (2021) Building blocks of protein structures: physics meets biology. Phys Rev E
82. Haas DJ (1970) Polypeptide alpha-carbon models: a new concept in protein model construction. Biopolymers 9(12):1547–1552
83. Jumper J et al (2021) Highly accurate protein structure prediction with AlphaFold. Nature 596(7873):583–589
84. Tunyasuvunakool K et al (2021) Highly accurate protein structure prediction for the human proteome. Nature 596(7873):590–596
85. Baek M et al (2021) Accurate prediction of protein structures and interactions using a three-track neural network. Science 373(6557):871–876
86. Ruff KM, Pappu RV (2021) AlphaFold and implications for intrinsically disordered proteins. J Mol Biol:167208
87. Chakravarty D, Porter LL (2022) AlphaFold2 fails to predict protein fold switching. Protein Sci 31(6):e4353
88. Varadi M et al (2022) AlphaFold protein structure database: massively expanding the structural coverage of protein-sequence space with high-accuracy models. Nucleic Acids Res 50 (D1):D439–D444
89. Bush V (1945) Science: the endless frontier. In: A report to the president
90. McCulloch W, Pitts W (1943) A logical calculus of the ideas immanent in nervous activity. Bull Math Biopys 5:115–133
91. Sejnowski TJ (2020) The unreasonable effectiveness of deep learning in artificial intelligence. Proc Natl Acad Sci USA 117(48): 30033–30038
92. Wolfram S (2023) What is ChatGPT doing and why does it work? Wolfram Media, Inc
93. Marks DS et al (2011) Protein 3D structure computed from evolutionary sequence variation. PLoS One 6(12):e28766
94. Wigner EP (1960) The unreasonable effectiveness of mathematics in the natural sciences. Commun Pure Appl Math 13:1–14
95. Chen SJ et al (2023) Opinion: protein folds vs. protein folding: differing questions, different challenges. Proc Natl Acad Sci USA 120(1):e2214423119

Chapter 2

Secondary Structure in Free and Assisted Modeling of Proteins with the Coarse-Grained UNRES Force Field

Emilia A. Lubecka, Cezary Czaplewski, Adam K. Sieradzan, Agnieszka G. Lipska, Łukasz Dziadek, and Adam Liwo

Abstract

Secondary structure is a solid scaffold on which the three-dimensional structure of a protein is built. Therefore, care must be taken to reproduce the secondary structure as accurately as possible in the simulations of protein systems. In this chapter, we summarize the physics-based energy terms that govern secondary-structure formation, the auxiliary restraints on secondary structure derived from bioinformatics and from the experimental data, and the role of those in the modeling of protein structures, dynamics, and thermodynamics with the physics-based coarse-grained UNRES force field. Examples illustrating the methodology discussed and further directions of development are presented.

Key words UNRES force field, Secondary structure, Physics-based model, Energy function, PSIPRED, Virtual-bond angles, NMR-based restraints

1 Introduction

The very first notion that proteins are three-dimensional objects that consist of regular building blocks assembled into purposeful architectures came with the discovery of helical and sheet structures by Pauling and Corey more than 70 years ago [1, 2]. This discovery truly revolutionized the field of protein science, since globular proteins were until then regarded as structureless objects, the ellipsoidal outline considered sufficient to describe their shapes and structures [3]. Early X-ray experiments suggested that regular structures were present only in fibrous proteins such as keratin [4] or fibroin [5].

Starting from the determination of the three-dimensional structure of hemoglobin [6], it became apparent that regular secondary-structure scaffold is present in globular proteins. Secondary-structure formation is also thought to initiate protein folding according to the diffusion and collision mechanism

Andrzej Kloczkowski et al. (eds.), *Prediction of Protein Secondary Structure*, Methods in Molecular Biology, vol. 2867,
https://doi.org/10.1007/978-1-0716-4196-5_2,

[7]. Therefore, the methods for the modeling of protein structure, as well as dynamics, thermodynamics, and interactions, must properly account for the formation and stability of secondary structure.

The formation of secondary structure is mainly a result of the interplay of backbone–hydrogen-bonding and backbone–local interactions. These can be modeled at the all-atom level with classical force fields, even though it took quite some time to tune the nonbonded and torsional potentials to get a good balance between the α-helical and β-structures [8–16]. While the coarse-grained models enable us to extend the time and size scales of the structures and events covered tremendously [17–23], they also pose a bigger challenge regarding the accounting for the formation of regular secondary structure compared to the all-atom models. A way around is to impose biasing potentials on the backbone dihedral angles [24] or restraints on the distances [19, 21]. On the other hand, very simple coarse-grained models, such as the tube model of protein backbone [25] or the model of coins threaded on the backbone [26], reproduce basic secondary-structure motifs surprisingly well. The excellent performance of the two models mentioned above seems to stem from the proper consideration of the geometric constraints inherent in polypeptide chains, which narrow down both local structure and possible hydrogen-bond geometries, thus largely pre-sculpting protein-energy landscape [25, 26].

In this chapter, we discuss the handling of protein secondary structure in the UNited RESidue (UNRES) physics-based model of polypeptide chains [27–29] developed in our laboratory, for a long time in collaboration with the late Harold A. Scheraga, Cornell University. In Subheading 2, we outline the model and the respective effective energy function. In Subheading 3, we discuss the terms of the UNRES energy function that are responsible for the formation of regular helical and sheet structures and point to the essential role of the energy terms that account for the coupling of the backbone–local and backbone–electrostatic interactions at the coarse-grained level in proper secondary-structure modeling. In Subheading 4, we discuss the secondary-structure restraints in UNRES that are obtained with the bioinformatics methods. In Subheading 5, we discuss the restraints on secondary structure from Nuclear Magnetic Resonance (NMR) experiments. Finally, in Subheading 6, we summarize the developments and outline future directions in secondary-structure modeling with UNRES.

2 UNRES Model of Polypeptide Chains

UNRES is a heavily coarse-grained model of polypeptide chains, in which each residue is reduced to the united peptide group (p) and united sidechain (SC). The geometry of the polypeptide backbone is defined by a sequence of α-carbon (C^{α}) atoms, which are not interaction sites [27–29]. Each peptide group is located in the

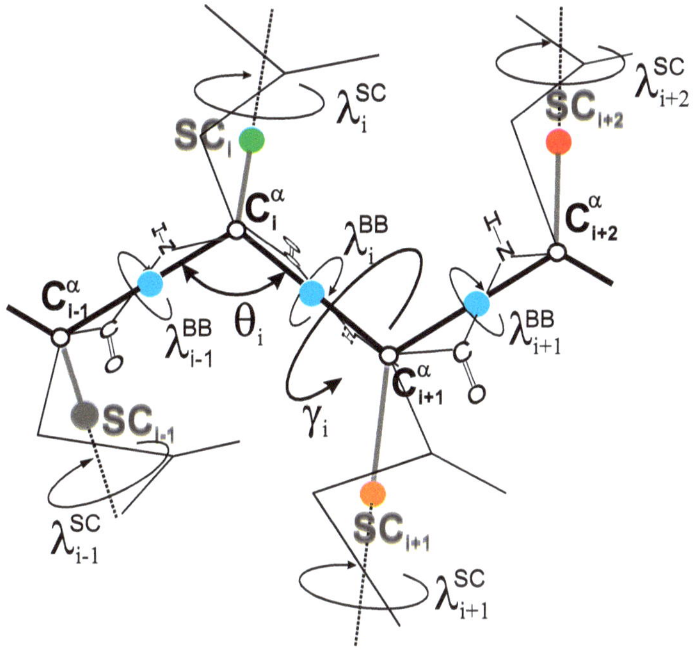

Fig. 1 Illustration of the UNRES coarse-graining scheme with a fragment of polypeptide chain from residue $i-1$ through $i+2$. The interaction sites are united peptide groups (p), each located halfway between the two consecutive C^α atoms, and united sidechains (SC), each attached to the corresponding C^α atom. The C^α atoms are not interaction sites. The geometry of the coarse-grained chain is defined by the C^α and SC coordinates. All peptide groups are assumed in the *trans* configuration, implying the equilibrium value of the $C^\alpha\cdots C^\alpha$ virtual-bond length of about 3.8 Å. The backbone geometry of an n-residue polypeptide chain can thus also be defined in terms of the virtual-bond angles $\theta_2, \theta_3, \ldots, \theta_{n-1}$ (each indexed after the respective C^α at its vertex) and virtual-bond dihedral angles $\gamma_2, \gamma_3, \ldots, \gamma_{n-2}$ (each indexed after the first C^α atom at its axis). The coarse-grained energy function of the chain is obtained by integrating the Boltzmann factors in the angles λ_i^{BB}, $i = 1, 2, \ldots, n-1$ and λ_i^{SC}, $i = 2, 3, \ldots, n-1$, which are the angles for the rotation of the peptide and the sidechain groups (whose chemical bonds are shown as thin lines) about the respective $C^\alpha\cdots C^\alpha$ or $C^\alpha\cdots$ SC virtual-bond axes, and over the solvent degrees of freedom (not shown)

middle between the two consecutive C^α atoms, and each sidechain is attached to the respective C^α atom. The united peptide groups represent only the carboxyamide-group atoms, and the united sidechain encompasses the C^α and H^α atoms, along with the sidechain atoms. The solvent is considered in an implicit manner. The model is depicted in Fig. 1.

The UNRES effective energy function has been derived by using a bottom-up approach. It originates from the potential of mean force (PMF) of polypeptide chains in water, in which the degrees of freedom not present in the model are integrated out

[30, 31]. These are the solvent degrees of freedom and the internal degrees of freedom of the chain restricted to a given coarse-grained geometry. Of the latter, the collective rotation angles λ of the peptide groups and those of the sidechains about the respective virtual-bond axes are the most important; these angles are indicated in Fig. 1. In our approach [30, 31], the PMF has been expressed by a truncated series of Kubo cluster-cumulant functions [32], each of which can be identified with a certain kind of energy term. A cluster-cumulant function is, in turn, expanded into the Kubo cluster cumulant series, the first nonvanishing term of the series constituting the analytical approximation to the respective energy term [30, 31]. These approximations are constructed based on the recently developed scale-consistent theory of coarse graining [31], which enables us to embed the atomic details in the effective energy function.

The UNRES effective energy function is expressed by Eq. (1).

$$\begin{aligned} U = {} & w_{SC}\sum_{i<j} U_{SC_iSC_j} + w_{SCp}\sum_{i\neq j} U_{SC_ip_j} + w_{pp}^{VDW}\sum_{i<j-1} U_{p_ip_j}^{VDW} \\ & + w_{pp}^{el} f_2(T)\sum_{i<j-1} U_{p_ip_j}^{el} + w_{tor} f_2(T)\sum_i U_{tor}(\gamma_i, \theta_i, \theta_{i+1}) \\ & + w_b\sum_i U_b(\theta_i) + w_{rot}\sum_i U_{rot}(\theta_i, \hat{\mathbf{r}}_{SC_i}) \\ & + w_{bond}\sum_i U_{bond}(d_i) + w_{ssbond}\sum_i U_{ssbond}(d_i^{SS}) \\ & + w_{corr}^{(3)} f_3(T)\sum_{i<j-1} U_{corr;\,ij}^{(3)} + w_{turn}^{(3)} f_3(T)\sum_i U_{turn;\,i}^{(3)}. \end{aligned} \tag{1}$$

The first four terms in UNRES correspond to pairwise site–site interactions. The terms $U_{SC_iSC_j}$ are sidechain–sidechain interaction potentials [33, 34]. They are represented by the modified Gay–Berne functional forms [35] and, therefore, have the axial and not the spherical symmetry. The terms $U_{SC_ip_j}$ are spherically symmetric excluded-volume potentials that prevent the collapse of the united sidechains on the backbone. The terms $U_{p_ip_j}^{VDW}$ and $U_{p_ip_j}^{el}$ are the nonbonded and mean-field–electrostatic interaction potentials, respectively, of united peptide groups, $U_{p_ip_j}^{VDW}$ having the spherical and $U_{p_ip_j}^{el}$ the axial symmetry, respectively.

The next three terms account for local interactions. The U_{bond} terms, d_i in Eq. (1) being the length of the ith virtual bond, are the bond-deformation potentials, U_b and U_{tor} are the backbone–virtual-bond angle and the backbone–virtual-bond torsional potentials, respectively, θ_i and γ_i in Eq. (1) denoting the backbone–virtual-bond and backbone–virtual-bond dihedral angles, respectively (Fig. 1). It should be noted that the U_{tor} terms depend not only on the virtual-bond dihedral angles but also on virtual-bond angles, which feature results from the application of the scale-consistent methodology to the derivation of these potentials [31, 36]. The U_{rot} terms are the sidechain-rotamer potentials, in which $\hat{\mathbf{r}}_{SC_i}$ denotes the local coordinates of the unit vector pointing from C_i^α to SC_i [37]. The additional bonding terms, U_{ssbond}, account for the energetics of disulfide bonds, including their formation and breaking [38, 39]. These bonding terms are nonlocal.

The last two terms, $U_{corr}^{(3)}$ and $U_{turn}^{(3)}$, respectively, are the multibody terms that account for the coupling of the backbone–local and backbone–electrostatic interactions [30, 31]; these terms are the major players in reproducing regular secondary structures. Although the $U_{corr}^{(3)}$ and $U_{turn}^{(3)}$ terms are technically pairwise, they encompass interactions not only between two peptide groups as the $U_{p_ip_j}^{el}$ terms do, but also include the local interactions with the surrounding backbone [30, 31]. The $U_{corr}^{(3)}$ and $U_{turn}^{(3)}$ terms have a lower than axial symmetry, because they not only depend on the distance between the peptide groups and the orientation of the respective $C^\alpha \cdots C^\alpha$ virtual-bond axes but also on the virtual-bond angles θ and virtual-bond dihedral angles γ of the two interacting backbone sections [30, 31].

The *w*s in Eq. (1) are the weights of the energy terms. These weights have been determined, along with some other parameters, by maximum-likelihood calibration of the force field [40], which is the final top-down step to complete UNRES force-field parameterization.

The factors $f_n(T)$ in Eq. (1) account for the dependence of those contributions to the UNRES effective energy function that correspond to higher-order terms in the Kubo cluster-cumulant expansion on temperature [41], as given by Eq. (2).

$$f_n(T) = \frac{\ln[\exp(1) + \exp(-1)]}{\ln\left\{\exp\left[(T/T_\circ)^{n-1}\right] + \exp\left[-(T/T_\circ)^{n-1}\right]\right\}}, \qquad (2)$$

where $T_\circ = 300$ K. It should be noted that the temperature dependence is a direct consequence of the fact that the UNRES energy function approximates the PMF (a temperature-dependent quantity) of polypeptide chains in water. The energy terms affected by the temperature factors correspond to the formation of ordered secondary structure. These terms decay with increasing temperature, in agreement with the experimental observation that secondary structure melts at higher temperatures. This point will be recalled in Subheading 3.

3 UNRES Energy Terms to Shape Secondary Structure

In the early versions of UNRES, backbone hydrogen bonding was represented by the mean-field $U_{p_ip_j}^{el}$ terms, while the torsional terms alone (U_{tor}) were responsible for local backbone structure [27]. The analytical formula for $U_{p_ip_j}^{el}$ was obtained by representing the peptide-group electrostatic interactions by dipole–dipole interactions. Based on this model, analytical expressions for the first and the second energy moments were derived by averaging over the angles of rotation λ of the peptide group about their axes (Fig. 2) [42]. These analytical expressions were subsequently parameterized [27] based on the energy surface of peptide-group interactions calculated with the ECEPP/3 all-atom force field [9]. Later, the parameterization was revised by using the ab initio-calculated energy surface [43]. The developed formulas reproduce very well the dependence of the mean-field energy (averaged over the rotations of the peptide group about the $C^{\alpha} \cdots C^{\alpha}$ axes) of two interacting peptide groups on orientation, having the global minimum when the peptide-bond axes are stacking onto each other (Fig. 2). The early UNRES (with the $U_{p_ip_j}^{el}$ terms) scored the best prediction of the structure of the HDEA protein (target T0061; an all-α-helical protein) in the Third Community Wide Experiment on the Critical Assessment of Techniques for Protein-Structure Prediction (CASP3) [44].

Even though peptide-group alignment is better for the peptide groups engaged in a β-sheet compared to those engaged in an α-helix, the number of peptide-group contacts in an α-helix is bigger compared to that in a β-sheet with the same chain length. Consequently, the early UNRES could only model α-helical proteins without imposing any restraints [27, 45, 46] (Fig. 3a). The deficiency of the force field could not be rectified by balancing U_{tor} and $U_{p_ip_j}^{el}$ or even by introducing the coupling terms between pairs of interacting peptide groups [48], which were similar to the terms introduced by Kolinski in his CABS statistical force field [49, 50], based on the statistics of the proteins structures derived from the Protein Data Bank (PDB) [51]. The only benefit from introducing the new coupling terms was better accuracy of the modeling of α-helical sections [48].

The difficulties of early UNRES in modeling β-sheet structures can be explained based on the Ramachandran maps [52, 53]. The left-handed extended-strand conformation of a single L-amino-acid residue forms a broad region in the map, while the α-helical conformation corresponds to a shallow region. Thus, locally the extended structures are preferable to α-helical ones. Moreover, the peptide groups of the extended strands are positioned nearly perpendicularly to the zigzag formed by a strand, which makes them ready to form hydrogen bonds with those of the neighboring

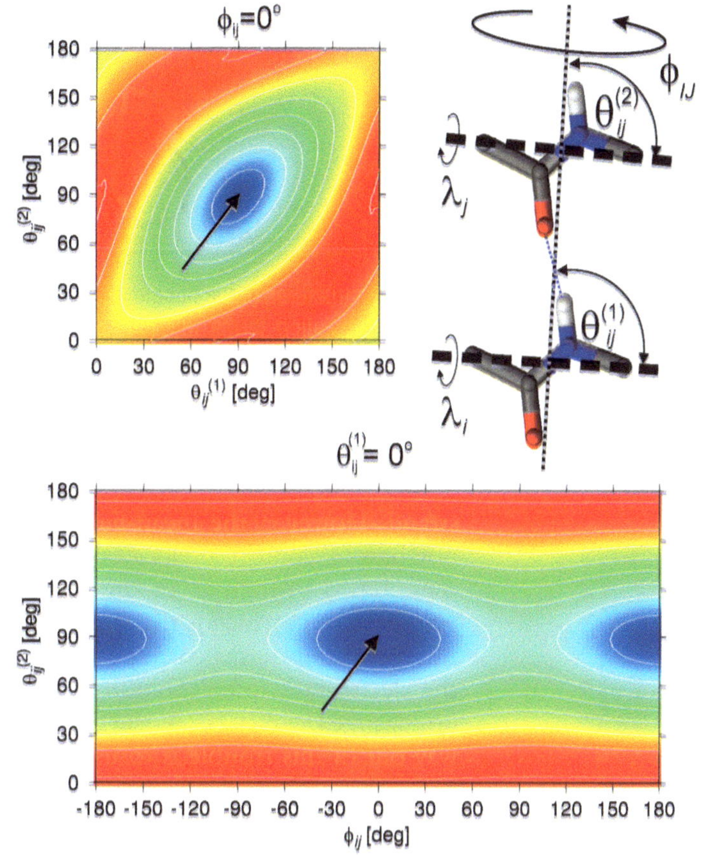

Fig. 2 Contour plots of the $U^{el}_{p_i p_j}$ UNRES energy term in the angles that define peptide-group orientation of the $C^{\alpha} \cdots C^{\alpha}$ virtual-bond axes (shown as thick dashed lines in the upper right part of the figure). The peptide-group centers are at the distance of 7.5 Å. The angle $\theta^{(1)}_{ij}$ is that of the axis of peptide group *i* with the line linking peptide-group centers. The angle $\theta^{(2)}_{ij}$ is that of the axis of peptide group *j* with the line linking peptide-group centers. The angle ϕ_{ij} is the rotation angle of peptide group *i* about the axis linking the peptide-group centers. The upper-left picture shows the section of $U^{el}_{p_i p_j}$ in the $(\theta^{(1)}_{ij}, \theta^{(2)}_{ij})$ plane for $\phi_{ij} = 0°$, while the lower panel shows the section of $U^{el}_{p_i p_j}$ in the $(\theta^{(2)}_{ij}, \phi_{ij})$ plane for $\theta^{(1)}_{ij} = 90°$. The λ_i and λ_j angles, which have been averaged over to get the mean-field $U^{el}_{p_i p_j}$ term, are also shown in the upper-right part of the figure. The energy scale runs from blue (− 1 kcal/mol) to red (− 0.2 kcal/mol) with the increment of 0.1 kcal/mol. The arrows point to the energy minima, which correspond to the stacking configuration of the peptide-groups axes, as observed in β-sheets. It can also be seen that concerted changing of the $\theta^{(1)}_{ij}$ and $\theta^{(2)}_{ij}$ angles or rotating the virtual-bond axes about the line linking peptide-group centers perpendicular to this line results in small energy increase, consistent with the experimental helical and sheet structures, in which such distortions are observed

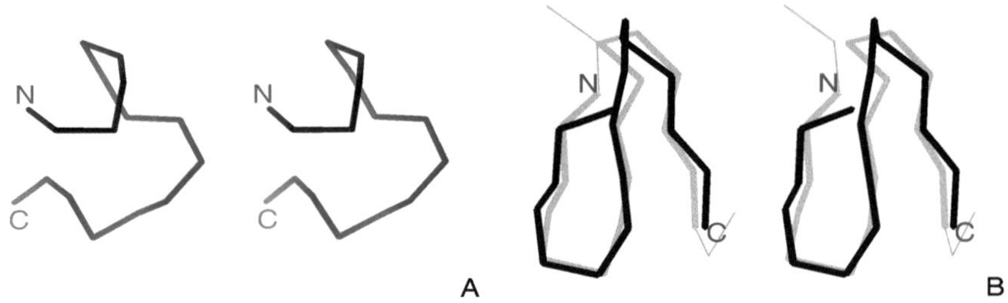

Fig. 3 Stereoview of the lowest-energy structures of the betanova miniprotein [47] calculated with the UNRES force field without inclusion of the correlation terms to couple the backbone–local with the backbone–electrostatic interaction (**a**) and stereoview of the superposition of the structure of betanova calculated with inclusion of these correlation terms (light gray lines) on the NMR structure (dark gray lines), the C^{α} RMS deviation being 1.2 Å (**b**). The N- and C-termini are indicated for tracing purposes (Reproduced from *J. Chem. Phys.* **115**, 2323 (2001), with the permission of AIP Publishing)

strand. However, the mean-field $U_{p_ip_j}^{el}$ UNRES energy terms do not have a context of local chain conformation. Consequently, these terms strongly favor the structures with more contacts per peptide group, which is the feature of α-helical structures. Adding the correlation terms corresponding just to the coupling between the electrostatic interactions, as we attempted initially [48], could not change the picture. At the first thought it appears that the only way out of this trouble is to use explicit polypeptide backbone in the coarse-grained model (as in the AWSEM coarse-grained force field [54]) or, at least, to place explicit dipoles [55] or oppositely charged sites [56] on the peptide groups. All these ways result in increasing the number of interaction sites and of the number of the degrees of freedom.

Looking for the improvement of the quality of secondary structure modeled with UNRES without making the backbone representation more detailed, we started to develop the UNRES force field systematically [30], by using the Kubo cluster-cumulant formalism [32]. As a result, we derived the analytical formulas the third-order correlation terms [$U_{corr}^{(3)}$ and $U_{turn}^{(3)}$ in Eq. (1)] that couple the backbone–local and backbone–electrostatic interactions [30]. We found that these terms can be expressed in terms of interactions of fictitious dipoles located on the peptide groups, with the magnitudes and directions of the dipoles dependent on the backbone–virtual-bond dihedral angles γ, the central axes of which are the axes of the interacting peptide groups (Fig. 4) [43]. The fictitious dipole moment of the *i*th peptide group is a vector sum of the components associated with the $C_{i-1}^{\alpha} \cdots C_i^{\alpha} \cdots C_{i+1}^{\alpha}$ frame and that associated with the $C_i^{\alpha} \cdots C_{i+1}^{\alpha} \cdots C_{i+2}^{\alpha}$ frame. The first frame corresponds to the residue ending with the peptide group and the second one to the residue beginning with it. The components of each dipole moment are determined by the Ramachandran maps of residue with indices i and $i + 1$, respectively [30]. Consequently,

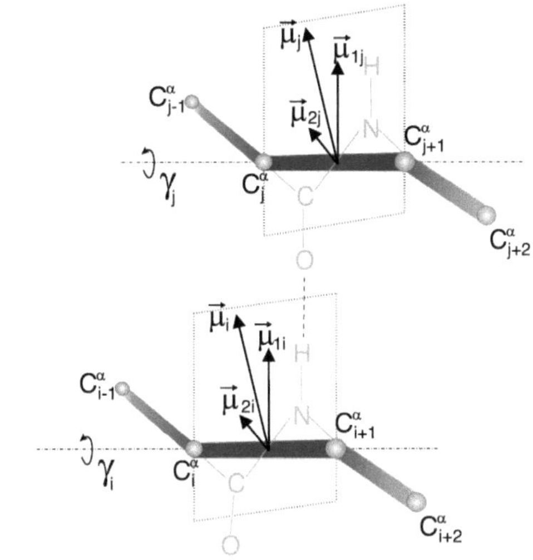

Fig. 4 Illustration of the $U^{(3)}_{corr}$ UNRES energy term that couples the electrostatic interactions of peptide groups p_i and p_j with the local interactions within the residues adjacent to them [30, 31]. The coupling term can be expressed as the energy of interaction of two fictitious dipoles $\boldsymbol{\mu}_i$ and $\boldsymbol{\mu}_j$ (shown as arrows along with their components), representing the peptide groups. The peptide groups, their planes aligned with the dipole moments, are shown with light-gray lines and atom symbols. The components $\boldsymbol{\mu}_{1i}$ and $\boldsymbol{\mu}_{2i}$, respectively, of the dipole moment $\boldsymbol{\mu}_i$, are defined in the C^{α}_{i-1}, C^{α}_i, C^{α}_{i+1} and C^{α}_i, C^{α}_{i+1}, C^{α}_{i+2} local frames, respectively, and those of $\boldsymbol{\mu}_j$ are defined likewise. The local frames encompass the residues preceding and following the peptide group p_i and p_j, respectively. The dipole-moment components depend on the average features of the Ramachandran maps of the residues preceding and following the interacting peptide groups [30, 31]. The dipole-moment components are rotated with respect to each other from their local-coordinate systems about the $C^{\alpha}_i \cdots C^{\alpha}_{i+1}$ and $C^{\alpha}_j \cdots C^{\alpha}_{j+1}$, axes, respectively, by the angles γ_i and γ_j, respectively. Their vector sums (after this rotation) constitute the dipole moments $\boldsymbol{\mu}_i$ and $\boldsymbol{\mu}_j$, respectively, thus making the dipole moments and, consequently, the $U^{(3)}_{corr}$ term dependent on local chain conformation (Adapted with permission from A. Liwo et al., *J. Phys. Chem. B* **108**, 9421–9438 (2004). Copyright 2004 American Chemical Society)

the $U^{(3)}_{corr}$ and $U^{(3)}_{turn}$ terms provide the coupling between the backbone–local and backbone–electrostatic interactions [30]. The cluster-cumulant formalism also enabled us to revise the parameterization of the torsional potentials to define them as the second-order correlation terms corresponding to the coupling of the Ramachandran maps of two adjacent residues [30], without having to calculate the energy map of the whole dipeptide and subtract the pairwise-interaction terms first, as in our early work on UNRES [27].

The new correlation terms, along with the torsional terms, were parameterized based on the ECEPP/3 [9] all-atom energy surfaces of model systems first [30] and, subsequently, on ab initio energy surfaces [43, 57]. In this intermediate stage of UNRES development, we also introduced the fourth-, fifth-, and sixth-order correlation terms [30, 43], which later turned out unnecessary. Therefore, these terms are not present in the present UNRES energy function given by Eq. (1). As shown in Fig. 3b, in which the UNRES-modeled structure of a three-stranded β-sheet miniprotein is superposed on the respective experimental structure [47], introducing the correlation terms enabled us to model the β structures [30].

With the third- and higher-order coupling terms, UNRES was able to model the structures of proteins with all structural classes, as demonstrated in the CASP4 experiment [58]. However, it was not possible to develop a single version of UNRES to cover all structural classes. One version was developed for all-α proteins [59], another one for all-β, and yet another for the $\alpha+\beta$-proteins [60]. Later, with the addition of double-torsional potentials [57, 61] and supplementary torsional potentials dependent on the dihedral angles involving united sidechains [62], as well as owing to the introduction of the maximum-likelihood method of force-field calibration [63], we managed to develop a version of UNRES to treat proteins with all structural classes [64]. This version of UNRES was successful in the CASP11 experiment [65]. However, the β-strands were too much compressed, which resulted in reduced modeling accuracy of the β-sheet sections.

The problems mentioned above were caused by the dependence of the correlation and torsional terms only on the backbone–virtual-bond dihedral angles γ and disregarding that on the backbone–virtual-bond angles θ (Fig. 1). The latter are close to 90° for α-helical and extended for the β-strand sections of the structure [66]. With the development of the scale-consistent theory of coarse graining [31], we derived the formulas for the $U_{corr}^{(3)}$, $U_{turn}^{(3)}$, and U_{tor} terms that include the dependence on the virtual-bond angles θ. Their most important feature is that the trigonometric functions of dihedral angles with multiplicity n are multiplied by the nth powers of the sines of the adjacent virtual-bond angles θ [31]. Consequently, if a fragment of the chain is nearly linear, the respective energy terms tend to zero instead of becoming numerically unstable or undefined, as happens without this dependence. The fictitious dipole moments shown in Fig. 4 now depend on the backbone–virtual-bond angles θ, thus enabling their alteration for α-helical and β-strand sections [31]. After maximum-likelihood calibration, the new variant of the UNRES force field reproduced the local geometry of both helices and strands

[40]. The force field proved to be predictive both with benchmark proteins of different structural classes [40] and in the CASP13 experiment [67].

With the recent improvements of the correlation and torsional potentials, UNRES reproduces the details of protein secondary structure reasonably well. With moderate restraints, as described in Subheadings 4 and 5, it can be used with confidence in modeling protein structure, dynamics, and thermodynamics. However, it can predict the structures of only comparatively small proteins in the ab initio mode. As suggested in our report of UNRES performance in the CASP14 experiment [68], this deficiency could be caused by the absence of long-range coupling terms. This observation once more points to the importance of properly accounting for the stability of secondary structure, which shapes the backbone scaffold, in correct modeling of the tertiary structure of proteins. Recently [69], we derived the formulas for the multitorsional potentials and found that these stabilize extended segments of the chain and determine the directions of the chain preceding or following helical segments. Subsequently [70], we revised the formulas for the torsional potentials to split them into the backbone-only potentials, those involving sidechains, and the improper torsional potentials. This new description covers the old torsional potentials, the supplementary sidechain torsional potentials introduced in our earlier work [62], and the sidechain local potentials [U_{rot} in Eq. (1)]. Both features are under implementation.

As follows from Eq. (1), all terms responsible for secondary-structure formation (i.e., $U^{el}_{p_ip_j}$, U_{tor}, $U^{(3)}_{corr}$, and $U^{(3)}_{turn}$) are multiplied by the temperature factors, which makes them decrease with increasing temperature. Before this dependence was introduced, the folding-transition temperatures obtained from coarse-grained molecular dynamics simulations with UNRES were nonphysically high [17]. Attempts at force-field optimization to get physical folding-transition temperatures were unsuccessful except for simple β-proteins, in which the formation of secondary and that of tertiary structure occur simultaneously [41]. For helical proteins, two clearly separated transitions were observed in simulations: a higher-temperature one to form helices and a lower-temperature one to pack them [41]. The introduction of the temperature factors enabled us to develop a force field, with which the folding–unfolding transition of simple proteins is a concerted process [41], as observed experimentally [71].

4 Bioinformatics-Based Restraints on Secondary Structure

Even with the highly coarse-grained UNRES model, full-blown ab initio folding simulations of the structure of big proteins and protein assemblies are impractical, because they would take ages to get to the folded structure. Moreover, none of the protein force fields, be it all-atom or coarse-grained, including UNRES, does yet have full ab initio folding capacity. Consequently, restraining at least a part of the system studied appears a necessity. In this section we summarize the bioinformatics-based restraints on secondary structure in UNRES.

The primary variables to restraint the secondary structure are the backbone–virtual-bond dihedral angles γ shown in Fig. 1. In our first attempt, we restrained the helical and extended strand sections by using a quartic penalty function on the virtual-bond dihedral angles, as expressed by Eq. (3) [65, 72].

$$V_s(\gamma) = \begin{cases} \frac{1}{4}\mathrm{mod}(\gamma - \gamma_l, 2\pi)^4 & \text{for } \mathrm{mod}(\gamma - \gamma_l, 2\pi) < 0 \\ 0 & \text{for } \mathrm{mod}(\gamma - \gamma_l, 2\pi) \geq 0 \\ & \text{and } \mathrm{mod}(\gamma - \gamma_u, 2\pi) \leq 0 \\ \frac{1}{4}\mathrm{mod}(\gamma - \gamma_u, 2\pi)^4 & \text{for } \mathrm{mod}(\gamma - \gamma_u, 2\pi) > 0, \end{cases} \tag{3}$$

where $\gamma_l = 40°$, $\gamma_u = 60°$ to restraint a virtual-bond dihedral angle to a helical conformation, and $\gamma_l = 140°$, $\gamma_u = 220°$ to restrain γ to an extended conformation.

The main source of the restraints is secondary-structure prediction; we use PSIPRED [73] for this purpose. The respective restraint is assigned if PSIPRED predicts a helical state for two consecutive residues or an extended state for two consecutive residues. The C^α atoms of these residues are then on the axis of the restrained virtual–bond dihedral angle γ. No restraints are imposed if each of the two consecutive residues is predicted to be in a different state or if both are predicted to be in the coil state.

The final output from PSIPRED, which is usually taken for further analysis, points to the most probable secondary-structure state of a residue. The probability of alternative states is usually low, but, sometimes, even all three states (helix, extended, and coil) have comparable PSIPRED-predicted probabilities. This more detailed information is available if the raw ASCII-format output from PSIPRED is used. Therefore, we subsequently developed a bimodal restraint function that takes into account ambiguous secondary-structure assignment, as expressed by Eq. (4) [67].

$$\begin{aligned} V_{dih}(\gamma_i) = & -\ln\left\{ P_i^{coil} P_{i+1}^{coil} + P_i^{hel} P_{i+1}^{hel} \exp\left[-\frac{1-\cos(\gamma_i - \overset{\circ}{\gamma}_{hel})}{\sigma_{hel}^2} \right] \right. \\ & \left. + P_i^{ext} P_{i+1}^{ext} \exp\left[-\frac{1-\cos(\gamma_i - \overset{\circ}{\gamma}_{hel})}{\sigma_{ext}^2} \right] \right\}, \end{aligned} \tag{4}$$

where P_i^X is the probability for a residue to occur in state X, where X is *ext* (for extended), *hel* (for helical), or *coil*, while γ_X° and σ_X are the center and the standard deviation of the cosine-Gaussian (which approaches the regular Gaussian for small $\gamma_i - \gamma_X$) corresponding to state X; we took $\gamma^\circ_{hel} = 50^\circ$, $\sigma_{hel} = 30^\circ$, $\gamma^\circ_{ext} = 180^\circ$, $\sigma_{ext} = 40^\circ$.

It should be noted that no cosine-Gaussian corresponds to the coil state; however, the minima corresponding to the extended and helical states become more shallow as the weight of the coil state increases. Two minima appear when PSIPRED assigns comparable weights to the helix and extended state. The potential presence of helix and extended-structure minima prevents the restraint function from fixing a chain fragment in a secondary structure which is only marginally more stable with respect to the alternative one.

Restraints on virtual-bond dihedral angles are sufficient for the α-helical but not for the β-sheet sections, which are nonlocal. Here the restraints from residue–residue contact prediction can be used. We have introduced a Lorentz-like bounded flat-bottom penalty function, which is defined by Eq. (5). Such restraints are imposed on the distances between the C^α atoms or sidechain centers of the residues for which contacts were predicted [72].

$$V_c(d, d_l, d_u, A) = \begin{cases} A\dfrac{(d-d_l)^4}{\sigma^4 + (d-d_l)^4} & \text{for } d < d_l \\ 0 & \text{for } d_l \leq d \leq d_u \\ A\dfrac{(d-d_u)^4}{\sigma^4 + (d-d_u)^4} & \text{for } d > d_u, \end{cases} \tag{5}$$

where d is the distance in the computed structure, d_l and d_u are the lower and upper contact-distance boundaries, respectively, σ is the thickness of the transition region between zero and maximum restraint height, and A is the height of the restraint well, which we assume to be equal to the confidence of a contact. The penalty function does have the upper boundary, which feature results in zero gradient if a restraint is grossly violated. This feature is important for the often encountered case of contradictory contact-distance restraints. We tested the conformational-search algorithm with simulated contact-distance restraints and found that even 50% of false restraints do not deteriorate its performance [72].

The introduction of secondary-structure restraints (both on virtual-bond dihedral angles and on residue–residue contacts), supplemented by other residue–residue contact restraints, improves the quality of the modeled structures [72]. This observation is illustrated in Fig. 5, in which the bar plots of the Global Distance Test Total Score (GDT_TS) [74] of the best models obtained with unrestrained UNRES simulations are compared with those of UNRES models obtained with dihedral-angle restraints, contact restraints, and combination thereof. GDT_TS is a measure of the fraction of a model that is similar to the experimental structure [74].

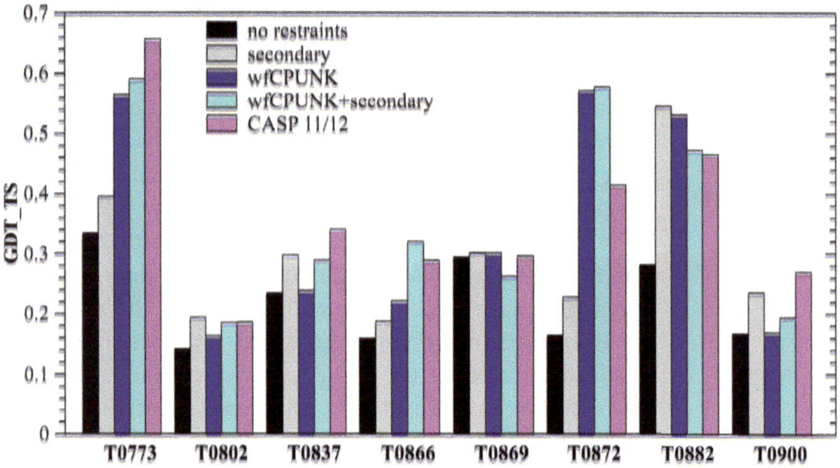

Fig. 5 Bar plots of the GDT_TS [74], for the best models obtained by the multiplexed replica exchange molecular dynamics (MREMD) simulations of the CASP11/CASP12 targets T0773, T0802, T0837, T0866, T0869, T0872, T0882, and T0900 with various sets of distance restraints: no restraints (black), restraints from secondary-structure prediction by PSIPRED (secondary; gray), distance restraints used by the wfCPUNK group (wfCPUNK; purple), and both distance and secondary-structure restraints (wfCPUNK+secondary; cyan). The results obtained by the wfCPUNK group in the CASP11 and CASP12 experiments are shown in pink (Reproduced with permission from E.A. Lubecka and A. Liwo, *J. Comput. Chem.*, **2019**, 40, 2164–2178. Copyright 2019 John Wiley & Sons)

The restraints on secondary structure can also be extracted from the models of a given protein obtained with various bioinformatics methods. This approach covers both the local (angular) and long-range restraints. We developed a consensus-fragment approach to protein-structure modeling [75, 76], in which restraints on the distances, virtual-bond angles, virtual-bond dihedral angles, and sidechain orientation with respect to the backbone derived from similar fragments found in bioinformatics-based models of a given protein are used. The restraints have the form of multi-log-Gaussians, as given by Eq. (6).

$$V_x = -\sum_i \ln\left\{ M_i^{excl} + \sum_{k=1}^{M_i^{incl}} \exp\left[-\frac{1}{2}\left(\frac{x_i - x_i^{(m_k)}}{\sigma_{x_i}^{(m_k)}} \right)^2 \right] \right\}, \tag{6}$$

where V_x is the penalty function imposed on a given set of geometric parameters ($C^\alpha \cdots C^\alpha$-distances, virtual-bond angles, virtual-bond dihedral angles or local sidechain coordinates), M_i^{incl} is the number of templates whose selected fragments contain the geometric parameter with index i, M_i^{excl} is the number of templates whose selected fragments do not contain the geometric parameter i, $M_i^{incl} + M_i^{excl} = M$, M being the total number of templates, and $m_1, m_2, \ldots m_{M_i^{incl}}$ are the indices of the templates which contribute to the restraints on the geometric parameter i. x_i and $x_i^{(m_k)}$ are the

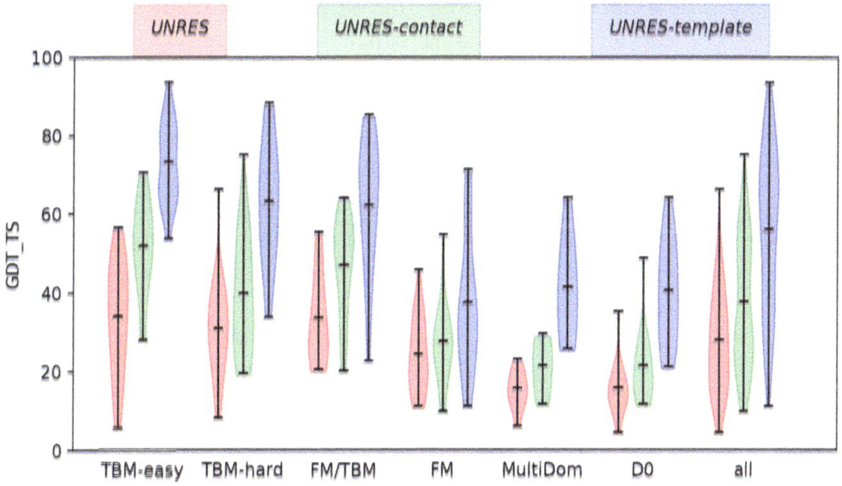

Fig. 6 Violin plots of the GDT_TS [74] of the first-ranked models of the respective CASP14 evaluation unit (EU) types and all EUs for the UNRES (that used only PSIPRED-based secondary-structure restraint), UNRES-contact (that used PSIPRED-based secondary-structure restrains and restraints from residue–residue contact prediction), and UNRES-template (that used consensus-fragment restraints derived from the server models submitted to CASP 72 hours after a target was released) groups. The mean, minimum, and the maximum values in the violin plots are shown as black horizontal lines, the GDT_TS ranges are shown as vertical lines, and approximate distributions are shown as color-filled shapes. The symbols of CASP categories are as follows: TMB-easy—easy template-based modeling targets (significant sequence homology to known structures); TMB-hard—hard template-based modeling targets (similar structures are present in the database, but sequence homology is weaker); FM/TBM—targets in the transition region between free and template-based modeling, with some resemblance to the structures present in the database; FM—free modeling targets for which no templates are present in the database; MultiDom— multi-domain targets; D0—"domain-0" targets comprising whole proteins (Reproduced with permission from A. Antoniak et al., *J. Mol. Graph. Model.*, **2021**, 108, 108008. Copyright 2021 Elsevier)

values of the *i*th geometric parameter of a given kind in the calculated and m_kth reference structure, respectively, and the σ's are the standard deviations of the respective Gaussians.

Including the template-based restraints results in a remarkable improvement of the quality of models, as demonstrated in Fig. 6, in which three UNRES-based methodologies tested in the CASP14 experiment are compared in terms of the GDT_TS measure. It can be seen from the figure that introducing restraints from the fragments of the bioinformatics-derived models gives better results that using the secondary-structure and contact restraints.

5 NMR-Based Restraints on Secondary Structure

The secondary-structure restraints available from NMR data are those imposed on the backbone ϕ and ψ angles (which are, in turn, determined from proton or paramagnetic carbon- or nitrogen-isotope chemical shifts or from the vicinal coupling constants) and those imposed on the proton–proton distances, the upper boundaries of which are determined from Nuclear Overhauser Effect (NOE) experiments [77]. Neither of those is for direct use in the UNRES model, because all-atom polypeptide chain coordinates are required to calculate them. A way around is to convert the coarse-grained chain into the all-atom chain, as done in the algorithm developed by Latek and Kolinski [78]. However, rebuilding the all-atom chain takes time and precludes the use of gradient algorithms for conformational search, such as molecular dynamics.

Because the backbone–virtual-bond angles θ and backbone–virtual-bond dihedral angles γ are analytical functions of the backbone angles ϕ and ψ given the standard valence geometry of polypeptide chains [66], it was straightforward to map the lower and upper boundaries of ϕ and ψ to those of θ and γ and, consequently, to use these restraints in UNRES [67, 79]. The penalty function imposed on the virtual-bond angles θ and virtual-bond dihedral angles γ is as that given by Eq. (3) except that the lower and upper boundaries are determined by NMR.

Using proton-distance restraints requires estimating interproton distances from the coarse-grained geometry. Based on our scale-consistent theory of coarse graining [31], we have recently developed the formulas for the distances involving the amide, H^{α} and H^{β} protons, parameterized them based on known protein structures, and validated against NMR-determined protein structures to create the ESCASA (EStimation of Coordinates of Atoms in proteins from coarse-grained geometry by Simple Analytical formulas) algorithm [80]. ESCASA was subsequently tested with a set of 13 benchmark proteins of the Montelione/NEF Benchmark Data Set [81] for which both NMR and X-ray data were available. The results of UNRES/ESCASA protein-structure modeling were compared with the respective NMR structures and with modeling using the CYANA software [82]. As shown in Fig. 7, in which the bar plots of the GDT_TS and those of root mean square deviation over the C^{α} atoms of the models from the respective X-ray structures (C^{α}-RMSD) are presented, the UNRES/ESCASA modeled structures are of comparable quality with the PDB-deposited or CYANA-modeled structures, despite the fact that UNRES is a highly coarse-grained model of polypeptide chains. The UNRES/ESCASA algorithm has also been tested in the CASP14 experiment, ranking #1 for the N1088-D1 target [83].

The ESCASA algorithm has been added [84] to the UNRES web server for the modeling of protein structure, dynamics, thermodynamics, and interactions [85, 86]. We are currently working on adding explicit chemical-shift restraints to the UNRES/ESCASA algorithm and on including the distances involving the ^{13}C and ^{15}N atoms.

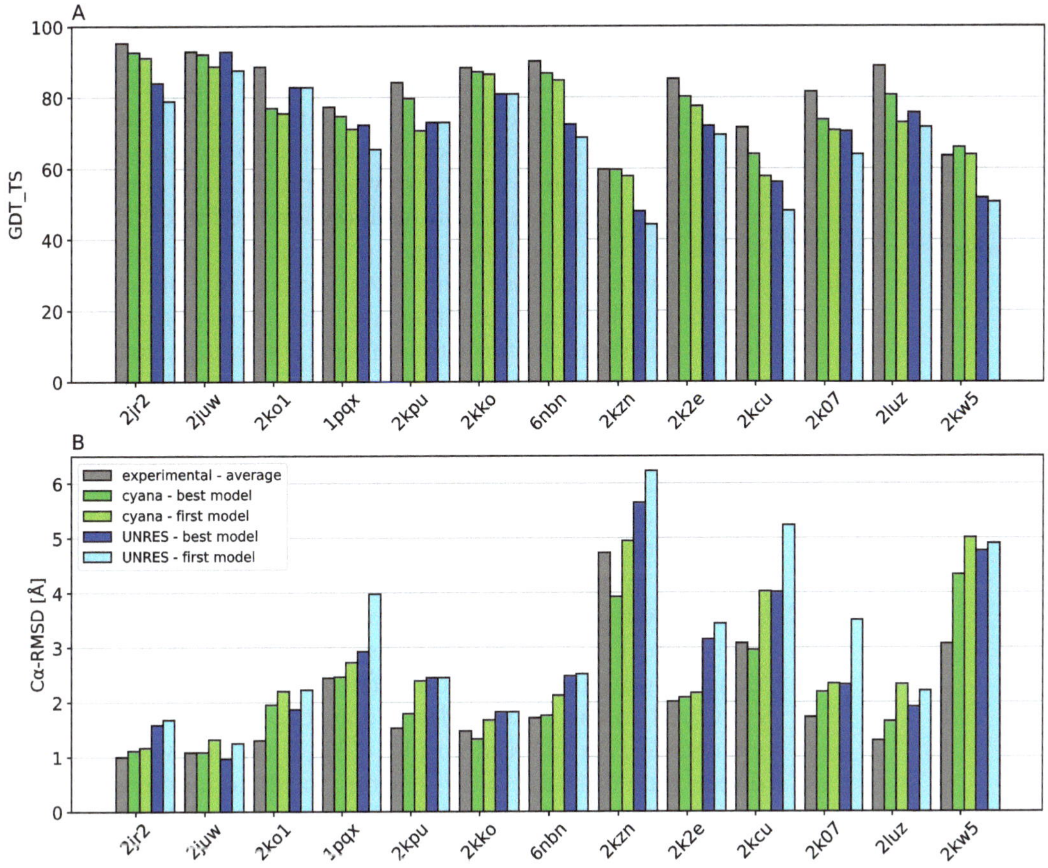

Fig. 7 Bar plots of (**a**) the GDT_TS and (**b**) C^{α}-RMSD for the experimental NMR structures (gray), CYANA [82] best models (green), CYANA first models (light green), UNRES best models (blue), and UNRES first models (light blue) of the 13 proteins selected from the Montelione/NEF Benchmark Data Set [81]. The X-ray structures were the reference structures in the computation of GDT_TS [74] and C^{α}-RMSD. Flexible ends and flexible loops were excluded from comparison (Reproduced with permission from E.A. Lubecka and A. Liwo, *J. Comput. Chem.*, **2022**, 43, 2047–2059. Copyright 2022 John Wiley & Sons)

6 Conclusions and Future Directions

The history of the development of UNRES presented in Subheading 3 points out that much of this process was devoted to proper modeling of secondary structure, which is the scaffold of structured proteins. With such a heavily coarse-grained model, a feature of clear advantage because of a tremendous extension of time and size scales of the systems treated, developing the physics-based energy terms to account for secondary-structure architecture and stability was particularly difficult. Even though, starting from the very early version, UNRES did include physics-based energy terms with axial and not spherical symmetry to account for the orientation of united peptide groups [27], this feature was insufficient for modeling β-sheet structures. It took further effort to realize that the coarse-grained model requires the terms that couple backbone–local and backbone–electrostatic interactions to reproduce all kinds of secondary structure and to develop and parameterize the respective physics-based terms [30, 31, 40]. As stated at the end of Subheading 3, still more development is required to splitting the main–chain–local terms into those involving only the backbone and those involving also the interactions with the attached sidechains [70] as well as introducing long-range correlations over the extended strands and α-helical segments [69]. The necessity of careful representation of hydrogen-bonding and local interactions so that they could orchestrate the formation of regular secondary structures is a take-home message for the developers of coarse-grained models of proteins.

Despite the advances in the development of physics-based coarse-grained energy terms, we will have to live with using restraints on secondary and tertiary structure to perform meaningful simulations of real-sized protein systems. The first reason for it is that it is impractical to attempt at ab initio simulations of big protein systems; those would take excessively long time even while using coarse-grained models. The second reason is that anything happens about 1000 times faster on the coarse-grained level compared to the all-atom level, due to averaging over the fine-grained degrees of freedom [17, 23]. Therefore, while a system diverts little from the initial structure in all-atom simulations with explicit solvent, the coarse-grained simulations with implicit solvent result in much larger deviations, unless the force field is perfect, which will require further development. Restraints on secondary structure, including those on the distances between the strands forming β-sheet sections, are the most obvious choice to keep the secondary structure in sections, where no major structural changes are expected. Using the restraints derived from templates is also a plausible option. On the other hand, the restraints derived from NMR experiments, which can be entered into coarse-grained modeling with the use of the ESCASA algorithm [80], can greatly aid the modeling of intrinsically disordered proteins (IDPs) or proteins with intrinsically disordered regions (IDRs), the structures of which are weakly defined.

Acknowledgements

This work was supported by the National Science Centre under grant UMO-2021/40/Q/ST4/00035. Computational resources were provided by (a) the Centre of Informatics—Tricity Academic Supercomputer & Network (CI TASK) in Gdańsk, (b) the Interdisciplinary Center of Mathematical and Computer Modeling (ICM) the University of Warsaw under grant No. GA71-23, and (c) our 796-processor Beowulf cluster at the Faculty of Chemistry, University of Gdańsk.

References

1. Pauling L, Corey RB, Branson HR (1951) The structure of proteins—2 hydrogen-bonded helical configurations of the polypeptide chain. Proc Natl Acad Sci USA 37:205–211
2. Pauling L, Corey RB (1951) The pleated sheet, a new layer configuration of polypeptide chains. Proc Natl Acad Sci USA 37:251–256
3. Polson A (1936) Determination of particle weight and shape from diffusion and viscosity data. Nature 137:740
4. Astbury WT, Woods HJ (1933) X-ray studies of the structure of hair, wool, and related fibers. II.–The molecular structure and elastic properties of hair keratin. Philos Trans Royal Soc A 232:333–394
5. Coleman D, Howitt FO (1947) Studies of silk proteins. I. The properties and constitution of fibroin. the conversion into a water-soluble form and its bearing on the phenomenon of denaturation. Proc Math Phys Eng Sci 190: 145–169
6. Muirhead H, Perutz M (1963) Structure of haemoglobin: a three-dimensional Fourier synthesis of reduced human haemoglobin at 5.5 Å resolution. Nature 199:633–638
7. Karplus M, Weaver DL (1994) Protein folding dynamics: the diffusion-collision model and experimental data. Prot Sci 3:650–668
8. Weiner SJ, Kollman PA, Nguyen DT, Case DA (1986) An all-atom force field for simulations of proteins and nucleic acids. J Comput Chem 7:230–252
9. Némethy G, Gibson KD, Palmer KA, Yoon CN, Paterlini G, Zagari A, Rumsey S, Scheraga HA (1992) Energy parameters in polypeptides. 10. Improved geometrical parameters and nonbonded interactions for use in the ECEPP/3 algorithm with application to proline-containing peptides. J Phys Chem 96:6472–6484
10. Ponder JW, Case DA (2003) Force fields for protein simulations. Adv Prot Chem 66:27–85
11. MacKerell Jr AD, Feig M, Brooks CL (2004) Improved treatment of the protein backbone in empirical force fields. J Am Chem Soc 126: 698–699
12. Brooks BR, Brooks III CL, MacKerell AD, Nilsson L, Petrella RJ, Roux B, Won Y, Archontis G, Bartels C, Boresch S, Caflisch A, Caves L, Cui Q, Dinner AR, Feig M, Fischer S, Gao J, Hodoscek M, Im W, Kuczera K, Lazaridis T, Ma J, Ovchinnikov V, Paci E, Pastor RW, Post CB, Pu JZ, Schaefer M, Tidor B, Venable RM, Woodcock HL, W. Yang XW, York DM, Karplus M (2009) CHARMM: the biomolecular simulation program. J Comput Chem 30:1545–1615
13. Lindorff-Larsen K, Piana S, Palmo K, Maragakis P, Klepeis JL, Dror RO, Shaw DE (2010) Improved side-chain torsion potentials for the Amber Ff99SB protein force field. Proteins 78:1950–1958
14. Nerenberg PS, Head Gordon T (2011) Optimizing protein-solvent force fields to reproduce intrinsic conformational preferences of model peptides. J Chem Theory Comput 7: 1220–1230
15. Robertson MJ, Tirado Rives J, Jorgensen WL (2015) Improved peptide and protein torsional energetics with the OPLS-AA force field. J Chem Theory Comput 11:3499–3509
16. Tang WS, Fawzi NL, Mittal J (2020) Refining all-atom protein force fields for polar-rich, prion-like, low-complexity intrinsically disordered proteins. J Phys Chem B 124:9505–9512
17. Liwo A, Khalili M, Scheraga HA (2005) Ab initio simulations of protein-folding pathways by molecular dynamics with the united-residue model of polypeptide chains. Proc Natl Acad Sci USA 102:2362–2367
18. Voth G (2008) In: Voth G (ed) Coarse-graining of condensed phase and biomolecular systems, chap 1, 1st edn. CRC Press/Taylor & Francis Group, Boca Raton/Milton Park, pp 1–4
19. Marrink SJ, Tieleman DP (2013) Perspective on the Martini model. Chem Soc Rev 42: 6801–6822
20. Kmiecik S, Gront D, Kolinski M, Wieteska L, Dawid AE, Kolinski A (2016) Coarse-grained protein models and their applications. Chem Rev 116:7898–7936
21. Souza PCT, Alessandri R, Barnoud J, Thallmair S, Faustino I, Grünewald F, Patmanidis I, Abdizadeh H, Bruininks BMH, Wassenaar TA, Kroon PC, Melcr J, Nieto V, Corradi V, Khan HM, Domański J, Javanainen M, Martinez Seara H, Reuter N, Best RB, Vattulainen I, Monticelli L, Periole X, Tieleman DP, Vries AH, Marrink SJ (2021) Martini 3: a general purpose force field for coarse-grained molecular dynamics. Nat Methods 18:382–388
22. Liwo A, Czaplewski C, Sieradzan AK, Lipska AG, Samsonov SA, Murarka RK (2021) Theory and practice of coarse-grained molecular dynamics of biologically important systems. Biomolecules 11:1347

23. Sieradzan AK, Sans-Duñó J, Lubecka EA, Czaplewski C, Lipska AG, Leszczynski H, Ocetkiewicz KM, Proficz J, Czarnul P, Krawczyk H, Liwo A (2023) Optimization of parallel implementation of UNRES package for coarse-grained simulations to treat large proteins. J Comput Chem 44:602–625
24. Brown S, Fawzi NJ, Head Gordon T (2003) Coarse-grained sequences for protein folding and design. Proc Natl Acad Sci USA 100: 10712–10717
25. Hoang TX, Trovato A, Seno F, Banavar JR, Maritan A (2004) Geometry and symmetry presculpt the free-energy landscape of proteins. Proc Natl Acad Sci USA 101:7960–7964
26. Banavar JR, Giacometti A, Hoang TX, Maritan A, Škrbić T. A geometrical framework for thinking about proteins. (2023) Proteins: Struct. Func. Bioinf. 1–15. https://doi.org/10.1002/prot.26567 (Early View)
27. Liwo A, Pincus MR, Wawak RJ, Rackovsky S, Scheraga HA (1993) Prediction of protein conformation on the basis of a search for compact structures; test on avian pancreatic polypeptide. Protein Sci 2:1715–1731
28. Liwo A, Baranowski M, Czaplewski C, Gołaś E, He Y, Jagieła D, Krupa P, Maciejczyk M, Makowski M, Mozolewska MA, Niadzvedtski A, Ołdziej S, Scheraga HA, Sieradzan AK, Ślusarz R, Wirecki T, Yin Y, Zaborowski B (2014) A unified coarse-grained model of biological macromolecules based on mean-field multipole-multipole interactions. J Mol Model 20:2306
29. Sieradzan AK, Czaplewski C, Krupa P, Mozolewska MA, Karczyńska AS, Lipska AG, Lubecka EA, Gołaś E, Wirecki T, Makowski M, Ołdziej S, Liwo A (2022) Modeling the structure, dynamics, and transformations of proteins with the UNRES force field. Springer US, New York, pp 399–416
30. Liwo A, Czaplewski C, Pillardy J, Scheraga HA (2001) Cumulant-based expressions for the multibody terms for the correlation between local and electrostatic interactions in the united-residue force field. J Chem Phys 115: 2323–2347
31. Sieradzan K, Makowski M, Augustynowicz A, Liwo A (2017) A general method for the derivation of the functional forms of the effective energy terms in coarse-grained energy functions of polymers. I. Backbone potentials of coarse-grained polypeptide chains. J Chem Phys 146:124106
32. Kubo R (1962) Generalized cumulant expansion method. J Phys Soc Jpn 17:1100–1120
33. Liwo A, Ołdziej S, Pincus MR, Wawak RJ, Rackovsky S, Scheraga HA (1997) A united-residue force field for off-lattice protein-structure simulations. I. Functional forms and parameters of long-range side-chain interaction potentials from protein crystal data. J Comput Chem 18:849–873
34. Makowski M, Liwo A, Scheraga HA (2017) Simple physics-based analytical formulas for the potentials of mean force of the interaction of amino acid side chains in water. VII. Charged-hydrophobic/polar and polar-hydrophobic/polar side chains. J Phys Chem B 121: 379–390
35. Gay JG, Berne BJ (1981) Modification of the overlap potential to mimic a linear site-site potential. J Chem Phys 74:3316–3319
36. Liwo A, Czaplewski C, Sieradzan AK, Lubecka EA, Lipska AG, Golon Ł, Karczyńska A, Krupa P, Mozolewska MA, Makowski M, Ganzynkowicz R, Giełdoń A, Maciejczyk M (2020) In: Strodel B, Barz B (eds) Progress in molecular biology and translational science. Computational approaches for understanding dynamical systems: protein folding and assembly, chap 2, vol 170. Academic, London, pp 73–122
37. Kozłowska U, Maisuradze GG, Liwo A, Scheraga HA (2010) Determination of side-chain-rotamer and side-chain and backbone virtual-bond-stretching potentials of mean force from AM1 energy surfaces of terminally-blocked amino-acid residues, for coarse-grained simulations of protein structure and folding. 2. Results, comparison with statistical potentials, and implementation in the UNRES force field. J Comput Chem 31:1154–1167
38. Chinchio M, Czaplewski C, Liwo A, Ołdziej S, Scheraga HA (2007) Dynamic formation and breaking of disulfide bonds in molecular dynamics simulations with the UNRES force field. J Chem Theory Comput 3:1236–1248
39. Krupa P, Sieradzan AK, Mozolewska MA, Li H, Liwo A, Scheraga HA (2017) Dynamics of disulfide-bond disruption and formation in the thermal unfolding of ribonuclease A. J Chem Theory Comput 13:5721–5730
40. Liwo A, Sieradzan AK, Lipska AG, Czaplewski C, Joung I, Żmudzińska W, Hałabis A, Ołdziej S (2019) A general method for the derivation of the functional forms of the effective energy terms in coarse-grained energy functions of polymers. III. Determination of scale-consistent backbone-local and correlation potentials in the UNRES force field and force-field calibration and validation. J Chem Phys 150:155104

41. Liwo A, Khalili M, Czaplewski C, Kalinowski S, Ołdziej S, Wachucik K, Scheraga HA (2007) Modification and optimization of the united-residue (UNRES) potential energy function for canonical simulations. I. Temperature dependence of the effective energy function and tests of the optimization method with single training proteins. J Phys Chem B 111:260–285
42. Liwo A, Pincus MR, Wawak RJ, Rackovsky S, Scheraga HA (1993) Calculation of protein backbone geometry from α-carbon coordinates based on peptide-group dipole alignment. Protein Sci 2:1697–1714
43. Liwo A, Ołdziej S, Czaplewski C, Kozłowska U, Scheraga HA (2004) Parameterization of backbone-electrostatic and multibody contributions to the UNRES force field for protein-structure prediction from ab initio energy surfaces of model systems. J Phys Chem B 108:9421–9438
44. Orengo CA, Bray JE, Hubbard T, LoConte L, Sillitoe I (1999) Analysis and assessment of ab initio three-dimensional prediction, secondary structure, and contacts prediction. Proteins Struct Funct Genet Suppl 3:149–170
45. Liwo A, Lee J, Ripoll DR, Pillardy J, Scheraga HA (1999) Protein structure prediction by global optimization of a potential energy function. Proc Natl Acad Sci USA 96:5482–5485
46. Lee J, Liwo A, Scheraga HA (1999) Energy-based *de novo* protein folding by conformational space annealing and an off-lattice united-residue force field: Application to the 10-55 fragment of staphylococcal protein A and to *apo*-calbindin D9K. Proc Natl Acad Sci USA 96:2025–2030
47. Kortemme T, Ramirez Alvarado M, Serrano L (1998) Design of a 20-amino acid, three-stranded β -sheet protein. Science 282:253–256
48. Liwo A, Kaźmierkiewicz R, Czaplewski C, Groth M, Ołdziej S, Wawak RJ, Rackovsky S, Pincus MR, Scheraga HA (1998) United-residue force field for off-lattice protein-structure simulations; III. Origin of backbone hydrogen-bonding cooperativity in united-residue potentials. J Comput Chem 19:259–276
49. Kolinski A, Skolnick J (1992) Discretized model of proteins. I. Monte Carlo study of cooperativity in homopolypeptides. J Chem Phys 97:9412–9426
50. Kolinski A (2004) Protein modeling and structure prediction with a reduced representation. Acta Biochim Pol 51:349–371
51. Berman HM, Westbrook J, Feng Z, Gilliland G, Bhat TN, Weissig H, Shindyalov IN, Bourne PE (2000) The protein data bank. Nucleic Acids Res 28:235–242
52. Ramachandran GN, Ramakrishnan C, Sasisekharan V (1963) Stereochemistry of polypeptide chain configurations. J Mol Biol 7:95–99
53. Lovell SC, Davis IW, Arendall III B, de Bakker PIW, Word JM, Prisant MG, Richardson JS, Richardson DC (2003) Structure validation by Cα geomtery; ϕ, ψ and Cβ deviation. Proteins 50:437–450
54. Davtyan A, Schafer NP, Zheng W, Clementi C, Wolynes PG, Papoian GA (2012) AWSEM-MD: protein structure prediction using coarse-grained physical potentials and bioinformatically based local structure biasing. J Phys Chem B 116:8494–8503
55. Alemani D, Collu F, Cascella M, Dal Perraro M (2010) A nonradial coarse-grained potential for proteins produces naturally stable secondary structure elements. J Chem Theory Comput 6:315–324
56. Ganesan SJ, Matysiak S (2014) Role of backbone dipole interactions in the formation of secondary and supersecondary structures of proteins. J Chem Theory Comput 10:2569–2576
57. Ołdziej S, Kozłowska U, Liwo A, Scheraga HA (2003) Determination of the potentials of mean force for rotation about $C^{\alpha} \cdots C^{\alpha}$ virtual bonds in polypeptides from the *ab initio* energy surfaces of terminally-blocked glycine, alanine, and proline. J Phys Chem A 107:8035–8046
58. Pillardy J, Czaplewski C, Liwo A, Lee J, Ripoll DR, Kaźmierkiewicz R, Oldziej S, Wedemeyer WJ, Gibson KD, Arnautova YA, Saunders J, Ye YJ, Scheraga HA (2001) Recent improvements in prediction of protein structure by global optimization of a potential energy function. Proc Natl Acad Sci USA 98:2329–2333
59. Lee J, Ripoll DR, Czaplewski C, Pillardy J, Wedemeyer WJ, Scheraga HA (2001) Optimization of parameters in macromolecular potential energy functions by conformational space annealing. J Phys Chem B 105:7291–7298
60. Pillardy J, Czaplewski C, Liwo A, Wedemeyer WJ, Lee J, Ripoll DR, Arłukowicz P, Ołdziej S, Arnautova YA, Scheraga HA (2001) Development of physics-based energy functions that predict medium-resolution structure for proteins of the α, β, and α/β structural classes. J Phys Chem B 105:7299–7311
61. Sieradzan AK, Hansmann UHE, Scheraga HA, Liwo A (2012) Extension of UNRES force field to treat polypeptide chains with D-amino acid residues. J Chem Theory Comput 8: 4746–4757

62. Sieradzan AK, Krupa P, Scheraga HA, Liwo A, Czaplewski C (2015) Physics-based potentials for the coupling between backbone- and side-chain-local conformational states in the united residue (UNRES) force field for protein simulations. J Chem Theory Comput 11:817–831
63. Zaborowski B, Jagieła D, Czaplewski C, Hałabis A, Lewandowska A, Żmudzińska W, Ołdziej S, Karczyńska A, Omieczynski C, Wirecki T, Liwo A (2015) A maximum-likelihood approach to force-field calibration. J Chem Inf Model 55:2050–2070
64. Krupa P, Hałabis A, Żmudzińska W, Ołdziej S, Scheraga HA, Liwo A (2017) Maximum likelihood calibration of the UNRES force field for simulation of protein structure and dynamics. J Chem Inf Model 57:2364–2377
65. Krupa P, Mozolewska MA, Wiśniewska M, Yanping Y, He Y, Sieradzan AK, Ganzynkowicz R, Lipska AG, Karczyńska A, Ślusarz M, Ślusarz R, Giełdoń A, Czaplewski C, Jagieła D, Zaborowski B, Scheraga HA, Liwo A (2016) Performance of protein-structure predictions with the physics-based UNRES force field in CASP11. Bioinformatics 32:3270–3278
66. Nishikawa K, Momany FA, Scheraga HA (1974) Low-energy structures of two dipeptides and their relationship to bend conformations. Macromolecules 7:797–806
67. Lubecka EA, Karczyńska AS, Lipska AG, Sieradzan AK, Zięba K, Sikorska C, Uciechowska U, Samsonov SA, Krupa P, Mozolewska MA, Golon Ł, Giełdoń A, Czaplewski C, Ślusarz R, Ślusarz M, Crivelli SN, Liwo A (2019) Evaluation of the scale-consistent UNRES force field in template-free prediction of protein structures in the CASP13 experiment. J Mol Graph Model 92:154–166
68. Antoniak A, Biskupek I, Bojarski KK, Czaplewski C, Giełdoń A, Kogut M, Kogut MM, Krupa P, Lipska AG, Liwo A, Lubecka EA, Marcisz M, Maszota Zieleniak M, Samsonov SA, Sieradzan A, Ślusarz MJ, Ślusarz R, Wesołowski PA, Zięba K (2021) Modeling protein structures with the coarse-grained UNRES force field in the CASP14 experiment. J Mol Graph Model 108:108008
69. Liwo A, Sikorska C (2022) Origin of correlations between local conformational states of consecutive amino-acid residues and their role in shaping protein structures and in allostery. J Phys Chem B 126:9493–9505. Correction: tJ Phys Chem B 127:425–426 (2023)
70. Lipska AG, Sieradzan AK, Atmaca S, Czaplewski C, Liwo A (2023) Towards consistent physics-based modeling of local backbone structures and chirality change of proteins in coarse-grained approaches. J Phys Chem Lett 14:9824–9833 Submitted
71. Kubelka J, Hofrichter J, Eaton WA (2004) The protein folding 'speed limit'. Curr Opin Struct Biol 14:76–88
72. Lubecka EA, Liwo A (2019) Introduction of a bounded penalty function in contact-assisted simulations of protein structures to omit false restraints. J Comput Chem 40:2164–2178
73. McGuffin LJ, Bryson K, Jones DT (2000) The PSIPRED protein structure prediction server. Bioinformatics 16:404
74. Zemla A (2003) LGA: a method for finding 3D similarities in protein structures. Nucleic Acids Res 13:3370–3374
75. Mozolewska MA, Krupa P, Zaborowski B, Liwo A, Lee J, Joo K, Czaplewski C (2016) Use of restraints from consensus fragments of multiple server models to enhance protein-structure prediction capability of the UNRES force field. J Chem Inf Model 56:2263–2279
76. Karczyńska A, Zięba K, Uciechowska U, Mozolewska MA, Krupa P, Lubecka EA, Lipska AG, Sikorska C, Samsonov SA, Sieradzan AK, Giełdoń A, Liwo A, Ślusarz R, Ślusarz M, Lee J, Joo K, Czaplewski C (2020) Improved consensus-fragment selection in template-assisted prediction of protein structures with the UNRES force field in CASP13. J Chem Inf Model 60:1844–1864
77. Wüthrich K (1986) NMR of proteins and nucleic acids. New York: Wiley
78. Latek D, Koliński A (2011) CABS-NMR—De Novo tool for rapid global fold determination from chemical shifts, residual dipolar couplings and sparse methyl-methyl NOEs. J Comput Chem 32:536–544
79. Lubecka E, Liwo A (2022) A coarse-grained approach to NMR-data-assisted modeling of protein structures. J Comput Chem 43:2047–2059
80. Lubecka E, Liwo A (2021) ESCASA: Analytical estimation of atomic coordinates from coarse-grained geometry for nuclear-magnetic-resonance-assisted protein structure modeling. I. Backbone and H^{β} protons. J Comput Chem 42:1579–1589
81. Everett JK, Tejero R, Murthy SBK, Acton TB, Aramini JM, Baran MC, Benach J, Cort JR, Eletsky A, Forouhar F, Guan R, Kuzin AP, Lee HW, Liu G, Mani R, Mao B, Mills JL, Montelione AF, Pederson K, Powers R, Ramelot T, Rossi P, Seetharaman J, Snyder D, Swapna GVT, Vorobiev SM, Wu Y, Xiao R, Yang Y, Arrowsmith CH, Hunt JF, Kennedy MA, Prestegard JH, Szyperski T, Tong L, Montelione GT (2016) A community resource

of experimental data for NMR/X-ray crystal structure pairs. Prot Sci 25:30–45

82. Güntert P, Mumenthaler C, Wüthrich K (1997) Torsion angle dynamics for NMR structure calculation with the new program DYANA. J Mol Biol 273:283–298
83. Huang YJ, Zhang N, Bersch B, Fidelis K, Inouye M, Ishida Y, Kryshtafovych A, Kobayashi N, Kuroda Y, Liu G, LiWang A, Swapna GVT, Wu N, Yamazaki T, Montelione GT (2021) Assessment of prediction methods for protein structures determined by NMR in CASP14: Impact of AlphaFold2. Proteins 89: 1959–1976
84. Ślusarz R, Lubecka E, Czaplewski C, Liwo A (2022) Improvements and new functionalities of UNRES server for coarse-grained modeling of protein structure, dynamics, and interactions. Front Biomol Sci 9:1071428
85. Czaplewski C, Karczyńska A, Sieradzan AK, Liwo A (2018) UNRES server for physics-based coarse-grained simulations and prediction of protein structure, dynamics and thermodynamics. Nucl Acids Res 46:W304–W309
86. Krupa P, Karczyńska AS, Mozolewska MA, Liwo A, Czaplewski C (2021) UNRES-Dock protein-protein and peptide-protein docking by coarse-grained replica-exchange MD simulations. Bioinformatics 37:1613–1615

Chapter 3

Improving Protein Secondary Structure Prediction by Deep Language Models and Transformer Networks

Tianqi Wu, Weihang Cheng, and Jianlin Cheng

Abstract

Protein secondary structure prediction is useful for many applications. It can be considered a language translation problem, that is, translating a sequence of 20 different amino acids into a sequence of secondary structure symbols (e.g., alpha helix, beta strand, and coil). Here, we develop a novel protein secondary structure predictor called TransPross based on the transformer network and attention mechanism widely used in natural language processing to directly extract the evolutionary information from the protein language (i.e., raw multiple sequence alignment [MSA] of a protein) to predict the secondary structure. The method is different from traditional methods that first generate a MSA and then calculate expert-curated statistical profiles from the MSA as input. The attention mechanism used by TransPross can effectively capture long-range residue–residue interactions in protein sequences to predict secondary structures. Benchmarked on several datasets, TransPross outperforms the state-of-art methods. Moreover, our experiment shows that the prediction accuracy of TransPross positively correlates with the depth of MSAs, and it is able to achieve the average prediction accuracy (i.e., Q3 score) above 80% for hard targets with few homologous sequences in their MSAs. TransPross is freely available at https://github.com/BioinfoMachineLearning/TransPro.

Key words Protein secondary structure prediction, Natural language processing, Attention, Transformer

1 Introduction

Studying protein structures and their functions is critical for addressing many biomedical problems such as protein engineering and drug design [1–4]. Predicting the protein secondary structure from sequence can provide useful information for studying tertiary and quaternary structures of proteins and their function [5–7].

Protein secondary structure (SS) can be defined as the local configuration of the polypeptide chain. The two regular local configuration types, alpha-helix (H) and beta-strand (E), determined by the hydrogen bonding patterns, were initially proposed in 1951 [8, 9]. The nonregular local configuration is often called coil or

Andrzej Kloczkowski et al. (eds.), *Prediction of Protein Secondary Structure*, Methods in Molecular Biology, vol. 2867,
https://doi.org/10.1007/978-1-0716-4196-5_3,

loop (C). The annotation of the protein secondary structure is further expanded to the eight types and is commonly computed from the tertiary structure of proteins by the standard annotation tools such as DSSP [10]. For simplification, the eight-state representation is usually reduced to the three-state representation (alpha-helix [H], beta-strand [E], and coil [C]). Since the tertiary structures of most proteins are unknown, there is a need to directly predict the protein secondary structure from sequence.

Many computational protein SS prediction methods have been developed over the last several decades. One major improvement in the protein SS prediction came from the innovation in enhancing the input information. Instead of using a single sequence of a target protein to predict its secondary structure, the sequence profile-based methods [11–19] compute the sequence profile from protein multiple sequence alignment (MSA) of the protein [20–23] to extract the evolutionary information for SS prediction. Using protein sequence profile for secondary structure prediction is substantially more accurate than using the single sequence of a target protein. The sequence profile is usually represented by expert-designed statistical models such as position-specific scoring matrices [21] and hidden Markov models [22]. As more and more protein sequences are produced, reasonable multiple sequence alignments can be generated for most proteins to construct good sequence profiles; the profile-based secondary structure prediction methods have reached a high accuracy (e.g., >80%) on most benchmarks. However, for some proteins that do not have many homologous sequences in the protein sequence databases for constructing reasonable multiple sequence alignments and profiles, the profile-based prediction methods cannot deliver optimal results. In this case, the single-sequence-based prediction method that only depends on the single sequence information itself [18, 24] has been developed to improve the prediction accuracy.

Since 1990s, neural networks had been shown to be the most successful methods for protein secondary structure prediction [11, 13]. When deep neural networks consisting of many more layers than traditional shallow neural networks started to achieve significant success in several computing domains such as image processing and computer vision in mid-2020s, different deep learning architectures such as deep belief networks, gated recurrent neural networks, long- and short-term memory networks, convolutional neural networks, residual networks, and inception networks have been applied to the secondary structure prediction problem [12, 25–27].

Recently, the language models in the natural language processing (NLP) were adapted for protein sequence analysis [28–31], creating a new avenue to improve protein secondary structure prediction. The language models that can interpret the meaning of the words in a long-range context is a promising technique to handle the long-range amino acid interactions in protein sequences that are needed for predicting some secondary structures in proteins, particularly beta-sheets that often involve the long-range interactions. In this work, we explore the application of language models in protein SS prediction and develop a predictor—TransPross—based on the transformer network and the attention mechanism to effectively obtain evolutionary information from the raw MSA directly to predict secondary structure. Tested on several different test sets including the TransPross test set, CASP13 free modeling (FM) test set, and CASP14 test set, TransPross achieves better performance than the state of the art of the sequence profile-based methods. The analysis of the performance of TransPross on hard targets with few homologous sequences in their MSAs also shows it often works well even when the MSAs are shallow. By taking a MSA as input, TransPross can be readily applied to many proteins that users may have obtained MSAs for. It can also be combined with any MSA generation tool to predict secondary structure prediction, which is more flexible than most traditional secondary structure predictions that use their own built-in MSA generation and profile generation programs to prepare the input. Therefore, TransPross is a useful tool that complements the existing protein secondary structure prediction tools.

2 Results

2.1 Comparison with Single-Sequence-Based Methods and Profile-Based Methods

We compare TransPross with PSIPRED, one of the most widely used state-of-the-art protein secondary structure prediction methods, on three different test sets: TransPross test set, CASP13-FM test set (which includes the free modeling [FM] targets of the 13th Critical Assessment of Techniques for Protein Structure Prediction [CASP13]), and CASP14 regular full-length targets with sequence length ≤ 500. FM targets were considered hard targets because they did not have similar structures in the Protein Data Bank (PDB) when they were released. PSIPRED can predict secondary structures from either sequence profile or a single protein sequence. PSIPRED-profile accepts the sequence profile calculated from multiple sequence alignments as input. PSIPRED-single predicts secondary structure from a protein single sequence. In addition to comparing TransPross with PSIPRED-profile, we also use PSIPRED-single and another single-sequence-based method SPOT-1D as the baseline to evaluate TransPross. The performance of the methods is evaluated by the Q3 accuracy score, which is the percent of residues of the proteins in a dataset that are correctly assigned into three categories (i.e., helix, beta-strand, and coil).

As the results shown in Table 1a, b, TransPross substantially outperforms the single sequence-based methods (SPOT-1D, PSIPRED-single) as expected. The Q3 accuracy score of

Table 1
Comparison of TransPross with other predictors in terms of the Q3 accuracy score on different test sets: (a) TransPross test set and CASP13 FM test set (b) CASP14 test set

Method	TransPross-test	CASP13
(a) Comparison of TransPross with SPOT-1D, PSIPRED-single and PSIPREDprofile on TransPross test set and CASP13 test set		
TransPross	*84.42*	*81.04*
SPOT-1D	76.97	73.21
PSIPRED-single	71.03	68.05
PSIPRED-profile	82.88	80.60

Method	CASP14
(b) Comparison of TransPross with PSIPRED-profile on CASP14 test set	
TransPross	*80.71*
PSIPRED-profile	79.62

TransPross for the three test sets are 84.42%, 81.04%, and 80.71%, respectively, which is also better than that of a state-of-the-art profile-based method—PSIPRED-profile.

2.2 Effect of the Quality of MSAs on Prediction Accuracy

To evaluate the impact of the quality of multiple sequence alignments (MSAs) on the performance of TransPross, we test TransPross on two different types of inputs: MSAs generated by using HHblits to search protein sequence against the big fantastic protein sequence database (BFD) [32], and MSAs generated by using DeepMSA [33] to search against Uniref30 and the mGnify databases. The quality of an MSA is determined by multiple factors such as the alignment accuracy, the number of homologous or nonhomologous sequences in the MSA, and the redundancy in the MSA. Here, we simply calculate the number of effective sequences (Neff) in the MSAs to approximate their quality [34]. In general, the higher the Neff, the more diverse and informative an MSA is. The average Neff for the two different inputs for 17 CASP13-FM targets is shown in Table 2; MSA BFD denotes the MSAs generated from the BFD database, whereas MSA DeepMSA represents the MSAs generated by DeepMSA. Because DeepMSA searches against several databases, MSA DeepMSA has a higher Neff than MSA BFD in most cases.

In Table 2, there is a positive relationship between the average Q3 accuracy score of TransPross and the average Neff of MSA on the 17 CASP13 FM targets. Figure 1 plots the Q3 score against Neff of the 17 CASP13 FM targets. The results indicate that increasing the number of effective sequences in MSAs is important

Table 2
The effect of Neff on the prediction accuracy of three-state secondary structure on 17 CASP13 FM domain targets

Input	Q3(%)	Neff
MSA BFD	76.09	202
MSA DeepMSA	81.04	382

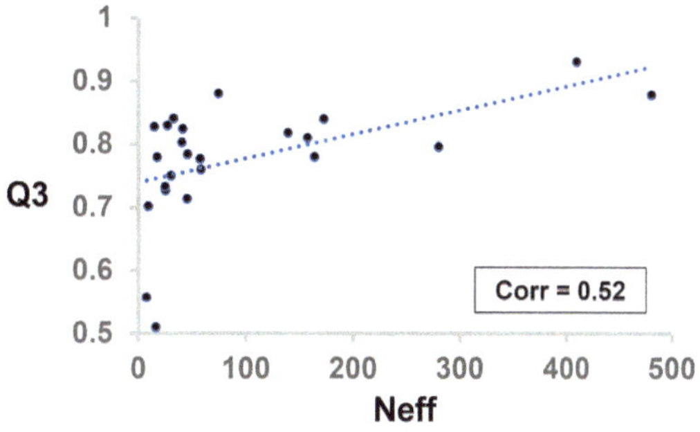

Fig. 1 The scatter plot of the Q3 accuracy score of secondary structure prediction versus Neff with linear fitting curve (Pearson's correlation coefficient = 0.52)

for improving secondary structure prediction, which is similar to protein contact prediction [34] and tertiary structure prediction [35]. The Pearson's correlation coefficient between the per-target Neff and the corresponding Q3 accuracy score on the 17 CASP13 FM targets is 0.52.

2.3 Secondary Structure Prediction for Hard Targets with Shallow MSAs

Targets with relatively few sequence homologs in their MSAs are generally defined as hard targets and the accurate predictions for them are likely more helpful for studying their structures and function. We assess the performance of TransPross on the 12 CASP13 FM targets with shallow MSAs (Neff <50). Interestingly, TransPross is able to achieve a pretty good average Q3 accuracy sore of 81.25%, slightly higher than PSIPRED-profile's 81.22%. Figure 2 illustrates a head-to-head comparison of TransPross predictions with the PSIPRED-profile predictions on 12 CASP13 FM targets with shallow alignments (Neff <50). In total, TransPross performed better in 8 of 12 cases, demonstrating its capability of accurately predicting secondary structure for hard targets with a small number of sequence homologs.

2.4 Running Time

We empirically calculated the running time of TransPross on the CASP13 FM targets with an average length of 148 residues. If the MSA of a target is provided, TransPross typically requires about 30 seconds on average to complete the prediction with a single Tesla V100 GPU. Because MSAs are often available for many

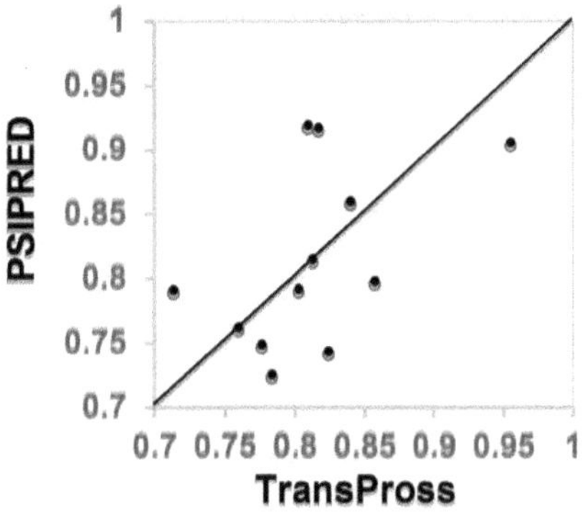

Fig. 2 Performance of TransPross versus PSIPRED-profile on 12 CASP13 FM domain targets with shallow MSAs (Neff <50) in terms of the Q3 accuracy score (denoted by a dot in the plot)

proteins in advance, TransPross can be applied to them to quickly generate secondary structure predictions. If MSAs are not available, users can use any MSA generation tool or the programs in the TransPross package to generate MSA first, which is generally much slower than predicting secondary structure from MSA with TransPross.

3 Materials and Methods

3.1 Datasets

We selected protein targets deposited into the Protein Data Bank (PDB) before May 2019 and extracted their true secondary structures. After filtering out the redundant sequences at the sequence identity cutoff of 90% by MMseqs2 [36] and setting the sequence length within the range [50, 500], there are 36,334 proteins left. We first randomly selected 5% of the targets to create the TransPross test set, and the rest of the targets were used for training and validation. Apart from the TransPross test set, the other two test sets are 17 CASP13 FM domains and 40 CASP14 regular targets with length $\leq$ 500. To ensure there was no overlap between the training data and test sets, we further filtered out the sequence redundancy between the training data and all the three test sets at the sequence identify cutoff 30% by using MMseqs2 [36]. We extracted the true three-state secondary structures of the proteins (i.e., strand E, helix H, and loop C) from their true tertiary structures as labels using DSSP [10]. The objective is to predict the secondary structure label for each residue of a protein from its sequence.

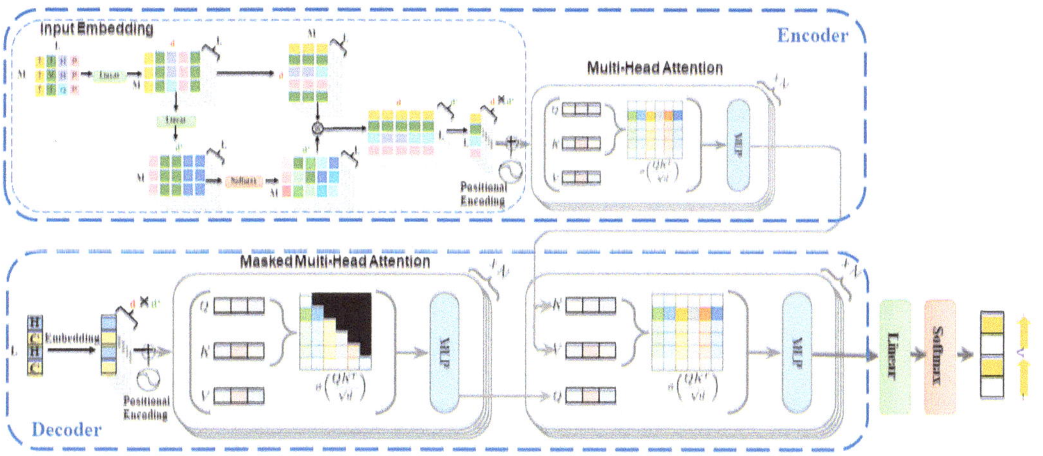

Fig. 3 The architecture of TransPross. It has two main modules: an encoder and a decoder. The input for the encoder is the MSA of a protein (M). The input for the decoder includes both the secondary structures prior to the current position under consideration and the output of the encoder. From the input, the decoder predicts the three-class secondary structure for each position in the protein sequence

3.2 Protein Sequence Language Model and Transformer Architecture

We apply a transformer network with the attention mechanism [37] that has achieved success in natural language translation to detect the relevant sequence context across the entire protein sequence for each amino acid position to predict its secondary structure type. The deep learning architecture is useful to capture long-range interaction between amino acids (like words in an English sentence) relevant to secondary structure prediction. The customized protein sequence-to-secondary-structure translation model of TransPross shown in Fig. 3 adopts an encoder-decoder structure. The encoder maps an input MSA of a protein in the symbolic representation to a sequence of continuous values (internal features) with the dimension of $L \times 512$(L: the protein sequence length). From the internal features, the decoder then generates an output sequence of three-state protein secondary structures one by one in an auto-regressive way [38].

Given the MSA for a target protein as the input, TransPross uses the encoder-decoder process to generate the secondary structure of the protein. The encoder consists of the input embedding component and self-attention component. The input embedding component first converts the input tokens to the value vector of the dimension $L \times M \times 64$ (M: the number of sequences in the MSA). In order to extract more important amino acids in each column of MSA, we also apply the column attention to the input MSA embedding vector by using the linear transformation and softmax function. The attention weight has the dimension of $L \times M \times 8$. The final output of the input embedding component is the multiplication of the column attention weights and the input MSA embedding vector with the dimension of $L \times 512$. In order to maintain the order of the sequence, we add the positional encoding to the final embeddings. The input embedding vector is further processed by the self-attention component composed of a stack of $N = 6$ identical subcomponents to generate the output of the encoder. Each subcomponent is composed of a multi-head self-attention layer (number of heads = 8) and a fully connected feed-forward layer. For each layer, there is a residual connection [39] around it, followed by the layer normalization [40] and one dropout operation [41] with dropout rate = 0.1.

Similar to the encoder, the decoder also consists of the embedding component and two attention components. It starts from the output tokens (i.e., true three-state secondary structure in the training phase or predicted secondary structure in the inference phase) of the previous positions in a target protein, which are converted to an embedding vector of the dimension $L \times 512$. A sinusoidal position encoding vector is added on top of the embedding vector, which is then fed into the first self-attention component, followed by the cross-attention component. The first self-attention component consists of a stack of $N = 6$ identical layers, similar to the attention component in the encoder except that it applies the masked self-attention mechanism to ensure that the prediction for the current position only depends on the output tokens prior to this position in the training phase. In the inference phase, the self-attention component in the decoder also considers only the previously generated secondary structure states when generating the next. Both the output of the self-attention component in the decoder and the output of the encoder stack are fed into the cross-attention decoder component composed of $N = 6$ identical subcomponents, each consisting of a multi-head self-attention layer (num of heads = 8) and a fully connected feed-forward layer. The output of the decoder stack is converted to the three-state protein secondary structure prediction through the linear transformation and softmax function.

We use Adam [42] as the optimizer with the following parameters: $\beta1 = 0.9$, $\beta2 = 0.999$, warmup steps = 4000, and weight decay = 0.001. We set the batch size to 30 and use the Kullback–Leibler divergence loss as the loss function to train TransPross. We set the number of training epochs to 100 with the early stopping to reduce overfitting. If there is no improvement in the validation loss for five consecutive epochs, the training stops. We train through tenfold cross validation. Five TransPross best models have been selected, and the final output is the ensemble of the five models.

3.3 Evaluation Metrics

We compare TransPross with a state-of-the-art profile-based protein secondary structure predictor PSIPRED and a single sequence predictor SPOT-1D. We evaluate the prediction performance in terms of the Q3 accuracy score, commonly used in the secondary structure prediction. The Q3 accuracy score is defined as the percentage of the correctly predicted secondary structures in three categories (strand E, helix H, and coil C) [43].

4 Conclusion

In this work, we introduce a novel transformer method based on the language model (TransPross) for protein secondary structure prediction. It takes a MSA of a protein as input and automatically encodes it as internal features for a decoder to predict three-class secondary structures of the protein. By using the column attention, TransPross is able to assign attention weights in each column corresponding to each residue position in the protein and extracts the evolutionary patterns to predict the per-residue secondary structure based on all the interdependencies over columns and rows across the whole MSA. In addition to achieving the state-of-the-art performance, TransPross can take a user-provided MSA from any source to generate secondary structure prediction quickly, making it a convenient tool for users to predict secondary structures at a large scale. TransPross can also be used by developers to create more deep-learning language and transformer models to improve protein secondary structure prediction.

Declarations

- *Funding: The work was partly supported by the National Science Foundation [grant no.: DBI1759934 and IIS1763246], National Institutes of Health [grant no.: R01GM093123], and the Thompson Missouri*

  Distinguished Professorship.
- *Conflict of interest/Competing interests: There is no conflict of interest.*
- *Consent for publication*

  All the authors approve the manuscript for publication.
- *Availability of data and materials*

  The data is available at https://zenodo.org/record/6762376#.Ys4St5PMKsA.
- *Code availability*

  The source code is available at https://github.com/BioinfoMachineLearning/TransPro.
- *Authors' contributions*

  JC conceived the project. TW and JC designed the method. TW implemented the method, performed the experiment, and collected the data. TW and JC analyzed the data. WC reviewed and summarized some related methods. TW and JC wrote the manuscript.

References

1. Verlinde CL, Hol WG (1994) Structure-based drug design: progress, results and challenges. Structure 2(7):577–587
2. Anderson AC (2003) The process of structure-based drug design. Chem Biol 10(9):787–797
3. Śledź P, Caflisch A (2018) Protein structure-based drug design: from docking to molecular dynamics. Curr Opin Struct Biol 48:93–102
4. Thornton JM, Laskowski RA, Borkakoti N (2021) AlphaFold heralds a data-driven revolution in biology and medicine. Nat Med 27(10):1666–1669
5. Adhikari B, Cheng J (2017) Improved protein structure reconstruction using secondary structures, contacts at higher distance thresholds, and noncontacts. BMC Bioinform 18(1): 1–13
6. Yang J, Zhang Y (2015) Protein structure and function prediction using I-TASSER. Curr Protoc Bioinformatics 52(1):5–8
7. Senior AW, Evans R, Jumper J et al (2020) Improved protein structure prediction using potentials from deep learning. Nature 577(7792):706–710
8. Pauling L, Corey RB, Branson HR (1951) The structure of proteins: two hydrogen-bonded helical configurations of the polypeptide chain. Proc Natl Acad Sci USA 37(4):205–211
9. Pauling L, Corey RB (1951) Configurations of polypeptide chains with favored orientations around single bonds: two new pleated sheets. Proc Natl Acad Sci USA 37(11):729–740
10. Kabsch W, Sander C (1983) Dictionary of protein secondary structure: pattern recognition of hydrogen-bonded and geometrical features. Biopolymers 22(12):2577–2637
11. Rost B, Sander C (1993) Prediction of protein secondary structure at better than 70% accuracy. J Mol Biol 232(2):584–599
12. Spencer M, Eickholt J, Cheng J (2014) A deep learning network approach to ab initio protein secondary structure prediction. IEEE/ACM Trans Comput Biol Bioinform 12(1):103–112
13. Jones DT (1999) Protein secondary structure prediction based on position-specific scoring matrices. J Mol Biol 292(2):195–202
14. Bettella F, Rasinski D, Knapp EW (2012) Protein secondary structure prediction with SPARROW. J Chem Inf Model 52(2):545–556
15. Magnan CN, Baldi P (2014) SSpro/ACCpro 5: almost perfect prediction of protein secondary structure and relative solvent accessibility using profiles, machine learning and structural similarity. Bioinformatics 30(18):2592–2597
16. Cuff JA, Barton GJ (2000) Application of multiple sequence alignment profiles to improve protein secondary structure prediction. Proteins 40(3):502–511
17. Wang S, Peng J, Ma J et al (2016) Protein secondary structure prediction using deep convolutional neural fields. Sci Rep 6(1):1–11
18. Heffernan R, Yang Y, Paliwal K et al (2017) Capturing non-local interactions by long short-term memory bidirectional recurrent neural networks for improving prediction of protein secondary structure, backbone angles, contact numbers and solvent accessibility. Bioinformatics 33(18):2842–2849
19. Fang C, Shang Y, Xu D (2018) MUFOLD-SS: new deep inception-inside-inception networks for protein secondary structure prediction. Proteins 86(5):592–598
20. Jeanmougin F, Thompson JD, Gouy M et al (1998) Multiple sequence alignment with Clustal X. Trends Biochem Sci 23(10): 403–405
21. Bhagwat M, Aravind L (2008) PSI-BLAST tutorial. In: Bergman NH (ed) Comparative genomics. Humana Press, Totowa, pp 177–186
22. Remmert M, Biegert A, Hauser A et al (2012) HHblits: lightning-fast iterative protein sequence searching by HMM-HMM alignment. Nat Methods 9(2):173–175
23. Johnson LS, Eddy SR, Portugaly E (2010) Hidden Markov model speed heuristic and iterative HMM search procedure. BMC Bioinform 11(1):431
24. Singh J, Litfin T, Paliwal K et al (2021) SPOT-1D-single: improving the single-sequence-based prediction of protein secondary structure, backbone angles, solvent accessibility and half-sphere exposures using a large training set and ensembled deep learning. Bioinformatics 37(20):3464–3472
25. Lyu Z, Wang Z, Luo F et al (2021) Protein secondary structure prediction with a reductive deep learning method. Front Bioeng Biotechnol 404
26. Zhao Y, Liu Y (2021) OCLSTM: optimized convolutional and long short-term memory neural network model for protein secondary structure prediction. PLoS One 16(2): e0245982
27. Guo Z, Hou J, Cheng J (2021) DNSS2: improved ab initio protein secondary structure prediction using advanced deep learning architectures. Proteins 89(2):207–217

28. Rao RM, Liu J, Verkuil R et al (2021) MSA transformer. Proceedings of the 38th International Conference on Machine Learning, PMLR 139:8844–8856
29. Rao R, Meier J, Sercu T et al (2020) Transformer protein language models are unsupervised structure learners. bioRxiv. https://doi.org/10.1101/2020.12.15.422761
30. Elnaggar A, Heinzinger M, Dallago C et al (2020) ProtTrans: towards cracking the language of life's code through self-supervised deep learning and high performance computing. arXiv:2007.06225
31. Nambiar A, Liu S, Hopkins M et al (2020) Transforming the language of life: transformer neural networks for protein prediction tasks. bioRxiv. https://doi.org/10.1101/2020.06.15.153643
32. Steinegger M, Mirdita M, Söding J (2019) Protein-level assembly increases protein sequence recovery from metagenomic samples manyfold. Nat Methods 16(7):603–606
33. Zhang C, Zheng W, Mortuza SM et al (2020) DeepMSA: constructing deep multiple sequence alignment to improve contact prediction and fold-recognition for distant-homology proteins. Bioinformatics 36(7): 2105–2112
34. Wu T, Hou J, Adhikari B et al (2020) Analysis of several key factors influencing deep learning-based inter-residue contact prediction. Bioinformatics 36(4):1091–1098
35. Jumper J, Evans R, Pritzel A et al (2021) Highly accurate protein structure prediction with AlphaFold. Nature 596(7873):583–589
36. Steinegger M, Söding J (2017) MMseqs2 enables sensitive protein sequence searching for the analysis of massive data sets. Nat Biotechnol 35(11):1026–1028
37. Vaswani A, Shazeer N, Parmar N et al (2017) Attention is all you need. Adv Neural Inf Proces Syst 30
38. Graves A (2013) Generating sequences with recurrent neural networks. arXiv:1308.0850
39. He K, Zhang X, Ren S et al (2016) Deep residual learning for image recognition. Proc IEEE Conf Comput Vis Pattern Recognit:770–778
40. Ba JL, Kiros JR, Hinton GE (2016) Layer normalization. arXiv:1607.06450
41. Srivastava N, Hinton G, Krizhevsky A et al (2014) Dropout: a simple way to prevent neural networks from overfitting. J Mach Learn Res 15(56):1929–1958
42. Kingma DP, Ba J (2014) Adam: a method for stochastic optimization. arXiv:1412.6980
43. Lee J (2006) Measures for the assessment of fuzzy predictions of protein secondary structure. Proteins 65(2):453–462

Chapter 4

Importance of Secondary Structure Data in Large Scale Protein Modeling Using Low-Resolution SURPASS Method

Aleksandra E. Badaczewska-Dawid and Andrzej Kolinski

Abstract

Secondary structure elements, such as alpha helices and beta strands, play a fundamental role in defining the overall fold of a protein. Leveraging secondary structure information is essential for encoding the structural features in coarse-grained protein models. Such models simplify the representation of amino acid residues, thereby reducing computational complexity. By incorporating accurate (even if only partial) secondary structure data, the models can efficiently search for the native conformation of proteins and preserve the core structural motifs across extended time frames. Here, the pivotal role of (predicted) secondary structure data in the coarse-grained modeling of protein tertiary and quaternary structures, along with their long-time dynamics, is investigated. Computational simulations of large protein systems using a low-resolution SURPASS model were performed. These case studies demonstrate the sufficiency of predicted secondary structure data in an accurate fold assembly. It leads to a realistic depiction of long-time dynamics in the recorded pseudo-trajectories by employing the Monte Carlo dynamics sampling schema, based on a long random sequence of local conformational modifications. This approach may provide a powerful tool for investigating the critical stages of protein folding. Future combination with knowledge-based potentials derived using machine learning techniques offers exciting opportunities to unravel the underlying mechanisms of biological processes in a variety of molecular complexes.

Key words Protein dynamics, Coarse-grained, Knowledge-based, Multiscale modeling, SURPASS

1 Introduction

Natural proteins are specific oligopolymers, built from twenty types of amino acid residues of the same chirality. The compositions and sequences of amino acids of these proteins may appear to be quite irregular, but they are definitely not random, and these sequences had been carefully edited during evolution processes responsible for the development of a plethora of living organisms. At specific conditions (in vitro or in silico), most natural proteins adopt well-defined three-dimensional structures, necessary for their biological functions. Similarly to many small molecules, many proteins can be crystallized, and quite frequently their crystal structures may

Andrzej Kloczkowski et al. (eds.), *Prediction of Protein Secondary Structure*, Methods in Molecular Biology, vol. 2867,
https://doi.org/10.1007/978-1-0716-4196-5_4,

maintain some of their biological functions. Therefore, it is believed that in many cases, protein crystallographic structures represent their functional forms in living organisms. Synthetic copolymers, composed from the same set of amino acids and with similar contents of particular residues, usually do not assemble unique structures.

Natural proteins are unique polymers for many other reasons. For instance, at appropriate conditions, they fold into their native structure quite rapidly. This phenomenon was the origin of the famous Levinthal paradox [1]. The conformational space of flexible oligomeric chains is immense, and a random search for a single lowest energy space may last ages of time. The explanation of fast folding is quite simple [2]. Namely, due to the specific local ordering and local stiffness of natural chains, the available conformational space becomes significantly narrowed. Secondary structure fragments emerge very rapidly, and the building of the entire native structure becomes reduced to the assembly from a few "prefabricated" blocks. Random (synthetic) oligopeptides usually do not exhibit appropriate secondary structure preferences (or accessibility) and therefore do not form dense, unique globules. These are the main reasons for the great importance of the secondary structure concept for protein structural biology. Of course, as for many other rules of structural biology, there are exceptions. Some natural proteins are structurally less regular, or even unstructured, and some synthetic proteins may be very similar to well-structured natural proteins [3].

Secondary structure assignments are based on the local shape of the main chains of polypeptides, which is mostly controlled by patterns of main chain hydrogen bonds. Assumed classifications of secondary structure may differ significantly, although most of them can be reduced to three terms, helices, beta-sheets, and less regular turns or loops. There are other regular types of secondary structure, more regular as beta-helices and less regular as various types of turns and loops, but they always adopt specific patterns of main chain hydrogen bonds. For the sake of clarity, we will limit our discussion to three types of secondary structure: α-helix, β-sheet, and loops, although the content of this work can be easily extended onto more specific classes of secondary structures. There were many methods for protein secondary structure prediction [4–7], although the most successful approaches were based on multiple sequence alignments drawing from the increasing size of structural data and from elaborated bioinformatics tools [8, 9]. With the growing size of protein data banks (PDB), the accuracy of the secondary structure prediction was permanently increasing [10].

1.1 Importance of Secondary Structure Data in Protein Modeling with Coarse-Grained Methods

Knowledge of secondary structure propensities can be very productive in structure prediction and modeling of protein dynamics and folding transitions [11], especially when the modeling methods are based on knowledge-based statistical models of interactions [12–16], although methods using physics-based modeling approaches can also benefit from secondary structure data in the form of distance biases [17].

The medium-resolution CABS coarse-grained modeling tool is an example of the knowledge-based approach that is strongly focused on the efficient use of general structural regularities that distinguish proteins from other polymers and are common for a majority (or a large subset) of natural proteins. The employed level of resolution of the CABS model (CA, CB, Sidechain) is a compromise between the simulations' speed and the necessity of dependable reconstruction of atomistic structures [18]. The force field is statistical, mimicking the effects of implicit solvent, interactions between side chains, and hydrogen-bond electrostatic interactions. In this context, the secondary structure data play a crucial role. The known, or predicted, secondary structure assignments (in the abovementioned 3-state convention) provide a template for very weak distance biases, preferring (but not enforcing) helix-like or beta-strand-like short-range shaping of protein chains. Such an approach allows productive simulations of the entire folding-unfolding process of not-too-large proteins. Local relaxation and flexibility can be very realistically simulated for very large protein systems.

Although the CABS model uses a very fast Monte-Carlo Dynamics (MCD) modeling strategy, providing realistic pictures of long-time system dynamics, it is still too slow for modeling large-scale transitions in big protein systems. That was a reason for designing a lower resolution and therefore faster coarse-grained modeling tool, SURPASS, used in this work (Fig. 1).

The low-resolution SURPASS coarse-grained modeling tool is another example of the knowledge-based approach that not only ensures computational feasibility and efficiency but also preserves crucial biophysical phenomena, contributing to the model's accuracy and practical applicability. SURPASS (Single United Residue per Pre-Averaged Secondary Structure fragment) uses a crude representation of protein chains. The SURPASS chains have the form of smoothened main chain traces. There are no specific properties of the united residues directly related to individual amino acids. The only sequence-specific information is encoded by the secondary structure assignments. The chain fragments treated as helical prefer characteristic for α-helices density of packing along the chain, and thick excluded volume spheres repulsing the united residues distant along the protein chain. The united residues representing β-strands are smaller, and directional attractive forces mimic the hydrogen-bond interactions between the parallel or anti-parallel strands. Of

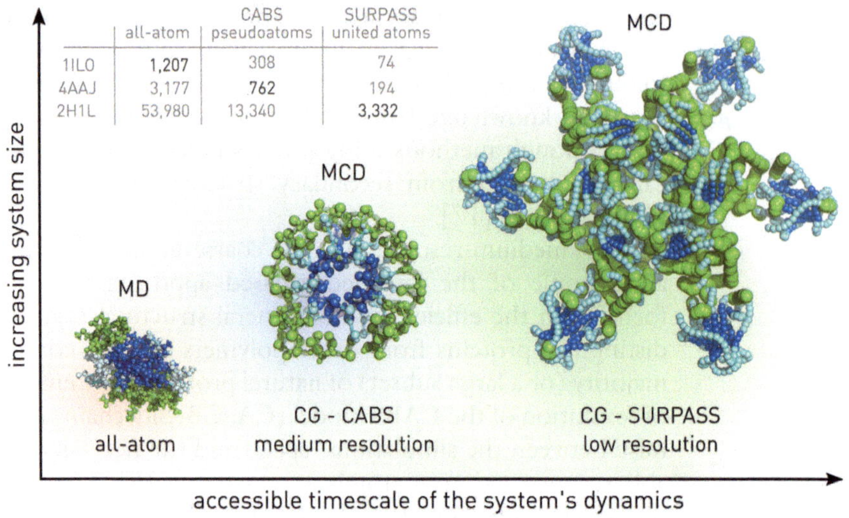

Fig. 1 Simulating larger systems requires more computational resources and often requires longer simulation times to obtain meaningful results. Therefore, highly accurate all-atom with Molecular Dynamics (MD) approaches are limited to smaller systems (roughly) up to 3000 atoms (PDB: 1ILO, 77 AA). Coarse-grained (CG) models simplify the representation of the system by grouping various atoms into a single pseudo atom. For example, the medium-resolution CABS model reduces the system size about 4 times (PDB: 4AAJ, 197 AA) while the low-resolution SURPASS model compresses it by 16 (PDB: 2H1L, 3335 AA). Combining CG representations with Monte Carlo Dynamics (MCD) sampling allows for longer simulations of larger systems

course, all united residues can change their character, although the energetically favorable states are defined by the experimental or predicted secondary structure assignments. Therefore, the secondary structure preferences, the preferable packing distances for the united residues, and the directional preferences of the main chain hydrogen bond counterparts are the only sequence-specific energetic contributions controlling the Monte Carlo dynamics sampling of the model chains. In our opinion, these features of the SURPASS model mimic the most fundamental, although in a very crude fashion, the essence of the global ordering of natural protein chains. As demonstrated already [19] the approximately correct secondary structure data are sufficient for the simulation of protein-like folding processes, thereby confirming the previously outlined explanations of Levinthal's paradox.

The versatility of the SURPASS model extends to its compatibility with multi-scale modeling, enabling a smooth transition from low coarse-grained resolution to more accurate (medium coarse-grained) CABS modeling, then to atomistic structures, and finally high-quality molecular dynamics (MD) refinements (Fig. 2). This interoperability facilitates a hierarchical modeling approach, where the initial broad strokes of protein structure dynamics established by SURPASS can be incrementally fine-tuned using CABS. Both CG models fit very well into integrative modeling protocols, using

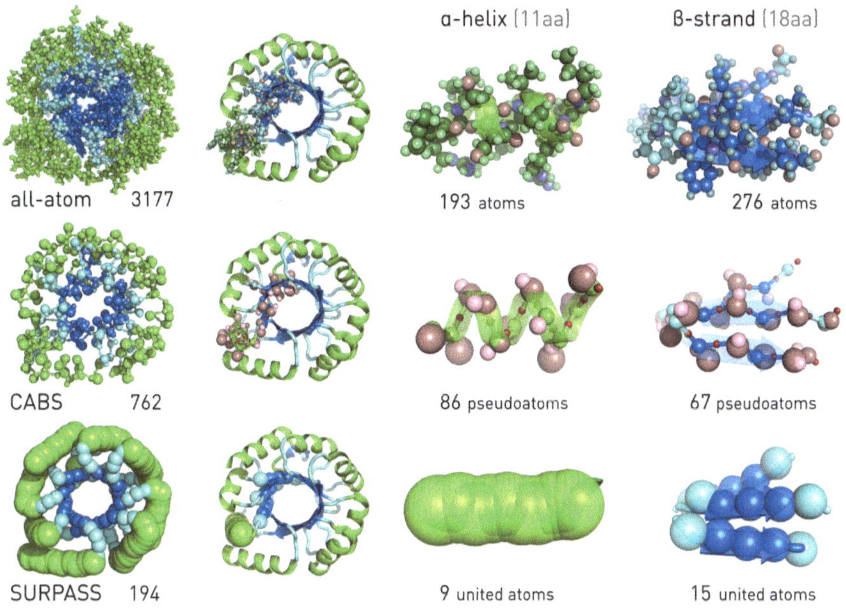

Fig. 2 The comparison of different representations of a protein structure (PDB: 4AAJ, composed of 197 amino acids) with the context of secondary structure (helix—green, extended—marine blue, loop—cyan). The system size decreases from all-atom representation (top row) of 3177 atoms to coarse-grained models: medium-resolution CABS (middle row) with 762 pseudo atoms (4 per each residue), and low-resolution SURPASS (bottom row) with 194 united atoms (~1 per residue). The two last columns show details of protein fragments shaped to α-helix and β-sheet, respectively

a variety of low-resolution and fragmentary experimental or bioinformatics data. It makes the emerging multiscale modeling scheme very efficient and easy to use, maximizing both computational efficiency and structural accuracy in the increasingly granular modeling workflow (Fig. 3).

1.2 The Next-Generation Knowledge-Based Force Fields Use AI-Powered Statistical Potentials

During the last few years, we witnessed huge progress in the development of AI methods in protein structural biology [17, 20–22]. For a large fraction of known proteins, the AlphaFold 2 is capable of predicting their three-dimensional structures with accuracy similar to the accuracy of experimental methods [23], supported by appropriate interpretation and computational refinement. Classical comparative modeling methods (or homology modeling) for in silico structure prediction are now to a large degree outdated [23]. At the same time for a fraction of large proteins and their complexes, the applicability of the AI methods for structure prediction is still limited [20]. Also, the prediction of dynamic properties of proteins and protein complexes remains a challenge of structural biology. On the other hand, the growing size of good quality structural data, now qualitatively accelerated, opens new possibilities for designing the statistical force fields for modeling protein structure and dynamics. Other

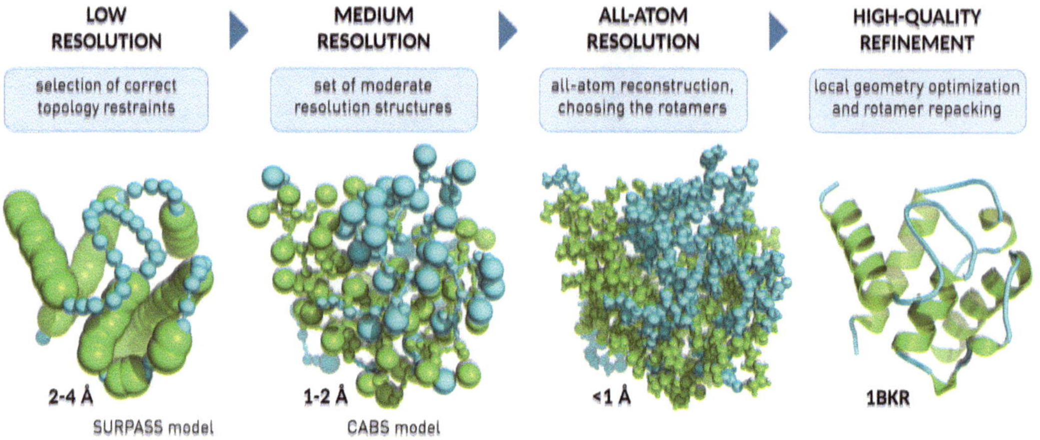

Fig. 3 Steps in the multiscale modeling pipeline of protein structure and dynamics. The approach begins with low-resolution SURPASS simulations that provide crude restraints (2–5 Å of resolution) for estimated correct topology. In the second step, moderate-resolution structures (1–2 Å) are generated using CABS, a more accurate coarse-grained model. A third step enhances the level of precision in the highest-ranked models by reconstructing atomic details. The final step aims to refine the high-resolution structures by considering energetics and fine-tuning their conformation and interactions

AI methods, especially those based on the protein language concept [24, 25] can be also very useful for the derivation of knowledge-based force fields on a new level of fidelity.

2 Methods

2.1 SURPASS Representation of Protein Structure

For this work, we employed the low-resolution SURPASS model for coarse-grained protein structure representation. The model utilizes a single united residue to average the four-residue long fragments along the protein chain, resulting in a compact and simplified shape. In terms of resolution, the unification of a few residues (all heavy atoms, including the peptide plate) goes below the accuracy of the CA trace. However, to ensure the inclusion of all residues, the merging frame moves by one position along the sequence. It guarantees that each amino acid (excluding the chain ends) contributes partially to four consecutive united pseudo atoms. Such a compact mapping, unique to the SURPASS representation, offers several advantageous features. (i) The number of beads in the coarse-grained chain corresponds to (n-3), where n is the amino acid sequence length. (ii) The packing pattern of the pseudo atoms captures the characteristic regularities of secondary structure elements, so we distinguish between 3 types of pseudo atoms: H (helical), E (extended), and C (coil). (iii) The overall representation resembles the topology of the protein in the commonly embraced "helix and ribbon" visualization (Fig. 4).

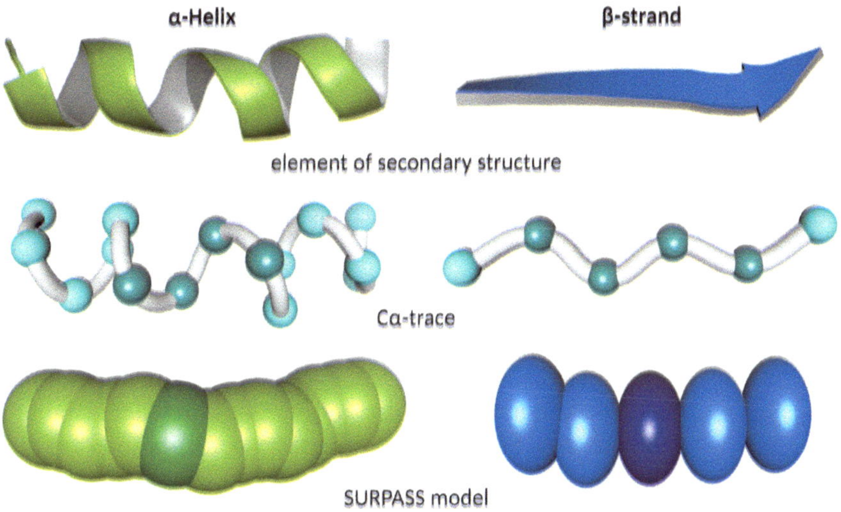

Fig. 4 Proteins form α-helices and β-sheets through hydrogen bonding (HB) interactions in the backbone (top row). α-Helices exhibit a repeating "helical wheel" pattern with HBs between residues (i) and (i+4) (middle row). The SURPASS model employs the same averaging frame, using a 4-residue-to-single-bead simplification of the system. It helps capture collective hydrogen bonding behavior and these implicit preferences in SURPASS support regular secondary structure formation. The shape and size of the SURPASS united atoms (H helix, E extended, C coil) determines the excluded volume preventing unphysical interactions

By reflecting secondary structure regularities in the model definition, SURPASS provides a context-rich view of protein structure while reducing the complexity of a system. It enables efficient exploration of protein folding, dynamics, and interactions at a coarse-grained level.

2.2 Knowledge-Based Forcefield

Unlike physics-based forcefields that rely on complex physical equations, knowledge-based approaches simplify the representation of protein interactions by leveraging empirical information. It makes them computationally more efficient while providing valuable insights into intricate protein interaction patterns. Knowledge-based force fields typically comprise several statistical potentials, which collectively describe the various types of preferences in protein structures. Each potential can represent a particular type of interaction or structural feature. For example, one statistical potential might represent short-range interactions between atoms, another—hydrogen bonding, the next one—secondary structural preferences, and so on. These potentials are derived from statistical analysis of the (experimental) data from the representative subset of high-resolution protein structures, approximating the observed regularities exhibited by different pairs of amino acids. The optimized weighted contributions from the various statistical potentials can serve as a potential of mean force (PMF) [26, 27]. The PMF essentially provides an effective, single-body potential that represents the net effect of many complex, multi-body interactions. By summing up the contributions, the PMF describes energetics and interactions of biomolecular systems as a rather simple function of several collective variables or coordinates. Consequently, the PMF delivers a comprehensive representation of the system's potential energy landscape, offering a computational advantage without compromising on detail.

By applying the PMF approach, the SURPASS force field describes the interactions in a broad range of protein systems. It incorporates secondary structure information to encode sequence effects and simplifies multi-body interactions enabling the implicit inclusion of solvent effects [28]. A few geometrical restrictions and parameters are implemented to enforce regularity and avoid non-physical interactions in the model. This includes distance and angle preferences for secondary structure elements and regular hydrogen bonds, as well as local repulsions to prevent excessive structure collapse. The interactions between pseudo residues are determined based on their secondary structure types, and specific contacts are rewarded with low soft-attraction energy. The optimized PMF converts the statistical frequencies of observed structural features in known protein structures into energy values, guiding the protein folding toward biologically realistic conformations.

Through the integration of the knowledge-based potentials, the SURPASS model achieves a comprehensive encoding of protein behavior, facilitating the exploration of protein structure, dynamics, and interactions at a surprisingly high level of accuracy as for the low-resolution model [19, 29].

2.3 Monte Carlo Dynamics Sampling

Monte Carlo (MC) sampling focuses on efficiently sampling discrete conformations of protein structure based on an energy function and probabilistic criteria, while molecular dynamics (MD) accurately simulates the time evolution of a system's dynamics through the solution of Newton's equations of motion. In contrast, Monte Carlo dynamics (MCD) combines elements from both approaches. It incorporates random moves or modifications of the protein chain, akin to MC, to explore the conformational space, and captures the dynamic behavior of the system over an extended time frame, similar to MD [30, 31].

MCD typically involves local conformational alterations, of which acceptance or rejection is determined by a probabilistic criterion and the corresponding change in energy. This approach enables more efficient and controlled exploration of the conformational space while yielding a registered pseudo-trajectory that provides a realistic depiction of the system's dynamics after a sufficiently long simulation.

In this study, we employed MCD with replica exchange (RE), leveraging 12 system copies to improve the exploration of different regions in the energy landscape (Fig. 5). The primary purpose of employing multiple replicas is to enhance the system's dynamics

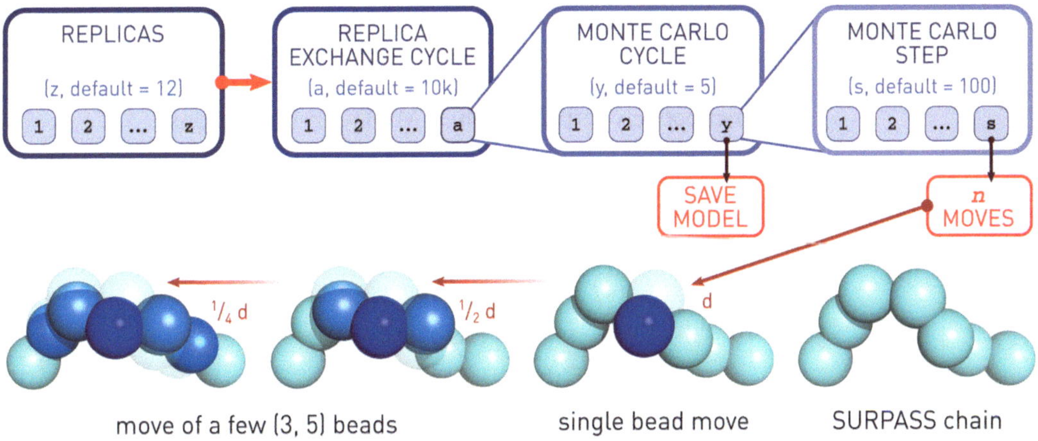

Fig. 5 The Monte Carlo Dynamics sampling scheme is employed in simulations using the SURPASS model. The pipeline consists of $z = 12$ replicas, initialized at different reduced temperature factors, and undergoes a total of $a = 10{,}000$ replica exchange events during the simulation time. Each replica exchange cycle incorporates $y = 5$ Monte Carlo (MC) cycles. After each cycle, a snapshot of the system conformation is recorded, resulting in a total of 50 k models per replica. Within each MC cycle, $s = 100$ MC steps are performed, attempting to make moves on $n = 3$ different types. In a multi-bead move, the distance range is sequentially halved for each pseudo atom at either end of the manipulated fragment. These moves are accepted or rejected based on the Metropolis criterion. This approach enables efficient exploration of the system's conformational landscape and the generation of a large number of models with realistic geometry

and avoid entrapment in local energetic minima. This is achieved by simulating each replica under slightly varied conditions, modulated by a reduced temperature factor. A higher factor correlates with increased system dynamics, enabling the system to overcome larger energetic barriers. This is particularly evident when neighboring replicas exchange reduced temperature factors, allowing them to bypass potential local minima.

We implemented short-range movers that specifically operated on 1, 3, and 5 beads within the SURPASS representation of the protein (Fig. 5). By focusing on the local modifications only, we targeted sampling of nearby conformational states, minimizing the disruption to the overall protein fold. The acceptance of the structural changes was determined using the well-established Metropolis criterion, maintaining the system's energy integrity throughout the sampling process.

2.4 Software Availability

SURPASS model, encompassing the protein chain representation, force field, and modeling framework, is readily accessible to the scientific community through the open-access repository hosted on GitHub, https://github.com/aedawid/surpass. The software, implemented in the C++ programming language, is available for download and requires compilation before utilization. Detailed instructions for downloading and compiling the software can be found in the accompanying documentation.

Upon successfully compiling the code, the "surpass" executable will be generated in the surpass/bin directory. To facilitate simulation scheduling, we provide an example syntax to guide users in effectively setting up their MCD-RE simulations:

```
./surpass \
 -verbose=FINE \
 -in:database=./ \
 -in:pdb=input.pdb -in:ss2=input.ss2 \
 -in:pdb:native=native-surpass.pdb \
 -out:pdb=tra.pdb \
 -sample:mc_outer_cycles=100
 -sample:mc_inner_cycles=10 \
 -sample:exchanges=10000 \
 -sample:replicas=1.0,1.1,1.2,1.3,1.4,1.5,1.6,1.7,1.8,
  1.9,2.0,2.2 \
 -sample:replicas:observation_mode=1
```

More tips for the necessary parameters and commands for specifying simulation details are outlined in the Notes section, enabling users to customize and execute simulations tailored to their specific research needs.

This study utilized the original version of the SURPASS model, which incorporates enhancements, updates, and improvements to the model published by Dawid et al., 2017–18 [19, 28].

3 Case Studies

In previous studies [19], we conducted systematic computational experiments by applying the SURPASS model to fold de novo a diverse array of single-domain globular proteins. The simulation settings and numerical outcomes were comprehensively detailed in the cited publications. From this extensive benchmark, we carefully selected representatives to demonstrate an in-depth case study, emphasizing the critical role of secondary structure in protein modeling. As a background story, we also provide a nuanced analysis of deeply coarse-grained results, illustrating the model's efficacy in balancing computational efficiency with structural accuracy.

The PDB entry 2W9Q reveals a single cystatin-like domain from Solanum tuberosum (potato) capable of assembling into larger multi-cystatin complexes, thereby enhancing its role as a cysteine protease inhibitor for plant defense against pests and pathogens [32, 33]. A single domain retains functionality and is well-folded independently; it exhibits a typical cystatin fold, composed of a tight 4-strand beta-sheet layered with a long alpha-helix, extended by a disordered N-terminal region hinting at regulatory functions or roles in complex assembly. Thus, overall secondary structure content is high, approximately 80%, constituting a compact $\alpha+\beta$ architecture. Notably, regions of increased hydrophobicity are present at both termini, contributing to structural stability, while a central hydrophilic region (spanning positions 43–47) features the critical conserved QXVXG motif, essential for protease interaction (https://www.rcsb.org/sequence/2W9Q). We selected this example for in-depth analysis anticipating it will effectively demonstrate the interplay between (primary and secondary) structural properties and protein fold assembly.

3.1 Identifying the Seed of Folding Events

The computational folding of a cystatin-like domain (PDB: 2W9Q) from a fully extended conformation involved using simultaneously 12 copies of the system, each sampling the conformational space independently. Consequently, we obtained 12 distinct pseudo-trajectories composed of system snapshots (models) registered periodically during simulation. These models were compared structurally to a known reference structure (in the SURPASS representation) by calculating the root-mean-square deviation (RMSD) allowing generation plots shown in Fig. 6. The RMSD values fluctuate over the course of the MC simulation for all replicas, indicating that the protein explored various conformations relative to the reference structure. Lower RMSD values (<6 Å in a SURPASS representation) indicate models that more closely resemble the known reference. It's important to note that this approach requires a reference structure, which may be a limitation in some applications. However, it remains valuable for validating newly proposed methods (like this one) or studying folding mechanisms and protein dynamics.

The color gradient in the background of each plot represents the reduced temperature factor. In Monte Carlo dynamics simulations this factor is a dimensionless quantity that governs the scale of conformational sampling and the dynamics of the protein. In practice, it affects the amplitude of atomic fluctuations in each attempt of move. Generally, a lower temperature factor (shown in blue shades) facilitates energy minimization leading to folding events or conformational freezing. Sampling at various temperature levels allows the system to escape local minima and potentially find a more stable conformation. Thus, higher temperatures (red shades) increase molecular motions and speed up exploration of the conformational space, facilitating transitions over energetic barriers, often through gradual unfolding.

Replica exchange occurs when the scaled temperature factor is swapped between two neighboring replicas upon meeting a specific

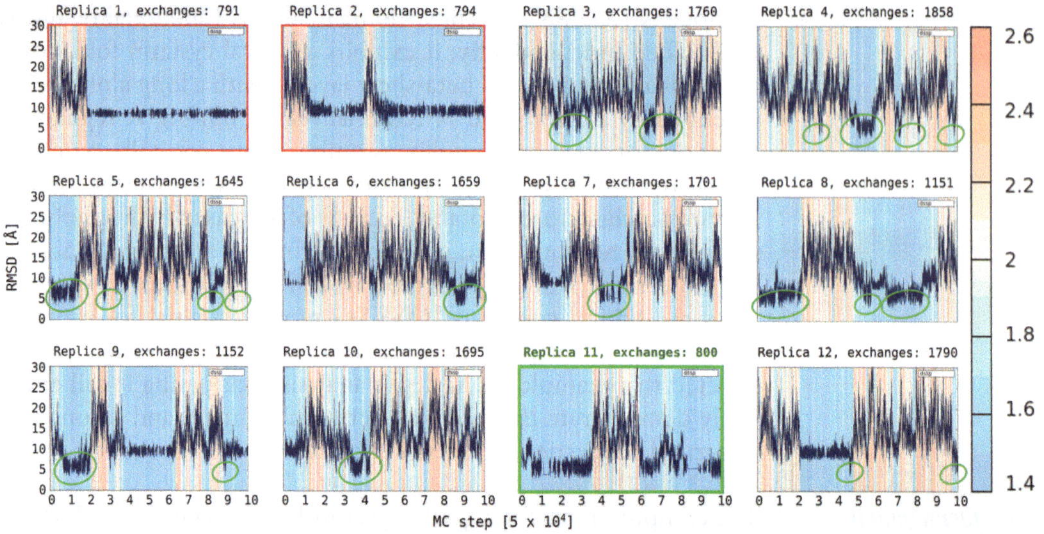

Fig. 6 The summary of SURPASS RE-MCD simulations for cystatin-like domain from Solanum tuberosum (PDB: 2W9Q). The computational experiment utilized 12 copies of the system. Neighboring replicas with an energy difference that meets the Metropolis criterion could exchange their dynamic conditions. The number of successful exchanges for each replica is provided. Each panel illustrates the evolution of RMSD over time [500,000 Monte Carlo steps], indicating the degree of similarity between models in each replica and reference structure, with high similarity marked with green ellipses. The background of each plot is determined by changes of the reduced temperature factor, a key indicator of protein dynamics during simulation. This factor influences the extent to which a protein system can explore its conformational space: lower values minimize conformational energy, whereas higher values enhance dynamics, enabling the system to overcome energetic barriers and escape local minima

energy criterion. The number of exchanges may vary between replicas, as not all exchange attempts are successful. As seen with the first two replicas (red-framed in Fig. 6), frequent rejections can trap a replica in specific conditions for longer periods, resulting in an inefficient sampling. While a higher exchange rate indicates a broader exploration of conformational space, too frequent exchanges (e.g., replica 6) can prevent the system from reaching equilibrium, hindering the approach to the global energy minimum. At both extremes, replicas contribute less effectively to the registered folding process.

Despite the aforementioned obstacles, the overall simulation is still valid as long as there are a few replicas with well-balanced sampling and a few folding-unfolding cycles. The background color-coding aids in estimating the number of these transitions. Ideally, observing several folding events in multiple replicas is expected. In this example, we have many of them, marked with green ellipses. The most valuable are transitions with sharp decreases in RMSD followed by a period of RMSD fluctuations around relatively low value, as seen in replicas 3–11, that may denote stable folded states closer to the reference structure.

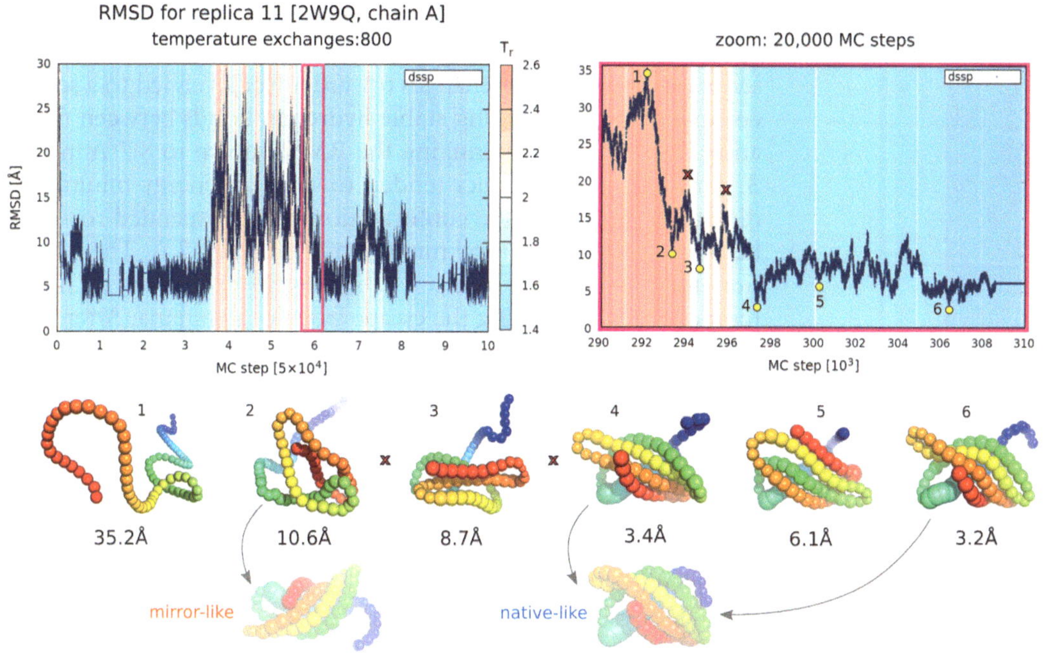

Fig. 7 The example of MC dynamics in a single replica for cystatin-like domain from Solanum tuberosum (PDB: 2W9Q). The main plot traces the RMSD values, demonstrating the conformational evolution over 50×10^4 MC steps with corresponding scaled temperature exchanges. A detailed zoom on 20,000 MC steps highlights key conformational states (numbered 1–6) and energetic barriers (indicated by red 'x' marks) in the selected unfolding-folding transition. Adjacent are structural snapshots capturing the folding progression at critical states, with RMSD values ranging from the initial expanded state (35.2 Å) to the final native-like conformation (3.2 Å)

To fully grasp protein folding mechanisms, a closer examination of multiple transitions is essential. Identifying and analyzing key conformational changes provides crucial insights into the folding process. In particular, the early nucleation and arrangement of secondary structure elements, such as alpha-helices and beta-sheets, are likely to play a significant role in steering the protein toward its final three-dimensional structure. As an illustrative demonstration of such in-depth analysis, we show an example of a rapid folding event leading to a stable state, identified in replica 11 around 30×10^4 MC steps. The chosen transition is highlighted with a pink frame on the full RMSD evolution plot, with the pertinent 20,000 MC steps magnified in a separate plot at the top of Fig. 7.

Upon closer examination of the initially rapid folding event, the zoomed-in view reveals a more intricate process. Starting from an expanded state with an RMSD of ~35 Å (state 1), the protein quickly collapsed to a partially folded conformation with an RMSD of ~10.6 Å (state 2) within 1000 MC steps at high temperatures. This transient state hinted at a potential for mirror-like structure formation under cooler conditions. However, under sustained high temperatures, it re-expanded to about 18 Å, encountering the first energetic barrier (marked with a red 'x'). A series of temperature exchanges toward lower levels led to a second collapse, this time forming stable hydrogen bonds between nucleating beta strands and bringing the RMSD down to 8.7 Å in state 3. Although the dynamics tended towards an energy-minimizing direction, this perishable conformation briefly expanded to ~15 Å before stabilizing at a local minimum of around 12 Å. The simulation then underwent a series of successful exchanges to higher temperatures, facilitating the escape from local minima. After overcoming another unfolding episode at ~17 Å (the second energetic barrier, marked with a red 'x'), a sequence of temperature decreases led to a rapid folding into a native-like topology (state 4) just 3.4 Å from the native structure, featuring a well-formed 4-strand beta sheet layered with an alpha helix. In low-temperature fluctuations, the protein experienced a temporary reorganization of the beta strands yet preserved a near-native topology (state 5) approximately 6 Å from the reference. It ultimately transitioned back into a more stable native-like topology with an RMSD of 3.2 Å (state 6).

The Monte Carlo dynamics with exchanges of the reduced temperature factor between replicas enabled the model chain to efficiently navigate through the conformational space and overcome energetic barriers, as evidenced by the correlation between temperature shifts and the protein's structural transitions. The formation and stabilization of secondary structure elements, particularly the precise spatial organization of beta strands, proved pivotal in directing the folding process toward a native-like topology.

3.2 Identifying the Native State Amidst Conformational Diversity

In a well-tuned Monte Carlo dynamics simulation, the pseudo trajectory exhibits a high conformational diversity due to efficient sampling of the energy landscape, encompassing extended chains, partially folded intermediates, molten globules, and more globular shapes. However, when the modeling is successful, it should also include well-packed structures that closely resemble the native state. Despite the prevalence of various conformations, the native state, being stable and energetically favorable, is more likely to recur multiple times during simulation, making its fraction detectable in the trajectory. By leveraging this assumption, it becomes feasible to identify the native state by focusing on the most densely populated structural clusters within the modeling results. This approach enables the identification of the native topology in scenarios where the true native structure is not known, such as in de novo protein folding simulations.

Clustering can be performed on selected replicas, but a more holistic representation of the conformational space involves using models from all replicas. We merged 6 million models from 12 replicas and downsized the set by selecting every tenth model. In the next step, we created two 20,000-model subsets based on (a) the

Table 1
Table summarizes the structural clustering of two 20 k protein model datasets, selected by lowest contact energy and radius of gyration, using five metrics CMO, GDT, MXS, RMS and TMS across three lintkage criteria: average, maximum, and single

Method	Fraction Average	RMSD	N-clust	Fraction Maximum	RMSD	N-clust	Fraction Single	RMSD	N-clust
Contact energy									
CMO	0.89 (1st)	9.95	2	0.93 (1st)	9.95	3	0.47 (2nd)	4.02	2
GDT	0.62 (1st)	6.23	3	0.64 (1st)	5.90	4	0.09 (6th)	4.01	3
MXS	0.66 (1st)	6.57	4	0.68 (1st)	5.90	5	0.53 (1st)	4.01	3
RMS	0.61 (1st)	4.04	4	**0.69 (1st)**	**4.04**	5	**0.70 (1st)**	**4.05**	3
TMS	0.67 (1st)	6.23	4	0.69 (1st)	5.90	5	0.53 (1st)	4.01	3
Radius of gyration									
CMO	0.08 (6th)	7.06	1	0.12 (4th)	5.94	1	–	–	0
GDT	0.25 (3rd)	4.62	3	0.24 (6th)	5.81	4	<0.01 (4th)	4.96	5
MXS	0.15 (5th)	5.38	4	0.36 (5th)	5.29	6	<0.01 (5th)	4.96	4
RMS	**0.43 (1st)**	5.82	1	**0.48 (1st)**	5.82	2	–	–	0
TMS	0.31 (3rd)	4.62	5	0.35 (4th)	5.15	5	<0.01 (7th)	4.96	3

The fraction column denotes the total proportion of models with native- or mirror-like topology, with N-clust indicating the number of such clusters (out of total 10 clusters). Bracketed indices mark the densest native/mirror cluster. The RMSD values pertain to the medoids of clusters that are closest to a native reference structure, with values ≤6 Å suggesting native topology and those >6 Å indicating mirror-like topology. This topology recognition is based on contact patterns between beta-sheet strands, serving as a basic yet effective fold estimation method

lowest contact energy and (b) the lowest radius of gyration. These subsets were subjected to structural clustering using the Clusco tool [34], employing five different structural similarity metrics (CMO, GDT, MXS, RMS, and TMS) and three linkage criteria (average, maximum, and single), resulting in a total of 15 clusterings per subset. Notably, the best clustering performance was achieved using the RMS (Root Mean Square Deviation) metric for both model selection criteria, as shown in Table 1 and corresponding Fig. 8a, b.

Upon analyzing the fractions and RMSD values associated with the densest clusters, it's evident that the dataset selected by contact energy tends to have a larger fraction of native- or mirror-like topologies compared to the dataset selected by radius of gyration. Contact energy is a more discriminating selection criterion for identifying models with correct topology because it directly measures the energetically favorable interactions that are indicative of protein folding. In contrast, the radius of gyration is a more general measure of size and does not necessarily correlate with the correct folding pattern, thus resulting in a lower fraction of densest clusters being native- or mirror-like.

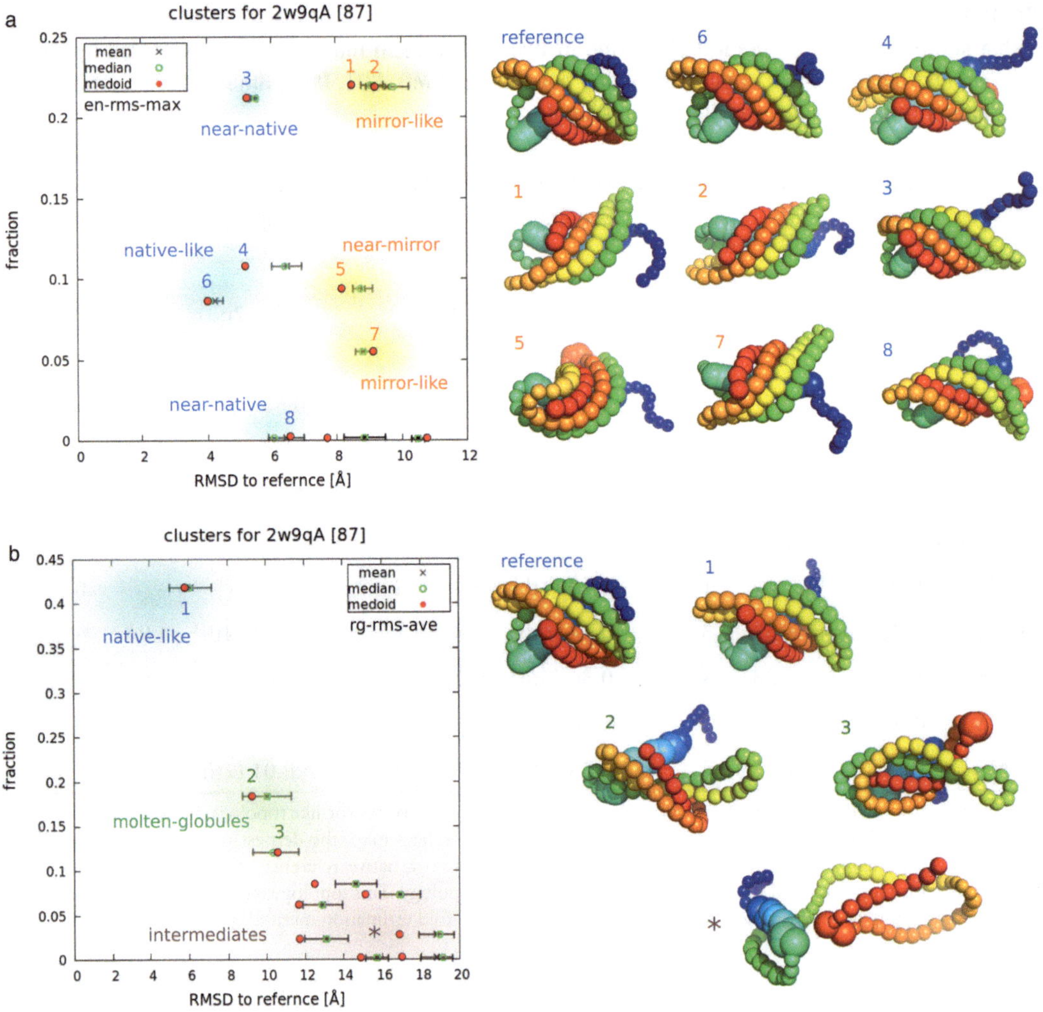

Fig. 8 (**a**) The results of structural clustering of SURPASS-generated protein models selected based on the Contact Energy criterion. The left panel depicts the clusters for protein 2w9qA, with the fraction of structures plotted against the root-mean-square deviation (RMSD) to a reference structure. Central tendencies within each cluster are marked as mean (black x), median (green circle), and medoid (red dot). On the right panel, the representative structures of each cluster are displayed. The structures are color-coded with blue for the N-terminal and red for the C-terminal. Clusters 3, 4, 6 and 8 are notable for their similarity to native conformation. (**b**) The results of structural clustering of SURPASS-generated protein models selected based on Radius of Gyration (rg). The left panel details the clusters identified for the protein model 2w9qA, with fractions of structures plotted against RMSD to a reference state. Central tendencies are indicated for each cluster: mean (black x), median (green circle), and medoid (red dot). The right panel visualizes the representative structures for each cluster, with the color gradient from blue (N-terminal) to red (C-terminal). The most dense cluster (1) is predominantly native-like, suggesting a conformation very similar to the functional fold. Clusters 2 and 3 are molten globules. Clusters marked with an asterisk are composed of partially folded intermediates, representing a variety of less structured, transitional conformations

Structural clustering of coarse-grained SURPASS models selected by the lowest contact energy criterion yielded 10 clusters. Each cluster's central tendencies (mean, median, and medoid) are shown in Fig. 8a. Analysis of cluster representatives (medoids) revealed that the two densest clusters exhibited mirror-like conformations, while the subsequent two were near-native (one strand replaced) or native-like. The majority of the remaining clusters also contain conformations that are either near-mirror or near-native, suggesting that the modeling approach is successful in capturing structurally relevant states. The observation of non-physiological mirror-like structures is an artifact of the current implementation of the knowledge-based forcefield. This is a minor limitation that does not detract significantly from the model's overall utility and can be addressed with future refinements.

Structural clustering of coarse-grained SURPASS models selected by the lowest radius of gyration resulted in the identification of 10 distinct clusters. The central tendencies of each cluster, including the mean, median, and medoid, are depicted in Fig. 8b. The densest cluster, representing about 43% of the models in the subset, exhibited native-like conformations. This illustrates the effectiveness of the SURPASS model in maintaining essential structural details, despite its coarse-grained nature. This fidelity enables a relatively straightforward reconstruction to an all-atom representation. The next two clusters are characterized as molten globules, indicating a state of partial folding where the proteins have attained a compact globular shape. These states are typically transitional. The remaining seven clusters, significantly smaller in fraction, consist of partially folded intermediates; these structures maintain some elements of secondary structure but have not achieved the precise tertiary organization.

The SURPASS model's ability to model protein structures is robust, often identifying conformations close to experimental data across various simulation trajectories, with accuracy slightly decreasing as protein size diminishes. It has the ability to reliably depict a broad spectrum of protein conformations, from highly ordered native-like structures to less ordered intermediate states, which opens an opportunity for many practical applications in protein dynamics studies.

3.3 The Role of Secondary Structure in Protein Folding

In the SURPASS model, akin to general protein folding phenomena, secondary structure nucleation is a key factor steering the folding process, as established in earlier sections. The current focus is on assessing the model's resilience to secondary structure inaccuracies and the extent to which gross errors affect its predictive capabilities.

The SURPASS model utilizes a knowledge-based force field to guide the protein dynamics during the simulation of biological processes such as folding. It uses amino acid sequence and

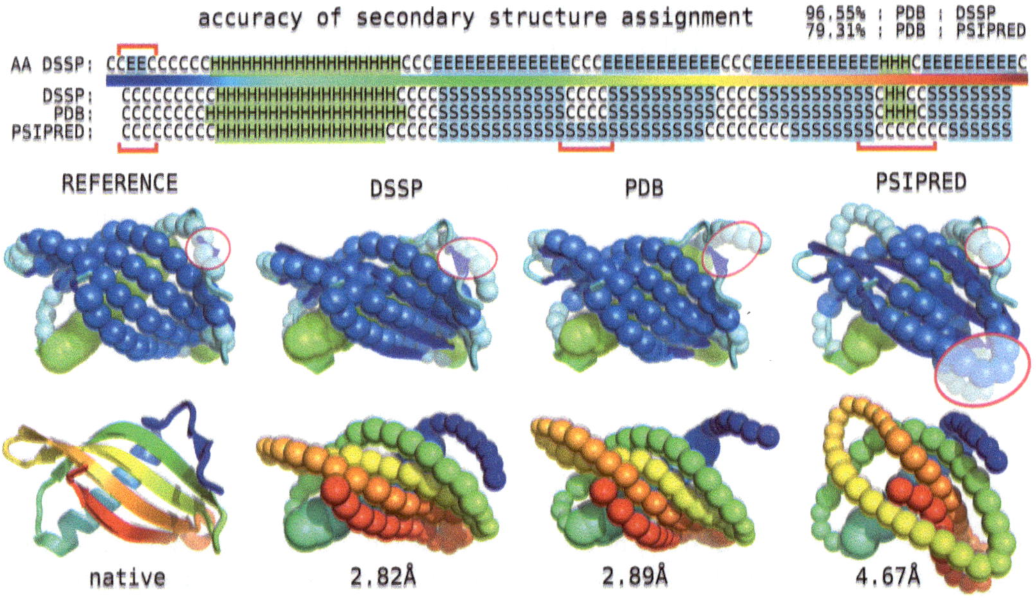

Fig. 9 The accuracy of secondary structure assignments from different sources: DSSP, PDB, and PSIPRED (prediction) for SURPASS representation aligned with DSSP assignment for all-atom (AA) representation. The lower part shows the 3D structures in the SURPASS representation of the best model with RMSD values indicating the deviation from the native structure

simplified secondary structure information as the only input. Notably, the source of secondary structure data has a marginal impact on model accuracy [19]; even predicted secondary structure can sometimes outperform assignments from the Protein Data Bank or DSSP. However, crude errors in secondary structure assignment pose challenges for the SURPASS model. These errors typically manifest as either the mis-assignment of long, continuous secondary structure elements instead of separate segments, or replacing elements of different types such as helices with β-strands, and vice versa. Such inaccuracies can lead to substantial distortions in the resulting models.

Figure 9 presents a comparison of the SURPASS model's performance using different sources of secondary structure data: DSSP, PDB assignments, and PSIPRED predictions. The top part of the figure shows the alignment of the secondary structure in a 3-letter notation (H, E, C) assigned per position in the protein sequence. The first row denotes all-atom representation, while the following three rows are in SURPASS representation. The nature of the SURPASS representation makes the system size trimmed to N-3, which in some cases may also affect the length of secondary structure elements.

In this example, the DSSP source has a minor discrepancy near the start of the sequence where it assigns coil (C) instead of extended strand (E), as indicated by the red brackets. The RMSD for the best model is 2.82 Å, which is close to the native structure, indicating a fairly accurate model despite pruning a short-extended fragment. PDB data seems to have a high degree of accuracy, with only a slight elongation of both helices. The RMSD for the best model is 2.89 Å, which still closely approximates the native structure. The structural difference is apparent mostly on the loose highly flexible N-termini. The PSIPRED predictions reveal several critical errors, especially at the C-terminal where a helix is erroneously omitted, and in the central region where two beta strands are incorrectly merged. These inaccuracies result in a pronounced RMSD of 4.67 Å, indicating a less accurate structure resulting from beta strands (green and yellow) that are incorrectly combined into a single rigid secondary structure element. Such a substantial error severely hampers the model's ability to replicate the native structure.

The SURPASS model demonstrates a robust tolerance from minor (element length, shift along the sequence) to moderate (presence of short elements) inaccuracies in secondary structure data. It maintains satisfactory accuracy in fold reproduction, as long as there are no major misclassifications, such as mistaking helices for beta-structures or merging several elements into non-physically long assignments. This resilience paves the way for successful de novo modeling of larger proteins, expanding protein dynamics studies beyond previous limitations.

4 Discussion

4.1 Applicability and Limitations of the SURPASS Model

The SURPASS model uses a single united atom representation of a protein residue and efficient Monte Carlo dynamics sampling schemes, controlled by a knowledge-based statistical force field. It has been shown that this simplified model of interactions is sufficient for surprisingly realistic reproduction of protein-like conformations, effectively controlling the fold assembly and dynamic properties of globular proteins. Crucially, it also enables the reconstruction of more realistic structures (and trajectories), thereby enhancing the accuracy to medium-resolution coarse-grained models like CABS or highly detailed all-atom structures for molecular dynamics refinement. The connection with finer granularity levels of protein representation opens the possibility for effective multiscale modeling of long-time (and large-scale) processes in larger protein systems. In essence, SURPASS stands out as a highly versatile tool, adept at bridging different scales and complexities in protein modeling.

The model of interactions built into the SURPASS force field is unique. Statistical potentials are not directly sequence-specific. Instead, secondary structure-based regularities define preferred interactions between united residues. Currently, the model supports a three-state classification (helix, strand, loop) though it is readily adaptable to accommodate different subsets of protein folds, such as those characteristic of beta helices. Previous studies, along with insights from this work, show that approximate secondary structure data free from crude errors is sufficient input for SURPASS to reliably trace folding trajectories and accurately identify correct folds using clustering procedures. This advancement paves the way for de novo modeling of larger proteins, extending beyond previous capabilities. Of course, achieving realistic models at atom-level resolution necessitates the integration of SURPASS simulations with higher-resolution methods.

4.2 Future Perspective

In the era of AlphaFold2 and similar methods, the task of predicting three-dimensional folds for single globular proteins appears outdated. However, despite these recent advances in Artificial Intelligence (AI) techniques, such approaches still fall short in predicting structurally heterogeneous proteins such as fold-switching proteins [35], large atypical proteins or complex cases like knotted proteins [36]. Additionally, AlphaFold2 tends to represent a single static structure, facing difficulties in predicting multiple conformations [37]. In these challenging scenarios, tools like SURPASS become invaluable, especially when integrated into multiscale modeling protocols. These approaches are essential for capturing the full complexity of protein behaviors, which often involve intricate processes occurring at different scales, from atomic to macromolecular levels. These methods continue to be critically important as they excel in offering a systematic framework for comprehending molecular mechanisms, enhancing interpretability, and gradually augmenting the level of detail to optimize accuracy and performance. The promising future perspective lies in the effective integration of AI-trained potentials into next-generation statistical force fields [38, 39] and coupling them with smartly projected coarse-grained representations such as SURPASS or CABS within a multiscale pipeline to combine the strengths of both approaches: the power of AI while maintaining realism and accuracy in the modeling process.

5 Notes

5.1 Technical Tips for Users

1. The SURPASS tool provides flexibility in configuring simulations through various options. The defaults are optimized on a large batch of simulations for diverse protein systems. However, many adjustable options allow users to customize their simulations based on their specific requirements and objectives to explore the conformational space efficiently.

./surpass	
-help	:print help message
-verbose	:set the verbosity level
-in:database	:path to bioshell parameters directory
-sample:seed	:sets random generator seed for MC sampling
-sample: mc_outer_cycles	:the number of large MC cycles (outer MC loop) to perform
-sample: mc_inner_cycles	:the number of small MC cycles (inner MC loop) to perform
-sample: mc_cycle_factor	:make each MC cycle N times longer
-in:pdb	:provide an input protein structure(s) in PDB format
-in:pdb:native	:provide the native (or reference) protein structure in PDB format
-out:pdb	:provide an output file to write structure in PDB format
-in:ss2	:provide an input secondary structure in PsiPred's SS2 format
-sample:t_start	:initial reduced temperature factor of the simulation
-sample:t_end	:final reduced temperature factor of the simulation
-sample:t_steps	:the reduced temperature factor increment between steps
-sample:replicas	:reduced temperature factor for replicas in REMC simulation (the number of temperature values defines the number of replicas)
-sample:replicas: observation_mode	:observation mode: 0 - same temperature; 1 - same conformation
-sample:exchanges	:the number of my_sampler exchanges

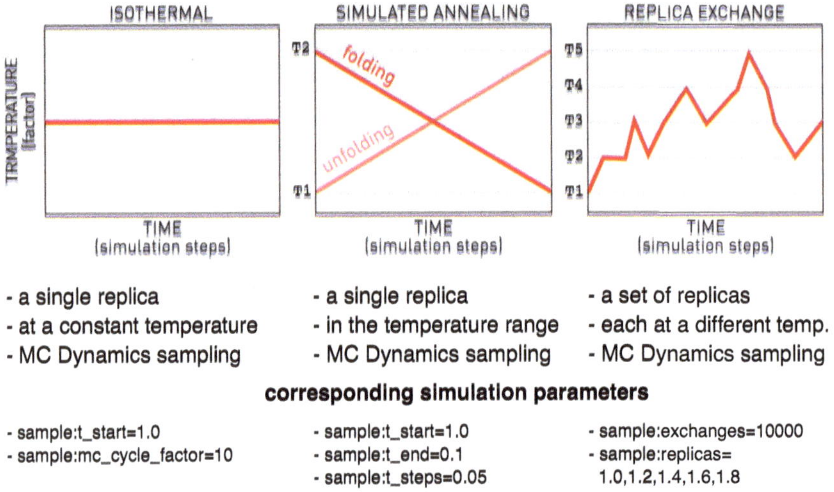

Fig. 10 Illustration of Monte Carlo with Replica Exchange method. Temperature is plotted vs the number of simulation time steps. The left plot shows an isothermal simulation, the central plot shows simulated annealing simulation, and the left plot the replica exchange simulation method

2. In SURPASS, the reduced temperature factor (Tr), which can be either constant (isothermal), gradually decreased (simulated annealing), or varied among replicas (replica exchange), determines the specific simulation variant used and influences the exploration of the conformational space. All simulation variants use Monte Carlo Dynamics as a sampling strategy:
   (a) Monte Carlo sampling on isothermal condition (MCI)
   (b) Monte Carlo with simulated annealing (MCA)
   (c) Monte Carlo sampling with replica exchange (REMC) (Fig. 10)
3. Reconstructed structures from SURPASS trajectory clusters (as demonstrated in the Subheading 3.2) can serve as valuable data for de novo prediction in challenging folding cases, or for a more precise energetic assessment of conformational transitions. The reconstruction can be effortlessly performed at a minimal level, focusing solely on the CA trace. A small set of key long-distance contacts, extracted from these structures, can then be utilized as restraints in higher-level simulations, such as those employing the medium coarse-grained CABS model (refer to Fig. 3 for details). This approach provides an efficient solution to overcome some limitations in current simulation techniques, thereby advancing our understanding and predictive capabilities in complex molecular systems.

Acknowledgments

A.K. acknowledges financial support from the National Science Centre, Grant number 2021/40/Q/NZ2/00078.

References

1. Levinthal C (1969) How to fold graciously. Mossbauer Spectrosc Biol Syst Proc 67:22–26. https://faculty.cc.gatech.edu/~turk/bio_sim/articles/proteins_levinthal_1969.pdf
2. Zwanzig R, Szabo A, Bagchi B (1992) Levinthal's paradox. Proc Natl Acad Sci USA 89:20–22. https://doi.org/10.1073/pnas.89.1.20
3. Bryson JW, Betz SF, Lu HS, Suich DJ, Zhou HX, O'Neil KT et al (1995) Protein design: a hierarchic approach. Science 270:935–941. https://doi.org/10.1126/science.270.5238.935
4. Smolarczyk T, Roterman-Konieczna I, Stapor K (2020) Protein secondary structure prediction: a review of progress and directions. Curr Bioinforma 15:90–107. https://doi.org/10.2174/1574893614666191017104639
5. Lyu Z, Wang Z, Luo F, Shuai J, Huang Y (2021) Protein secondary structure prediction with a reductive deep learning method. Front Bioeng Biotechnol 9:687426. https://doi.org/10.3389/fbioe.2021.687426
6. Kouza M, Faraggi E, Kolinski A, Kloczkowski A (2017) The GOR method of protein secondary structure prediction and its application as a protein aggregation prediction tool. Methods Mol Biol 1484:7–24. https://doi.org/10.1007/978-1-4939-6406-2_2
7. Sen TZ, Jernigan RL, Garnier J, Kloczkowski A (2005) GOR V server for protein secondary structure prediction. Bioinformatics 21:2787–2788. https://doi.org/10.1093/bioinformatics/bti408
8. Rost B, Sander C (1993) Prediction of protein secondary structure at better than 70% accuracy. J Mol Biol 232:584–599. https://doi.org/10.1006/jmbi.1993.1413
9. Rost B (1996) PHD: predicting one-dimensional protein structure by profile-based neural networks. Methods Enzymol 266:525–539. https://doi.org/10.1016/s0076-6879(96)66033-9
10. Rost B (2001) Review: protein secondary structure prediction continues to rise. J Struct Biol 134:204–218. https://doi.org/10.1006/jsbi.2001.4336
11. Kmiecik S, Gront D, Kolinski M, Wieteska L, Dawid AE, Kolinski A (2016) Coarse-grained protein models and their applications. Chem Rev 116:7898–7936. https://doi.org/10.1021/acs.chemrev.6b00163
12. Leman JK, Weitzner BD, Lewis SM, Adolf-Bryfogle J, Alam N, Alford RF et al (2020) Macromolecular modeling and design in Rosetta: recent methods and frameworks. Nat Methods 17(7):665–680. https://doi.org/10.1038/s41592-020-0848-2
13. Zhang Y (2008) I-TASSER server for protein 3D structure prediction. BMC Bioinformatics 9:40. https://doi.org/10.1186/1471-2105-9-40
14. Blaszczyk M, Jamroz M, Kmiecik S, Kolinski A (2013) CABS-fold: server for the de novo and consensus-based prediction of protein structure. Nucleic Acids Res 41:W406–W411. https://doi.org/10.1093/nar/gkt462
15. Jamroz M, Kolinski A, Kmiecik S (2013) CABS-flex: server for fast simulation of protein structure fluctuations. Nucleic Acids Res 41:W427–W431. https://doi.org/10.1093/nar/gkt332
16. Kurcinski M, Jamroz M, Blaszczyk M, Kolinski A, Kmiecik S (2015) CABS-dock web server for the flexible docking of peptides to proteins without prior knowledge of the binding site. Nucleic Acids Res 43:W419–W424. https://doi.org/10.1093/nar/gkv456
17. Liwo A, editor. (2019) Computational methods to study the structure and dynamics of biomolecules and biomolecular processes: from bioinformatics to molecular quantum mechanics, 2nd edn. Springer International Publishing, Cham. http://link.springer.com/10.1007/978-3-319-95843-9
18. Kolinski A (2004) Protein modeling and structure prediction with a reduced representation. Acta Biochim Pol 51:349–371. https://www.ncbi.nlm.nih.gov/pubmed/15218533
19. Dawid AE, Gront D, Kolinski A (2018) Coarse-grained modeling of the interplay between secondary structure propensities and protein fold assembly. J Chem Theory Comput 14:2277–2287. https://doi.org/10.1021/acs.jctc.7b01242

20. Bertoline LMF, Lima AN, Krieger JE, Teixeira SK (2023) Before and after AlphaFold2: an overview of protein structure prediction. Front Bioinform 3:1120370. https://doi.org/10.3389/fbinf.2023.1120370
21. Callaway E (2022) What's next for AlphaFold and the AI protein-folding revolution. Nature 604:234–238. https://doi.org/10.1038/d41586-022-00997-5
22. Senior AW, Evans R, Jumper J, Kirkpatrick J, Sifre L, Green T et al (2020) Improved protein structure prediction using potentials from deep learning. Nature 577:706–710. https://doi.org/10.1038/s41586-019-1923-7
23. Jumper J, Evans R, Pritzel A, Green T, Figurnov M, Ronneberger O et al (2021) Highly accurate protein structure prediction with AlphaFold. Nature 596:583–589. https://doi.org/10.1038/s41586-021-03819-2
24. Weissenow K, Heinzinger M, Rost B (2022) Protein language-model embeddings for fast, accurate, and alignment-free protein structure prediction. Structure 30:1169–77.e4. https://doi.org/10.1016/j.str.2022.05.001
25. Weissenow K, Heinzinger M, Steinegger M, Rost B (2022) Ultra-fast protein structure prediction to capture effects of sequence variation in mutation movies. bioRxiv. https://doi.org/10.1101/2022.11.14.516473
26. Anfinsen CB, Scheraga HA (1975) Experimental and theoretical aspects of protein folding. Adv Protein Chem 29:205–300. https://doi.org/10.1016/s0065-3233(08)60413-1
27. Larriva M, de Sancho D, Rey A (2006) Evaluation of a mean field potential for protein folding with different interaction centers. Physica A Stat Mech Appl 371:449–462. https://www.sciencedirect.com/science/article/pii/S0378437106004729
28. Dawid AE, Gront D, Kolinski A (2017) SURPASS low-resolution coarse-grained protein modeling. J Chem Theory Comput 13:5766–5779. https://doi.org/10.1021/acs.jctc.7b00642
29. Badaczewska-Dawid AE, Kolinski A, Kmiecik S (2020) Protocols for fast simulations of protein structure flexibility using CABS-flex and SURPASS. Methods Mol Biol 2165:337–353. https://doi.org/10.1007/978-1-0716-0708-4_20
30. Jamroz M, Orozco M, Kolinski A, Kmiecik S (2013) Consistent view of protein fluctuations from all-atom molecular dynamics and coarse-grained dynamics with knowledge-based force-field. J Chem Theory Comput 9:119–125. https://doi.org/10.1021/ct300854w
31. Jamroz M, Kolinski A, Kmiecik S (2014) CABS-flex predictions of protein flexibility compared with NMR ensembles. Bioinformatics 30:2150–2154. https://doi.org/10.1093/bioinformatics/btu184
32. Green AR, Nissen MS, Kumar GNM, Knowles NR, Kang C (2013) Characterization of Solanum tuberosum multicystatin and the significance of core domains. Plant Cell 25:5043–5052. https://doi.org/10.1105/tpc.113.121004
33. Nissen MS, Kumar GNM, Youn B, Knowles DB, Lam KS, Ballinger WJ et al (2009) Characterization of Solanum tuberosum multicystatin and its structural comparison with other cystatins. Plant Cell 21:861–875. https://doi.org/10.1105/tpc.108.064717
34. Jamroz M, Kolinski A (2013) ClusCo: clustering and comparison of protein models. BMC Bioinformatics 14:62. https://doi.org/10.1186/1471-2105-14-62
35. Chakravarty D, Porter LL (2022) AlphaFold2 fails to predict protein fold switching. Protein Sci 31:e4353. https://doi.org/10.1002/pro.4353
36. Dabrowski-Tumanski P, Stasiak A (2023) AlphaFold blindness to topological barriers affects its ability to correctly predict proteins' topology. Molecules 28:7462. https://doi.org/10.3390/molecules28227462
37. Hou M, Jin S, Cui X, Peng C, Zhao K, Song L et al (2024) Protein multiple conformation prediction using multi-objective evolution algorithm. Interdiscip Sci. https://doi.org/10.1007/s12539-023-00597-5
38. Greener JG, Jones DT (2021) Differentiable molecular simulation can learn all the parameters in a coarse-grained force field for proteins. PLoS One 16:e0256990. https://doi.org/10.1371/journal.pone.0256990
39. Baek M, Baker D (2022) Deep learning and protein structure modeling. Nat Methods 19:13–14. https://doi.org/10.1038/s41592-021-01360-8

Chapter 5

Machine Learning Techniques to Infer Protein Structure and Function from Sequences: A Comprehensive Review

Gopal Srivastava, Mengmeng Liu, Xialong Ni, Limeng Pu, and Michal Brylinski

Abstract

The elucidation of protein structure and function plays a pivotal role in understanding biological processes and facilitating drug discovery. With the exponential growth of protein sequence data, machine learning techniques have emerged as powerful tools for predicting protein characteristics from sequences alone. This review provides a comprehensive overview of the importance and application of machine learning in inferring protein structure and function. We discuss various machine learning approaches, primarily focusing on convolutional neural networks and natural language processing, and their utilization in predicting protein secondary and tertiary structures, residue-residue contacts, protein function, and subcellular localization. Furthermore, we highlight the challenges associated with using machine learning techniques in this context, such as the availability of high-quality training datasets and the interpretability of models. We also delve into the latest progress in the field concerning the advancements made in the development of intricate deep learning architectures. Overall, this review underscores the significance of machine learning in advancing our understanding of protein structure and function, and its potential to revolutionize drug discovery and personalized medicine.

Key words Protein structure, Protein function, Machine learning techniques, Protein sequence data, Convolutional neural networks, Natural language processing

1 Introduction

The majority of globular proteins conform to the "physiological conditions" rule [1], which implies that factors like pH, cofactor inclusion, or alterations in the protein sequence can impact its structure in a manner relevant to its function. For example, the addition of a cofactor can assist the protein in adopting its functional form [2]. Conversely, there has been a growing recognition of intrinsically disordered proteins and regions [3, 4]. Typically lacking a fixed structure, these proteins and regions assume a definite structure only under specific conditions. It was a longstanding challenge for the scientific community to determine if all the

Andrzej Kloczkowski et al. (eds.), *Prediction of Protein Secondary Structure*, Methods in Molecular Biology, vol. 2867, https://doi.org/10.1007/978-1-0716-4196-5_5,

necessary information for accurately predicting protein structures could be extracted from its sequence. To address this question, two types of methods have been employed, first principles- and knowledge-based approaches. Methods based on first principles focus on comprehending and utilizing the molecular forces that govern protein thermodynamics and kinetics. Physics-based techniques have been utilized to ascertain the secondary structures of proteins due to their evolutionary conservation. Identifying the secondary structure also aids in understanding the three-dimensional arrangement of proteins. Although physics-based approaches are appealing, they have proven to be challenging and computationally demanding, particularly for generating accurate models of larger proteins, often encountering difficulties in attaining the desired level of accuracy [5].

In parallel to physics-based approaches, knowledge-based techniques have been utilized to predict protein structures [6]. However, these algorithms faced challenges in achieving good performance due to their heavy reliance on high-quality data. As more structural data became available, the performance of knowledge-based algorithms improved significantly. Advancements in experimental techniques, such as X-ray crystallography, cryo-EM, and NMR, have led to a substantial increase in the number of protein structures used to train knowledge-based algorithms. As of June 2023, the Protein Data Bank (PDB) [7] contains 202,301 protein structures. This wealth of structural data shifted the field of protein structure and function prediction from first-generation methods, which focused on the amino acid propensity in the sequence, to second-generation techniques incorporating neighboring information to predict various protein features. Subsequently, third-generation methods emerged, combining evolutionary information and multi-sequence alignments (MSAs) with machine learning (ML) techniques, such as support vector machines (SVM), Bayesian models, hidden semi-Markov networks, and neural networks (NNs). This review delves into the latest ML-based tools for predicting protein structure and function using amino acid sequences. We first explore these contemporary tools and then focus our discussion on current ML architectures, specifically convolutional neural networks (CNN) and natural language processing (NLP) techniques.

2 Methods to Infer Structure and Function from Amino Acid Sequences

ML has emerged as a powerful tool for predicting protein structure and function directly from amino acid sequences. By training on annotated protein data, ML algorithms can extract meaningful patterns and relationships to generate predictive models. These models enable the inference of protein structure and function for novel sequences, providing valuable insights into protein behavior and facilitating drug discovery and enzyme design. Leveraging diverse features, ML offers a promising approach to accelerate and enhance our understanding of protein structures and functions solely from the sequence information.

2.1 Secondary Structure Prediction

Protein secondary structure prediction holds great importance in understanding the fundamental organization of proteins. The secondary structure refers to the local folding patterns of amino acid residues within a protein chain, including elements like α-helices, β-sheets, and turns. Accurate prediction of protein secondary structure provides insights into the protein overall fold, stability, and functional properties. It aids in identifying critical regions, such as active sites and binding interfaces, and facilitates the analysis of protein-protein interactions (PPIs) and molecular recognition events. Additionally, protein secondary structure prediction plays a vital role in annotating newly sequenced proteins, as well as in guiding the prediction of tertiary structures. It serves as a valuable tool in various fields, including structural biology, drug design, and protein engineering, enabling researchers to better understand protein behavior and design experiments to unravel their functional mechanisms.

Significant progress in predicting protein secondary structure has been made by incorporating evolutionary information derived from protein sequences [8]. Early methods primarily focused on individual residue features, but later approaches expanded their scope by incorporating information from neighboring residues and homologous sequences through MSAs [9]. A notable example is an NN-based method called SPINE, which achieved an impressive 80% accuracy by training on 2640 non-redundant proteins [10]. SPINE utilized position-specific scoring matrix (PSSM) and physiochemical structural properties as inputs. Furthermore, many recently developed tools have leveraged MSAs to extract evolutionary information and predict protein structures using advanced ML techniques. These include deep NNs, deep CNNs, deep graph neural networks (GNNs), and NLP.

Several tools, including PSIPRED [11], JPred [12], and DESTRUCT [13], employ feedforward neural networks (FFNN) to predict protein secondary structure. DESTRUCT goes beyond by also predicting torsion angles [13]. On the other hand, SPIDER3 utilizes a three-layered, window-based FFNN architecture with 150 hidden units and incorporates solvent-accessible surface area into its predictions [14]. Notably, the SPIDER3 algorithm combines a bidirectional recurrent neural network (RNN) with long short-term memory (LSTM) cells, followed by a FFNN. This comprehensive approach enables the prediction of four protein structural annotations, encompassing secondary structure, solvent accessibility, torsion angles, and structural disorder.

Additionally, SSPro and ACCPro are tools specifically designed for predicting protein secondary structure and solvent accessible surface areas [15]. SSPro utilizes a combination of bidirectional RNN and 1D-CNN. Its current version also includes a PSI-BLAST-based refinement step to further enhance accuracy.

CNNs are a specialized type of NNs renowned for their proficiency in image recognition [16]. However, their application extends beyond that domain, encompassing various other tasks such as NLP, speech recognition, and protein secondary structure prediction. CNNs leverage convolutional layers that perform specific operations on the input data, followed by an activation function. These layers automatically learn spatial hierarchical features from the input data. In the context of protein secondary structure prediction, CNNs analyze features in the protein sequence to make predictions. NetSurfP-2.0 is an example of a method that employs a deep CNN with two layers, each equipped with a varying number of filters and units [17]. This approach utilizes HHblits [18] and MMseqs2 [19] algorithms to extract features from the input protein sequences. Another method called OCLSTM combines the strengths of both CNNs and LSTMs [20]. It starts by utilizing an optimized CNN to extract amino acid features in the form of a PSSM matrix from the input sequences. Subsequently, bidirectional LSTM is employed to capture remote contacts between residues and predict protein secondary structure. OCLSTM has the ability to capture both spatial and temporal features from the input data, resulting in exceptional performance scores of 84.68%, 82.36%, 82.91%, 84.21%, and 85.08% on the Critical Assessment of Techniques for Protein Structure Prediction (CASP)10, CASP11, CASP12, CB513, and 25PDB datasets, respectively [20].

2.2 Residue Contact Prediction

Protein residue-residue (RR) contact prediction plays a crucial role in deciphering 3D structures and functional properties of proteins. RR contacts refer to the spatial proximity of amino acid residues within a protein, and accurately predicting these contacts can provide valuable insights into protein folding, stability, and interaction networks. By identifying RR contacts, we gain crucial information about the spatial arrangement of amino acids, which can guide the prediction of protein tertiary structures. Furthermore, RR contact prediction enables the characterization of PPIs, ligand binding sites, and important regions involved in molecular recognition. This information is invaluable for understanding protein function, designing novel therapeutics, and exploring biological processes at the molecular level. Overall, accurate prediction of RR contacts is essential for advancing our knowledge of protein structure and function, and it has far-reaching implications in drug discovery, structural biology, and systems biology.

In recent years, several highly accurate methods leveraging ML techniques have been developed for the prediction of RR contacts in proteins. These methods have significantly advanced our ability to extract valuable structural information from protein sequences. DeepContact is a prominent method that excels at identifying co-evolutionary patterns and determining RR contact probabilities in proteins [21]. Ranked among the top two CNN-based approaches in CASP12, it utilizes raw evolutionary couplings from the ASTRAL SCOPe 2.06 dataset, generated by CCMPred [22]. DNCON2, another method, employs CNNs to predict RR contacts [23]. It follows a two-stage prediction process, training five CNNs to predict contact maps at different distance thresholds (6, 7.5, 8, 8.5, and 10 Å). These maps are then utilized as 2D features for a final CNN that predicts short-, medium-, and long-range contacts within a single map. In DNCON2, long-range contacts refer to residues separated by at least 24 residues, medium-range contacts to those separated by 12 to 23 residues, and short-range contacts to those separated by 6 to 11 residues.

DeepCov is a CNN-based algorithm that employs fully connected layers and utilizes the frequencies of residue pairs along with their covariance as input features [24]. It leverages $441 \times m \times m$ features, where m represents the length of the protein sequence. These features are calculated based on the 20 amino acids, including a gap. DeepCov defines a contact between two residues when their $C\beta$ atoms (or $C\alpha$ atoms for glycine) are within 8 Å of each other. Compared to CCMpred [22] and MetaPSI-COV2 [25], DeepCov has demonstrated superior precision. On the other hand, RaptorX-Contact is a technique that predicts inter-residue distances by employing a deep, fully convolutional residual neural network (ResNet) [26]. It emerged as a winner in the CASP13 competition. The ResNet employed by this method predicts the contact/distance score of a pair of residues by considering their correlation with other residue pairs at different distances [27].

SPOT-Contact, an acronym for Sequence-based Prediction Online Tools for Contact map prediction, is an ensemble method comprising two bidirectional RNNs [28]. It leverages several 1D features to make predictions, including PSSM generated from the UniRef90 sequence database, HMM profile obtained from HHblits, and predicted structural probabilities from SPIDER3. Predictions from SPIDER3 encompass various aspects such as relative solvent-accessible surface area, half-sphere exposures based on the $C\alpha$ atom, eight sine/cosine backbone torsion angles, and the probabilities of the predicted secondary structure. Additionally, SPOT-Contact incorporates seven physicochemical properties [29]. As for 2D features, it utilizes outputs from CMMPred and mutual and direct-coupling information from DCA. Overall, SPOT-Contact incorporates a total of 71 1D features and three 2D features [28].

TripletRes is an advanced deep residual NN designed for the precise prediction of RR contacts in proteins leveraging discretized distance profiles and end-to-end training [30]. To enhance accuracy, TripletRes incorporates triplets of co-evolutionary matrices, including those from the Potts model, precision matrix, and covariance matrix. Notably, TripletRes surpasses the performance of top models from CASP11 and CASP12, as well as CAMEO targets [31], by impressive margins of 58.4% and 44.4%, respectively [30]. Furthermore, TripletRes outperforms DeepContact in predicting contacts within 50 non-redundant protein domains from CASP11 and CASP12 [30]. The average prediction of long-range contacts $L/10$, $L/5$, $L/2$, and L (where L represents the protein sequence length) achieved by TripletRes is 55.1%, 53.2%, 57.1%, and 58.4%, respectively, surpassing the corresponding predictions by DeepContact [30].

2.3 Tertiary Structure Prediction

Protein tertiary structure prediction plays a crucial role in understanding the functional aspects of proteins and their behavior within biological systems. The tertiary structure refers to the 3D arrangement of atoms within a protein, which directly influences its function and interactions with other molecules. The accurate prediction of protein tertiary structure enables the identification of critical regions, such as active sites and binding pockets, which are essential for rational drug design and targeted therapeutics. Additionally, it provides insights into protein folding, stability, and dynamics, shedding light on PPIs, enzyme catalysis, molecular recognition, and signaling pathways. Overall, the accurate prediction of protein tertiary structure is vital for advancing our understanding of complex biological processes and has significant implications for fields such as biomedicine, biotechnology, and molecular biology.

In recent years, significant progress has been made in the development of highly accurate methods that leverage ML techniques to predict protein tertiary structures. These methods have revolutionized our ability to infer the 3D arrangement of atoms within proteins solely from their amino acid sequences. By training on large datasets of experimentally determined protein structures, ML algorithms have become adept at recognizing patterns and capturing complex relationships that govern protein folding. Notably, AlphaFold, created by DeepMind, is an advanced computational tool designed to predict the 3D configuration of proteins based on their amino acid sequences [6]. By harnessing ML methodologies, it thoroughly examines sequence information and provides predictions for the most probable structure. AlphaFold has exhibited remarkable accuracy and has successfully predicted the structures of numerous previously unknown proteins. This groundbreaking technology has the potential to reshape the landscape of protein structure prediction and significantly contribute to our comprehension of protein functions.

AlphaFold combines evolutionary, physical, and geometrical considerations to improve protein structure prediction. It takes protein sequences and MSAs as input to predict 3D coordinates of heavy atoms, and then uses an evolutionary algorithm to refine the structure by repeatedly altering and assessing conformations until the most probable structure is found. The training network consists of two modules, an Evoformer module and an end-to-end structure prediction module. The Evoformer module converts the protein structure prediction problem into a graph in 3D space, where the edges of the graph are defined by residues in proximity. The relationships between residues are encoded in a residue pair matrix, while the MSA representation encodes the residues of the input sequences, with each row representing a sequence. The structure module uses a 3D backbone configuration represented by independent rotations and translations, each defined relative to the global frame.

The rotations and translations of the dihedral angles in the backbone structure prioritize the orientation of the main chain, constraining the location of the side chain of each residue within that frame. However, the geometry of the peptide bonds is not constrained, and the network may violate this constraint during the application of the structure module. This allows for local refinement of all parts of the chain without the need to solve complex loop closure problems. During the fine-tuning process, a violation loss term encourages satisfaction of peptide bond geometry. The final predicted structure is then relaxed using gradient descent in the Amber force field, which enforces the peptide bond geometry constraints. During CASP14, AlphaFold outperformed other methods with its highly accurate protein structures. The median backbone accuracy of AlphaFold structures in terms of the root-mean-square deviation (RMSD) [32] of Cα atoms was as low as 0.96 Å with a 95% confidence interval between 0.85 and 1.16 Å [6].

RoseTTAFold is another method that harnesses the power of ML technology to predict the tertiary structures of proteins, drawing inspiration from the impressive achievements of AlphaFold [33]. RoseTTAFold employs a variety of methodologies that are akin to those utilized in AlphaFold2. These approaches include shifting the starting point of the process from processed features like inverse covariance matrices to directly utilizing MSAs. The conventional 2D convolution approach is replaced with an attention mechanism that captures interactions between residues positioned far apart along the protein sequence. Furthermore, a two-track network architecture is employed to facilitate the iterative transformation and exchange of information between the 1D sequence level and the 2D distance map level. Additionally, an SE (3)-equivariant transformer network is introduced to refine the atomic coordinates, deviating from previous methods that focused on refining 2D distance maps. Lastly, an end-to-end learning process is implemented, optimizing all network parameters through backpropagation. The learning process begins with the final generated 3D coordinates and propagates through all network layers back to the input sequence.

To achieve a high level of prediction accuracy, RoseTTAFold enhances the two-track model with a third track trained on 3D backbone coordinates to better mimic connections between sequence, RR distances, orientation, and atomic coordinates. In contrast to AlphaFold, within this architectural framework, a dynamic exchange of information takes place among the 1D amino acid sequence information, the 2D distance map, and the 3D coordinates. This bidirectional flow enables the network to holistically analyze and infer relationships within and between the sequence, distances, and coordinates. By collectively reasoning about these interconnected elements, the network gains a comprehensive understanding of the underlying structural intricacies of proteins. Due to hardware limitations, RoseTTAFold was trained on disjointed sequences comprising two segments, resulting in a total of 260 residues. To derive optimal models, a combination and averaging procedures were applied to the 1D features and predictions of 2D distances and orientations for each subset. This was followed by employing two distinct approaches to generate the final 3D structures.

In the first approach, the projected distributions of RR distances and orientations were utilized in pyRosetta [34], enabling the generation of all-atom models. In the second approach, the averaged 1D and 2D features were inputted into a final SE (3)-equivariant layer. Through end-to-end training, commencing from the amino acid sequence and extending to 3D coordinates, the network directly generated backbone coordinates. RoseTTAFold generates high accuracy predictions at the residue level, surpassing all other tools in CASP14 except for AlphaFold2 [33]. The decrease in accuracy, in comparison to AlphaFold2, was attributed to the hardware limitations that constrained the model size. However, once these hardware limitations are overcome, RoseTTAFold is expected to exhibit accuracy on par with or even surpassing that of AlphaFold2.

The majority of end-to-end sequence-to-structure tools for the prediction of the protein structure have predominantly concentrated on modeling individual domains. These domains serve as the fundamental folding units of proteins, operating autonomously with distinct folding patterns and functionalities. In fact, a large number of prokaryotic and eukaryotic proteins contain two or more domains that perform specific cellular functions through cooperative domain interactions [35]. The development of I-TASSER-MTD aims to streamline the assembly process of multi-domain protein structures while also enabling structure-based function annotation using solely sequence information [36]. At its core, the I-TASSER-MTD tool draws upon the well-established I-TASSER [37] and DEMO [38] tools, known for their outstanding performance in multiple CASP challenges. The I-TASSER-MTD pipeline consists of several essential components, all guided by sequence information. These components include sequence-based domain boundary prediction, deep learning-based spatial restraint prediction, folding of single-domain structures, assembly of multi-domain structures, and structure-based function annotation. By integrating these components, I-TASSER-MTD provides a comprehensive framework for accurately predicting protein structures and annotating their functions solely based on sequence information.

The I-TASSER-MTD pipeline starts with the query sequence and employs various methods to collect threading templates from the PDB [7]. Domain boundaries are predicted using FUpred [39] and ThreaDom [40], while RR spatial restraints are generated by DeepPotential [41]. If the query is identified as a multi-domain protein and no template alignments can cover all domains, the pipeline initiates the "multi-domain assembly" mode. In this mode, individual domain models are constructed using D-I-TASSER and then assembled into a full-length model by DEMO. If the query is a single-domain protein or top template alignments cover all domains, the "multi-domain assembly" mode is turned off, and the full-length structure is directly modeled by D-I-TASSER. Finally, COFACTOR [42] is utilized to predict protein function annotations, including Enzyme Commission (EC) numbers, Gene Ontology (GO) terms, and ligand binding sites.

The accuracy of I-TASSER-MTD was tested using estimated TM-score (eTM-score) and estimated RMSD (eRMSD) on the DEMO benchmark set comprising 356 multi-domain proteins with various domain types. Notably, the eTM-score exhibits a strong correlation with the actual TM-score, as indicated by a high Pearson correlation coefficient (PCC) of 0.85. Additionally, the eTM-score demonstrates a low average error of 0.07. On the other hand, the PCC between eRMSD and RMSD is 0.82 and the average error between eRMSD and RMSD is 2.2 Å [36]. By utilizing an eTM-score cutoff of 0.5 to identify models with precise global topologies, both the false-negative and false-positive rates are consistently maintained below 0.15. This indicates that the fold-level predictions are accurate in more than 85% of cases, demonstrating that the performance of I-TASSER-MTD is highly reliable [36].

2.4 Protein Function Prediction

Proteins play a crucial role in the functioning of organisms, serving as the fundamental components of functional domains. Because of their remarkable diversity, proteins exhibit a wide array of functions, including enzymatic catalysis, energy provision, DNA binding for transcriptional control, DNA damage repair, and regulation of membrane transport, among others [43]. Nonetheless, the functions of the majority of proteins remain elusive. Among the staggering 179 million proteins in the UniProt database, only a small fraction, less than 0.1%, have been experimentally annotated [44]. Conducting traditional biological experiments to determine protein functions is laborious and expensive, making it impractical to experimentally annotate all unknown proteins. Consequently, computational methods have become pivotal in inferring protein functions. Recent advancements in computer science have revolutionized the conversion of protein sequences into computationally processable vectors. This breakthrough has led to a surge in the utilization of ML and deep learning (DL) techniques within the realm of biology and biomedicine.

Predicting protein functions from their sequences involves various computational approaches. Homology-based methods leverage the observation that proteins with similar sequences often have similar functions. By comparing the sequence of a protein with an unknown function to a database of known protein sequences, a match can be identified, enabling inference of the unknown protein's function. ML-based methods employ large datasets to train algorithms that recognize patterns and relationships between protein sequences and their functions. Subsequently, these trained algorithms can predict the functions of novel proteins based on their sequences. Structure-based methods, on the other hand, first predict the 3D structure of a protein from its sequence and then utilize this structural information to infer its function. This approach allows the identification of key functional regions and interactions with other molecules. Functional site prediction involves identifying specific residues or sequences within a protein that likely contribute to a particular function, such as ligand binding or catalytic activity. These methods can be used individually or in combination to predict protein function, with prediction accuracy influenced by multiple factors such as the quality and availability of training data, the complexity of the protein's function, and the limitations inherent in the computational methods that are employed.

In the field of protein function prediction, significant research has been dedicated to different categories. Presently, ML methods have emerged as highly accurate in inferring protein function. A notable example is DeepFam, an alignment-free protein family prediction approach that operates on protein sequences as input [45]. DeepFam utilizes convolutional and 1-max pooling layers to calculate the existence score of conserved regions. Its hidden layers detect conserved regions by activating units for k-mers associated with a particular protein family. DeepFam has demonstrated superior performance compared to other methods trained on the G protein-coupled receptor (GPCR) dataset. It achieved a remarkable accuracy of 97.2% at family, 86.8% at sub-family, and 81.2% at sub-subfamily levels. When assessing the performance against the Clusters of Orthologous Groups (COG) dataset [46, 47], Deep-Fam outperformed pHMM, 3-mer LR, and Protvec LR with accuracies of 95.4%, 91.8%, 85.6%, and 47.3%, respectively [45].

DeepCpf1 was developed with the aim of predicting the activity and specificity of CRISPR-Cpf1, a type of CRISPR-Cas protein, using CNNs [48]. CRISPR-Cpf1 holds immense potential for genome editing, but its activity and specificity are not fully understood. DeepCpf1 leverages DL techniques to analyze the relationship between the sequence of CRISPR-Cpf1 proteins and their activity and specificity. By capturing this information, it enables predictions regarding the activity and specificity of newly discovered CRISPR-Cpf1 proteins. The performance of CNN-based DeepCpf1 was assessed through a five-fold cross-validation and compared to other ML techniques such as gradient boosting trees, L2-regularized linear regression (LR), random forest, L1L2-regularized LR, L1-regularized LR, k-nearest neighbors, SVM, and NN. The CNN method outperformed the alternative approaches, demonstrating superior performance with AUC scores of 0.83, 0.81, 0.81, 0.81, 0.79, 0.79, 0.78, 0.78, and 0.76, respectively, for the methods mentioned in that order [48].

DeepLoc tackles the crucial task of accurately predicting the subcellular localization of proteins by harnessing the power of deep NNs and relying solely on sequence information [49]. The method utilizes an RNN that comprehensively processes the entire protein sequence and incorporates an attention mechanism to identify key protein regions essential for subcellular localization. The effectiveness of DeepLoc was validated by training and testing it on a dataset of experimentally annotated proteins from the UniProt database [44, 49]. Remarkably, it achieved an impressive accuracy of 78% across 10 functional categories and a remarkable accuracy of 92% for membrane-bound and soluble proteins [49]. By outperforming alternative algorithms, including those dependent on homology information, DeepLoc has established itself as an exceptionally accurate and valuable tool for predicting protein subcellular localization.

A recent study introduced a method leveraging CNN to predict protein family classification while considering the precise location of amino acid residues [50]. By utilizing a CNN-based model, this innovative approach automates the crucial task of protein family classification, extracting significant features from protein sequences and successfully categorizing them into their respective families. Importantly, the model was trained entirely from scratch, eliminating the need for pretrained weights and enhancing the flexibility of the approach. To assess its performance, the proposed method was rigorously evaluated using a benchmark dataset. The results showcased the method's exceptional capability, demonstrating its potential to achieve high accuracy in protein family classification. The tool exhibited an impressive average precision score of 0.98 [49], further highlighting the effectiveness of DL models in this domain. This research paves the way for more efficient and reliable protein family classification, contributing to advancements in computational biology and protein analysis.

SDN2GO is an advanced method that harnesses the power of DL to predict protein function [51]. This cutting-edge algorithm takes advantage of various data sources, including protein sequences, PPIs, and GO annotations, to accurately predict protein functions. To evaluate its performance, the model underwent rigorous testing using a benchmark dataset specifically curated to include proteins lacking GO annotations. Remarkably, SDN2GO outperformed existing methods for protein function inference, showcasing its superior capabilities. Specifically, the algorithm achieved an impressive area under the curve (AUC) value of 0.92 [51], underscoring its accuracy and reliability in protein function prediction. Importantly, the integration of diverse information sources in SDN2GO significantly enhances the predictive power, offering valuable insights into protein functions. SDN2GO holds great promise for a highly accurate and comprehensive protein function annotation.

3 Machine Learning Techniques to Annotate Protein Sequences

ML algorithms, such as CNNs and NLP, have emerged as powerful tools for predicting protein structure and function directly from amino acid sequences. CNNs are particularly well-suited for extracting meaningful patterns and features from sequential data, making them ideal for analyzing protein sequences. By training on annotated protein data, CNN models can learn the intricate relationships between amino acid residues and their corresponding structural and functional properties. This enables the inference of protein structure and function for novel sequences, providing valuable insights into protein behavior. NLP techniques, on the other hand, offer the ability to process and understand textual information related to proteins, such as scientific literature and protein databases. NLP methods can be used to extract relevant information, identify key features, and enhance the accuracy of predictions. The integration of CNNs and NLP in protein structure and function prediction represents a synergistic approach that combines the power of DL and language processing to unlock new possibilities in the field of bioinformatics and accelerate our understanding of proteins.

3.1 Convolutional Neural Networks

The CNN is a powerful class of artificial NNs that leverages the spatial and temporal correlations present in data. Its origins can be traced back to the neocognitron [52], a model inspired by the hierarchical organization of the animal visual cortex [53]. However, it was not until 1989 that the modern framework of CNNs was established [54], which significantly improved their architecture and performance. One notable milestone was the development of LeNet-5, a multilayer NN specifically designed for handwritten digit recognition [55]. This breakthrough demonstrated the effectiveness of CNNs in handling complex pattern recognition tasks. The success of NNs, including CNNs, heavily relies on the availability of large-scale training data and computational power. With the advent of the big data era and the advancements in computing capabilities, training deep NNs has become feasible, opening up new opportunities for CNNs in various fields and greatly expanding their application scope.

CNNs find wide application in the field of computer vision (CV). Within CV, CNNs are trained to comprehend and interpret visual data by excelling in face recognition [56, 57], human pose estimation [58, 59], and action recognition [60, 61]. By processing digital images captured from videos and cameras, well-trained CNN models can accurately identify and classify objects. Key tasks carried out by CNNs in CV include image classification [62, 63], object detection and segmentation [64, 65], video processing [60, 66], and super-resolution imaging [67, 68]. CNNs also play a crucial role in textual information processing. In text detection and recognition, CNN models are deployed to detect and identify text in various backgrounds, accounting for appearance variations [69, 70]. In speech recognition, CNNs are employed to classify spoken language at the word level using pooled datasets [71]. Within NLP, CNNs find extensive application in tasks such as language modeling [72] and sentence modeling [73]. The inherent ability of CNNs to extract local relations within spatiotemporal structures empowers them to capture long-range information and learn hierarchical representations of input words [74]. Remarkably, CNNs demonstrate exceptional performance even with 1D data. Utilizing 1D CNNs, time series prediction tasks, such as weather forecasting [75], traffic flow prediction [76], network intrusion detection, and signal identification, can effectively be tackled [77].

Although CNN architectures come in various forms, they share fundamental components. A CNN model comprises three key elements, the convolutional layer, pooling layer, and activation function, working together to extract meaningful features from the input data. As an example, the architecture of LeNet-5 is presented in Fig. 1. Convolutional layers perform convolutions, applying filters to the input and generating feature maps that capture local patterns. Pooling layers reduce the spatial dimensions of the feature maps, summarizing the information and enhancing computational

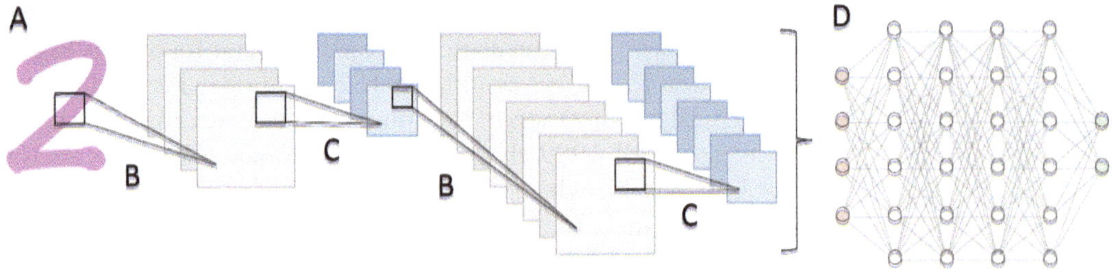

Fig. 1 Architecture of LeNet-5. The network receives visual data as input (**a**), with each unit in a layer extracting feature by convolving a small neighborhood of units from the preceding layer (**b**). Planes represent feature maps, and subsampling is employed to mitigate sensitivity to shifts and distortions in the output (**c**). Following the convolution layers, fully connected layers are utilized for global information fusion and label predictions (**d**)

efficiency. Finally, the activation function introduces non-linearity to the network, allowing it to learn complex relationships between features. Together, these components form the foundation of CNNs and enable their remarkable ability to capture and process spatial information in various applications.

3.1.1 Convolutional Layer

Convolutional layers in CNN serve as the fundamental components for feature extraction. Their primary goal is to capture high-level features from input data. A convolutional layer comprises a collection of kernels or filters, whose parameters are learned during training. Typically, these kernels have smaller dimensions compared to the input images and operate by sliding over the image as a window. At each position, the region of the image covered by the kernel is known as the receptive field. The kernel convolves with the image by element-wise multiplication between the kernel and the receptive field [78]. The kernel movement across the image is determined by the "stride," which represents the number of pixels it shifts at each step. To ensure consistent handling of border pixels, "padding" may be applied to the image matrix, depending on the kernel size and stride. Through this convolutional operation, CNNs efficiently extract relevant features from the input data. Figure 2 visually depicts the convolution process, which can be mathematically described by the following equation:

$$f_l^m(p, q) = \sum_c \sum_{x,y} i_c(x, y) \cdot e_l^m(u, v) \quad (1)$$

where each element $i_c(x, y)$ of the input image matrix is multiplied by the corresponding element $e_l^m(u, v)$ in the kernel (m denotes the m-th kernel and l denotes the l-th layer) to get the output feature $f_l^m(p, q)$. Variables p, q, x, y, u, and v represent row and column indices.

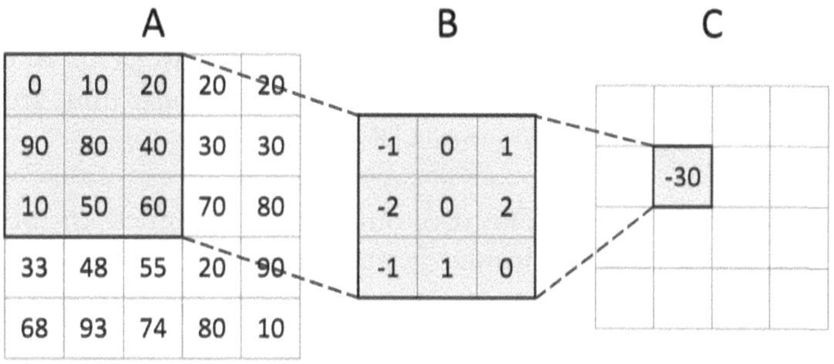

Fig. 2 Convolution operations for local pattern extraction. The kernel slides over the input feature matrix (**a**), and at each position, it performs element-wise multiplications between the kernel (**b**) and the corresponding masked region of the input. These multiplications are then summed up to produce a single value in the output feature matrix (**c**). This process captures local patterns and spatial relationships in the data

3.1.2 Pooling Layer

Once features are extracted, their precise location becomes less crucial compared to their relative positions to other features. To capture the dominant features within local regions, a pooling layer, also known as a down-sampling layer, is employed [79]. Similar to a convolutional layer, the pooling layer operates with a stride equal to the kernel size. This layer summarizes and condenses information from the receptive field, resulting in a significant reduction in the spatial dimensions of the input volume. By reducing the complexity of the network, the pooling layer helps mitigate overfitting. Furthermore, it can extract combination features that are resilient to small distortions and translational shifts. Several pooling methods have been devised, including max pooling, mixed pooling, average pooling, stochastic pooling, spatial pyramid pooling, and others [80, 81]. The general mathematical representation of this layer can be expressed as:

$$\boldsymbol{Z}_l^m = O_p\left(\boldsymbol{F}_l^m\right) \tag{2}$$

where $\boldsymbol{F}_l^m$ is the feature matrix from a convolutional layer, $O_p(\cdot)$ denotes the pooling operation, and $\boldsymbol{Z}_l^m$ represents the pooled feature map.

3.1.3 Activation Function

In order to introduce nonlinearity into the model, an activation function is typically applied immediately after the convolutional layer, as the convolution operation itself is a linear operation involving element-wise multiplication. This activation function plays a crucial role in enabling the model to learn complex patterns from the extracted features. Several commonly used activation functions include tanh, sigmoid, maxout, and the rectified linear unit (ReLU). Among these, ReLU has shown significant advantages, as it allows for faster training of the network while maintaining the same level of accuracy. Additionally, ReLU helps address the issue of vanishing gradients. As a result of these benefits, ReLU and its variants, such as leaky ReLU, parametric ReLU, randomized ReLU, and exponential linear unit, are widely utilized as activation functions [82, 83].

3.1.4 Dropout Layer

The dropout layer is designed to introduce randomness by randomly deactivating certain units or connections within the network. This deliberate skipping helps prevent the network from excessively relying on specific neurons or small combinations of neurons. By doing so, this operation compels the network to produce accurate outputs even when certain information is missing. Incorporating the dropout layer adds regularization to the network, leading to a reduction in overfitting and an enhancement in generalization. Different approaches can be employed within the dropout layer, including standard dropout, fast dropout [84], adaptive dropout [85], and spatial dropout [86].

3.1.5 Fully Connected Layer

Fully connected layers in CNN models are employed at the end to obtain the final output. While the convolutional layer works with local receptive fields in each instance, the fully connected layer performs a global operation, connecting every neuron in one layer to every neuron in the preceding layer. The fully connected layer initially applies a linear transformation to the output features derived from the previous layer. Subsequently, it combines these extracted features in a non-linear manner. The resulting outputs are utilized based on the specific objective of the CNN model, such as image classification [78].

3.2 Natural Language Processing

NLP combines linguistic and computer science to process and analyze human languages. It started out in the 1950s as the symbolic NLP, which utilizes a series of rules provided by the researchers to emulate the understanding of human languages [87]. Starting from the 1980s, the "statistical revolution" marked a significant breakthrough in the field, as researchers began delving into the application of machine learning algorithms for various NLP tasks [88]. With the increasing of computational power, the development of the theories of linguistics, and the enormous amount of data, ML algorithms using statistical inferences have gained popularity. Many models have been adopted to solve different NLP tasks, including the decision tree [89] and the hidden Markov model [90]. The former replaces human-written rules with a series of if-then decision nodes, and the latter uses the probabilistic framework of the hidden Markov process to make statistical inferences. Statistical models make predictions based on real-valued weights assigned to features in the input data, highlighting the importance of well-engineered features for accurate predictions. However, this reliance on feature quality presents a notable limitation of statistical models. In the early 2000s, the rise of NNs prompted researchers to shift their focus towards them, primarily due to the introduction of word embedding techniques. These techniques allowed for the conversion of individual words into real-valued vectors within a given corpus, thereby enhancing the appeal and interest in NNs. Such approach well captured the semantic properties of each word without the need to devise sophisticated feature engineering. It has been applied to many high-level NLP tasks, such as human-machine interaction, machine translation, and image annotation.

Following the significant achievements of NNs in NLP research, the field continued to progress, and in the 2010s, DL emerged as the forefront technology in the domain of NLP. Many different models have been employed for various NLP tasks, including gated recurrent neural networks (GRU) [91], the long-short-term memory (LSTM) [92], character-level CNNs [93], and transformers [94]. These models have improved the performance of all sorts of NLP tasks by a huge margin compared to their predecessors. Some of the most advanced state-of-the-art DL-based language models, BERT [95] and GPT3 [96], have already been successfully deployed to everyday commercial applications such as Siri, Alexa, and many website translation services. NLP models have also been employed to biomedical-related data for information extraction, such as in BioBERT [97] and PubMed-BERT [98]. Nonetheless, these approaches involved replacing the original corpus with biomedical-related corpus sourced from biomedical publications, which, while relevant and beneficial to the community, often lack the crucial scientific insights necessary for advancing domain sciences.

NLP algorithms have been widely adopted in biology and biomedicine since there are many biological data represented by sequences, e.g., protein sequences and DNA sequences. Thus, it is natural to employ NLP algorithms for biological applications ranging from protein classification to genomic analyses [99]. The underlying mechanisms between the biological sequences processing and the NLP are similar. They both consists of preprocessing, annotation, tokenization, and model learning. The disparity in data between traditional NLP and biological sequence processing primarily manifests in the higher occurrence of stop words (e.g., "the," "a," "that," etc.) in the former. Consequently, part-of-speech (POS) tagging holds greater significance in NLP, whereas it carries less importance in biological sequence processing due to the pre-existing curation of the data compared to NLP. Figure 3 shows examples of NLP algorithms that can be directly applied to the biological sequence processing. Classification tasks in NLP, such as sentiment classification and genre classification, share the same methods as in biological sequence processing. Compared to document clustering, the prediction of PPIs involves more intricate methods, as it goes beyond clustering and necessitates the accurate

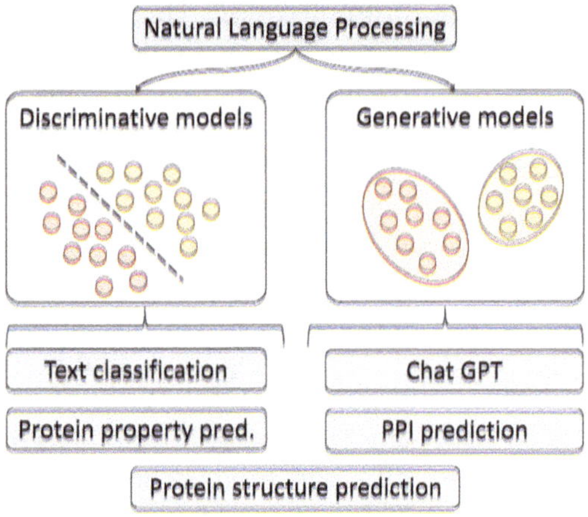

Fig. 3 Discriminative and generative NLP models. Discriminative models are primarily employed in tasks like text classification and summarization, resembling protein property prediction and folding problems. Generative models, extensively used in Natural Language Processing (NLP), have proven integral to applications like ChatGPT and can be applied to protein folding and protein-protein interaction (PPI) prediction tasks

prediction of specific relationships between different proteins. The most complicated task is arguably the protein folding problem. It is mathematically an NP complete problem, which means the time complexity scales non-polynomially as the size of the problem increases. However, it has been successfully tackled by the AlphaFold using a combination of NLP and other advanced algorithms [6]. In addition to protein sequence processing, many major breakthroughs have been made with DNA sequences as well. For instance, the N4-methylcytosine was successfully identified using NLP algorithms with word embeddings of the DNA sequences [100]. An integrated platform was developed to analyze common biological sequences, including DNA, RNA, and protein, based on NLP models [101].

4 Recent Advances in Machine Learning Architectures

Numerous advancements have taken place in ML architectures since their initial establishment in 1989. These progressions encompass regularization, parameter optimization, and structural reformulation [78, 102]. Notable performance enhancements were attained by reorganizing processing units and creating novel blocks. Recent architectural innovations, categorized based on the types of modifications, are summarized in Table 1.

Table 1
Categorization of deep CNN architectures

Spatial exploitation	Depth, width, multipaths	Attention mechanisms	Feature maps
AlexNet	Highway Networks	Residual Attention Network	SE-network
ZfNet	ResNet		
GoogleNet	DenseNet		
	Inception		
	Xception		

Depending on the type of architecture modifications, CNNs can be broadly categorized into four classes, namely, spatial exploitation, depth/width/multipath, attention, and feature-map exploitation-based CNNs. Each category represents a different approach to leveraging the power of CNNs for various tasks. The representative architectures within each category highlight the diverse range of design choices and techniques employed in modern CNN architectures

4.1 Spatial Exploitation-Based Architectures

AlexNet revolutionized the field of CNNs by expanding their applications beyond simple handwritten digit recognition to more complex image recognition and classification tasks [103]. To enhance the learning capability of CNNs, AlexNet increased the number of layers to eight, surpassing the original five layers in LeNet. In addition, parameter optimization strategies were implemented, and techniques such as dropout and ReLU were employed to address overfitting issues and achieve optimal performance. Another significant improvement introduced by AlexNet was training on the graphics processing unit (GPU) instead of the central processing unit (CPU), which greatly accelerated the learning process. The success of efficient learning with AlexNet ushered in a new era of research focused on advancing CNN architectures.

Prior to 2013, improvements in CNN learning mechanisms were primarily driven by trial and error, lacking an understanding of the underlying reasons. To gain better insights into network activity, ZfNet was proposed. It allowed for the monitoring of CNN performance by interpreting the activation of individual neurons. This analysis revealed that only a small number of neurons were active in the first two layers, while the rest remained inactive. Furthermore, the features learned by the second layer exhibited aliasing artifacts. These observations were used to adjust the network topology, resulting in significant performance improvements. Finally, GoogleNet, also known as Inception-V1, introduced the concept of the inception block in CNNs. An inception block integrates multi-scale convolutional transformations using split, transform, and merge techniques. By incorporating kernels of different sizes, it captures spatial information at various scales. GoogleNet addressed the issue of redundant information and reduced computation cost by employing sparse connections, omitting irrelevant feature maps. Additionally, it proposed the concept of an auxiliary learner to expedite the convergence process.

4.2 Architectures Utilizing Depth, Width, and Multipaths

Highway networks achieved enriched feature representations by leveraging high depth and effectively trained the deep network through the utilization of a multipath, cross-layer connectivity mechanism [104]. This architecture allows for the training of networks with over 100 layers, and it has been demonstrated that even with a 900-layer highway network, the convergence rate remains satisfactory [104]. ResNet is another architecture introducing the concept of residual learning [105]. It adopted the idea of a bypass pathway in the highway network and applied it to the designing of ResNet blocks. A block in ResNet is a conventional FFNN with an additional residual connection enabling cross-layer connectivity. This shortcut connection is data-independent and parameter-free, which can potentially prevent a vanishing gradient problem and accelerate the network convergence. The number of layers in ResNet varies from 34 to 1202 based on the type of the residual network architecture. DenseNet addressed the issue in ResNet that many layers contribute little or no information because the information is explicitly preserved through additive identity transformations [78]. In DenseNet, features from the previous layers are concatenated instead of added, as a result, the network might be able to explicitly distinguish between information that is preserved and information that is added to the network. Architectures belonging to the inception family, including Inception-V3, Inception-V4, Inception-ResNet, and Xception [106–109], are characterized by the usage of inception blocks, which leverage kernels of multiple sizes operating at the same level. This approach allows the network to expand its width rather than its depth. The variation in kernel sizes aids in extracting significant features from images that may exhibit substantial variations in size and spatial locality.

4.3 Attention-Based Architectures

The attention mechanism helps ML models focus on features that are relevant to the context, so that these models can identify objects from complex scenes and clustered backgrounds. Residual attention network (RAN) is constructed by stacking residual blocks with attention mechanism [110]. The attention module consists of two branches, namely the trunk branch and the mask branch, both of which employ a bottom-up, top-down strategy. The bottom-up architecture provides low-resolution features with rich semantic information, and the top-down structure generates dense features to infer each pixel. Through stacking multiple attention modules, RAN can efficiently reorganize noisy, cluttered, and complicated images.

4.4 Architectures Exploiting Feature Maps

Feature selection is a critical aspect of model performance, involving the adjustment of weights associated with kernels to mitigate noise and prevent network overfitting caused by a large set of feature maps. In the squeeze-and-excitation network (SE-network), SE-blocks are introduced as processing units to assign high weights to important feature maps and suppress less important feature maps [111]. The SE-block incorporates two operations, namely squeeze and excitation, where squeeze extracts a global view of feature maps by computing statistics, and excitation assigns weights to feature maps using a two-layer FFNN. These SE-blocks can be seamlessly integrated into any model prior to the convolutional layers.

5 Conclusions

ML techniques play a crucial role in elucidating protein structure and function, enabling a deeper understanding of biological processes and facilitating drug discovery. The rapid growth of protein sequence data has necessitated the use of DL as a powerful tool for predicting structure and function solely from sequence information. CNNs and NLP are prominent ML approaches extensively employed in predicting protein secondary and tertiary structures, residue-residue contacts, protein function, and subcellular localization. However, challenges remain, including the need for high-quality training datasets and the interpretability of models. Nevertheless, the development of intricate deep learning architectures demonstrates ongoing progress in the field. Overall, this review highlights the profound impact of machine learning on advancing our knowledge of protein structure and function, with the potential to revolutionize drug discovery and personalized medicine.

References

1. Anfinsen CB (1973) Principles that govern the folding of protein chains. Science 181(4096):223–230. https://doi.org/10.1126/science.181.4096.223
2. Wittung-Stafshede P (2002) Role of cofactors in protein folding. Acc Chem Res 35(4): 201–208. https://doi.org/10.1021/ar010106e
3. Lindorff-Larsen K, Kragelund BB (2021) On the potential of machine learning to examine the relationship between sequence, structure, dynamics and function of intrinsically disordered proteins. J Mol Biol 433(20):167196. https://doi.org/10.1016/j.jmb.2021.167196
4. Ramanathan A, Ma H, Parvatikar A et al (2021) Artificial intelligence techniques for integrative structural biology of intrinsically disordered proteins. Curr Opin Struct Biol 66:216–224. https://doi.org/10.1016/j.sbi.2020.12.001
5. Zhang Y (2008) Progress and challenges in protein structure prediction. Curr Opin Struct Biol 18(3):342–348. https://doi.org/10.1016/j.sbi.2008.02.004
6. Jumper J, Evans R, Pritzel A et al (2021) Highly accurate protein structure prediction with AlphaFold. Nature 596(7873): 583–589. https://doi.org/10.1038/s41586-021-03819-2
7. Berman HM, Westbrook J, Feng Z et al (2000) The Protein Data Bank. Nucleic Acids Res 28(1):235–242. https://doi.org/10.1093/nar/28.1.235
8. Yoo DP, Zhou BB, Zomaya YA (2008) Machine learning techniques for protein secondary structure prediction: an overview and evaluation. Curr Bioinforma 3(2):74–86. https://doi.org/10.2174/157489308784340676
9. Yang Y, Gao J, Wang J et al (2018) Sixty-five years of the long march in protein secondary structure prediction: the final stretch? Brief

Bioinform 19(3):482–494. https://doi.org/10.1093/bib/bbw129

10. Dor O, Zhou Y (2007) Achieving 80% ten-fold cross-validated accuracy for secondary structure prediction by large-scale training. Proteins 66(4):838–845. https://doi.org/10.1002/prot.21298
11. McGuffin LJ, Bryson K, Jones DT (2000) The PSIPRED protein structure prediction server. Bioinformatics 16(4):404–405. https://doi.org/10.1093/bioinformatics/16.4.404
12. Cuff JA, Clamp ME, Siddiqui AS et al (1998) JPred: a consensus secondary structure prediction server. Bioinformatics 14(10):892–893. https://doi.org/10.1093/bioinformatics/14.10.892
13. Wood MJ, Hirst JD (2005) Protein secondary structure prediction with dihedral angles. Proteins 59(3):476–481. https://doi.org/10.1002/prot.20435
14. Heffernan R, Yang Y, Paliwal K et al (2017) Capturing non-local interactions by long short-term memory bidirectional recurrent neural networks for improving prediction of protein secondary structure, backbone angles, contact numbers and solvent accessibility. Bioinformatics 33(18):2842–2849. https://doi.org/10.1093/bioinformatics/btx218
15. Magnan CN, Baldi P (2014) SSpro/ACCpro 5: almost perfect prediction of protein secondary structure and relative solvent accessibility using profiles, machine learning and structural similarity. Bioinformatics 30(18):2592–2597. https://doi.org/10.1093/bioinformatics/btu352
16. Valueva MV, Nagornov NN, Lyakhov PA et al (2020) Application of the residue number system to reduce hardware costs of the convolutional neural network implementation. Math Comput Simul 177:232–243. https://doi.org/10.1016/j.matcom.2020.04.031
17. Klausen MS, Jespersen MC, Nielsen H et al (2019) NetSurfP-2.0: improved prediction of protein structural features by integrated deep learning. Proteins 87(6):520–527. https://doi.org/10.1002/prot.25674
18. Heinzinger M, Elnaggar A, Wang Y et al (2019) Modeling aspects of the language of life through transfer-learning protein sequences. BMC Bioinformatics 20(1):723. https://doi.org/10.1186/s12859-019-3220-8
19. Steinegger M, Söding J (2017) MMseqs2 enables sensitive protein sequence searching for the analysis of massive data sets. Nat Biotechnol 35(11):1026–1028. https://doi.org/10.1038/nbt.3988
20. Zhao Y, Liu Y (2021) OCLSTM: optimized convolutional and long short-term memory neural network model for protein secondary structure prediction. PLoS One 16(2):e0245982. https://doi.org/10.1371/journal.pone.0245982
21. Liu L, Yang S, Liu Y et al (2022) DeepContact: high-throughput quantification of membrane contact sites based on electron microscopy imaging. J Cell Biol 221(9). https://doi.org/10.1083/jcb.202106190
22. Seemayer S, Gruber M, Soding J (2014) CCMpred – fast and precise prediction of protein residue-residue contacts from correlated mutations. Bioinformatics 30(21):3128–3130. https://doi.org/10.1093/bioinformatics/btu500
23. Adhikari B, Hou J, Cheng J (2018) DNCON2: improved protein contact prediction using two-level deep convolutional neural networks. Bioinformatics 34(9):1466–1472. https://doi.org/10.1093/bioinformatics/btx781
24. Jones DT, Kandathil SM (2018) High precision in protein contact prediction using fully convolutional neural networks and minimal sequence features. Bioinformatics 34(19):3308–3315. https://doi.org/10.1093/bioinformatics/bty341
25. Buchan DWA, Jones DT (2018) Improved protein contact predictions with the MetaPSICOV2 server in CASP12. Proteins 86(Suppl 1):78–83. https://doi.org/10.1002/prot.25379
26. Wang S, Sun S, Li Z et al (2017) Accurate de novo prediction of protein contact map by ultra-deep learning model. PLoS Comput Biol 13(1):e1005324. https://doi.org/10.1371/journal.pcbi.1005324
27. Luttrell J, Liu T, Zhang C et al (2019) Predicting protein residue-residue contacts using random forests and deep networks. BMC Bioinformatics 20(Suppl 2):100. https://doi.org/10.1186/s12859-019-2627-6
28. Hanson J, Paliwal K, Litfin T et al (2018) Accurate prediction of protein contact maps by coupling residual two-dimensional bidirectional long short-term memory with convolutional neural networks. Bioinformatics 34(23):4039–4045. https://doi.org/10.1093/bioinformatics/bty481
29. Meiler J, Müller M, Zeidler A et al (2001) Generation and evaluation of dimension-reduced amino acid parameter representations by artificial neural networks. J Mol Model

7(9):360–369. https://doi.org/10.1007/s008940100038
30. Li Y, Zhang C, Bell EW et al (2021) Deducing high-accuracy protein contact-maps from a triplet of coevolutionary matrices through deep residual convolutional networks. PLoS Comput Biol 17(3):e1008865. https://doi.org/10.1371/journal.pcbi.1008865
31. Robin X, Haas J, Gumienny R et al (2021) Continuous automated model EvaluatiOn (CAMEO)-perspectives on the future of fully automated evaluation of structure prediction methods. Proteins 89(12):1977–1986. https://doi.org/10.1002/prot.26213
32. Kabsch W (1976) A solution for the best rotation to relate two sets of vectors. Acta Crystallogr A 32(5):922–923. https://doi.org/10.1107/S0567739476001873
33. Baek M, DiMaio F, Anishchenko I et al (2021) Accurate prediction of protein structures and interactions using a three-track neural network. Science 373(6557):871–876. https://doi.org/10.1126/science.abj8754
34. Chaudhury S, Lyskov S, Gray JJ (2010) PyRosetta: a script-based interface for implementing molecular modeling algorithms using Rosetta. Bioinformatics 26(5):689–691. https://doi.org/10.1093/bioinformatics/btq007
35. Doolittle RF (1995) The multiplicity of domains in proteins. Annu Rev Biochem 64: 287–314. https://doi.org/10.1146/annurev.bi.64.070195.001443
36. Zhou X, Zheng W, Li Y et al (2022) I-TASSER-MTD: a deep-learning-based platform for multi-domain protein structure and function prediction. Nat Protoc 17(10): 2326–2353. https://doi.org/10.1038/s41596-022-00728-0
37. Roy A, Kucukural A, Zhang Y (2010) I-TASSER: a unified platform for automated protein structure and function prediction. Nat Protoc 5(4):725–738. https://doi.org/10.1038/nprot.2010.5
38. Zhou X, Hu J, Zhang C et al (2019) Assembling multidomain protein structures through analogous global structural alignments. Proc Natl Acad Sci USA 116(32):15930–15938. https://doi.org/10.1073/pnas.1905068116
39. Zheng W, Zhou X, Wuyun Q et al (2020) FUpred: detecting protein domains through deep-learning-based contact map prediction. Bioinformatics 36(12):3749–3757. https://doi.org/10.1093/bioinformatics/btaa217
40. Xue Z, Xu D, Wang Y et al (2013) ThreaDom: extracting protein domain boundary information from multiple threading alignments. Bioinformatics 29(13):i247–i256. https://doi.org/10.1093/bioinformatics/btt209
41. Li Y, Zhang C, Yu DJ et al (2022) Deep learning geometrical potential for high-accuracy ab initio protein structure prediction. iScience 25(6):104425. https://doi.org/10.1016/j.isci.2022.104425
42. Roy A, Yang J, Zhang Y (2012) COFACTOR: an accurate comparative algorithm for structure-based protein function annotation. Nucleic Acids Res 40(Web Server issue): W471–W477. https://doi.org/10.1093/nar/gks372
43. Alberts B, Johnson A, Lewis J et al (2002) Molecular biology of the cell, 4th edn. Garland Science, New York
44. UniProt C (2019) UniProt: a worldwide hub of protein knowledge. Nucleic Acids Res 47 (D1):D506–D515. https://doi.org/10.1093/nar/gky1049
45. Seo S, Oh M, Park Y et al (2018) DeepFam: deep learning based alignment-free method for protein family modeling and prediction. Bioinformatics 34(13):i254–i262. https://doi.org/10.1093/bioinformatics/bty275
46. Tatusov RL, Koonin EV, Lipman DJ (1997) A genomic perspective on protein families. Science 278(5338):631–637. https://doi.org/10.1126/science.278.5338.631
47. Galperin MY, Makarova KS, Wolf YI et al (2015) Expanded microbial genome coverage and improved protein family annotation in the COG database. Nucleic Acids Res 43(Database issue):D261–D269. https://doi.org/10.1093/nar/gku1223
48. Luo J, Chen W, Xue L et al (2019) Prediction of activity and specificity of CRISPR-Cpf1 using convolutional deep learning neural networks. BMC Bioinformatics 20(1):332. https://doi.org/10.1186/s12859-019-2939-6
49. Almagro Armenteros JJ, Sønderby CK, Sønderby SK et al (2017) DeepLoc: prediction of protein subcellular localization using deep learning. Bioinformatics 33(21):3387–3395. https://doi.org/10.1093/bioinformatics/btx431
50. Zhang D, Kabuka MR (2021) Protein family classification from scratch: a CNN based deep learning approach. IEEE/ACM Trans Comput Biol Bioinform 18(5):1996–2007. https://doi.org/10.1109/TCBB.2020.2966633
51. Cai Y, Wang J, Deng L (2020) SDN2GO: an integrated deep learning model for protein

function prediction. Front Bioeng Biotechnol 8:391. https://doi.org/10.3389/fbioe.2020.00391

52. Fukushima K, Miyake S (1982) Neocognitron: a self-organizing neural network model for a mechanism of visual pattern recognition. In: Competition and cooperation in neural nets. Springer, Berlin/Heidelberg, pp 267–285
53. Hubel DH, Wiesel TN (1968) Receptive fields and functional architecture of monkey striate cortex. J Physiol 195(1):215–243. https://doi.org/10.1113/jphysiol.1968.sp008455
54. Lecun Y, Boser BE, Denker JS et al (1992) Handwritten digit recognition with a back-propagation network. In: Lisboa PGJ (ed) Neural networks, current applications. Chapman Hall/CRC Publishers
55. Lecun Y, Bottou L, Bengio Y et al (1998) Gradient-based learning applied to document recognition. Proc IEEE 86(11):2278–2324. https://doi.org/10.1109/5.726791
56. Farfade S S, Saberian M J, Li L-J (2015) Multi-view face detection using deep convolutional neural networks. Paper presented at the proceedings of the 5th ACM on international conference on multimedia retrieval, Shanghai, China
57. Li H, Lin Z, Shen X, et al (2015) A convolutional neural network cascade for face detection. In: 2015 IEEE conference on Computer Vision and Pattern Recognition (CVPR), 7–12 June 2015, pp 5325–5334. https://doi.org/10.1109/CVPR.2015.7299170
58. Li S, Liu Z-Q, Chan AB (2014) Heterogeneous multi-task learning for human pose estimation with deep convolutional neural network. Paper presented at the 2014 IEEE conference on Computer Vision and Pattern Recognition (CVPR)
59. Bulat A, Tzimiropoulos G (2016) Human pose estimation via convolutional part heatmap regression. In: Computer vision—ECCV 2016. Springer International Publishing, Cham, pp 717–732
60. Wang X, Gao L, Song J et al (2017) Beyond frame-level CNN: saliency-aware 3-D CNN with LSTM for video action recognition. IEEE Signal Process Lett 24:510–514
61. Donahue J, Hendricks L A, Guadarrama S, et al (2015) Long-term recurrent convolutional networks for visual recognition and description. In: 2015 IEEE conference on Computer Vision and Pattern Recognition (CVPR), 7–12 June 2015, pp 2625–2634. https://doi.org/10.1109/CVPR.2015.7298878
62. Sermanet P, Chintala S, Lecun Y (2012) Convolutional neural networks applied to house numbers digit classification. 21st international conference on pattern recognition, ICPR 2012
63. Ciregan D, Meier U, Schmidhuber J (2012) Multi-column deep neural networks for image classification. In: 2012 IEEE conference on computer vision and pattern recognition, 16–21 June 2012, pp 3642–3649. https://doi.org/10.1109/CVPR.2012.6248110
64. Ronneberger O, Fischer P, Brox T (2015) U-Net: convolutional networks for biomedical image segmentation. In: Medical image computing and computer-assisted intervention—MICCAI 2015. Springer International Publishing, Cham, pp 234–241
65. Zhang Y, Qiu Z, Yao T, et al (2018) Fully convolutional adaptation networks for semantic segmentation. 2018 IEEE conference on computer vision and pattern recognition (CVPR). https://doi.org/10.1109/CVPR.2018.00712
66. Ullah A, Ahmad J, Muhammad K et al (2018) Action recognition in video sequences using deep bi-directional LSTM with CNN features. IEEE Access 6:1155–1166. https://doi.org/10.1109/ACCESS.2017.2778011
67. Dong C, Loy CC, He K et al (2016) Image super-resolution using deep convolutional networks. IEEE Trans Pattern Anal Mach Intell 38(2):295–307. https://doi.org/10.1109/TPAMI.2015.2439281
68. Lim B, Son S, Kim H, et al (2017) Enhanced deep residual networks for single image super-resolution. In: 2017 IEEE conference on Computer Vision and Pattern Recognition Workshops (CVPRW), 21–26 July 2017, pp 1132–1140. https://doi.org/10.1109/CVPRW.2017.151
69. Bai X, Shi B, Zhang C et al (2017) Text/non-text image classification in the wild with convolutional neural networks. Pattern Recogn 66:437–446. https://doi.org/10.1016/j.patcog.2016.12.005
70. Gomez L, Nicolaou A, Karatzas D (2017) Improving patch-based scene text script identification with ensembles of conjoined networks. Pattern Recogn 67(C):85–96. https://doi.org/10.1016/j.patcog.2017.01.032
71. Abdel-Hamid O, Mohamed A, Jiang H et al (2014) Convolutional neural networks for speech recognition. IEEE/ACM Trans Audio Speech Lang Process 22(10):

1533–1545. https://doi.org/10.1109/TASLP.2014.2339736
72. Kim Y, Jernite Y, Sontag D et al (2016) Character-aware neural language models. Proc AAAI Conf Artif Intell 30(1). https://doi.org/10.1609/aaai.v30i1.10362
73. Kalchbrenner N, Grefenstette E, Blunsom PA (2014) Convolutional neural network for modelling sentences. In: Proceedings of the 52nd annual meeting of the Association for Computational Linguistics (Volume 1: Long Papers), Baltimore, Maryland, June 2014. Association for Computational Linguistics, pp 655–665. https://doi.org/10.3115/v1/P14-1062
74. Gu J, Wang Z, Kuen J et al (2018) Recent advances in convolutional neural networks. Pattern Recogn 77:354–377. https://doi.org/10.1016/j.patcog.2017.10.013
75. Harbola S, Coors V (2019) One dimensional convolutional neural network architectures for wind prediction. Energy Convers Manag 195:70–75. https://doi.org/10.1016/j.enconman.2019.05.007
76. Han D, Chen J, Sun J (2019) A parallel spatiotemporal deep learning network for highway traffic flow forecasting. Int J Distrib Sens Netw 15(2):1550147719832792. https://doi.org/10.1177/1550147719832792
77. Li Z, Liu F, Yang W et al (2022) A survey of convolutional neural networks: analysis, applications, and prospects. IEEE Trans Neural Netw Learn Syst 33(12):6999–7019. https://doi.org/10.1109/TNNLS.2021.3084827
78. Khan A, Sohail A, Zahoora U et al (2020) A survey of the recent architectures of deep convolutional neural networks. Artif Intell Rev 53(8):5455–5516
79. Lee CY, Gallagher P, Tu Z (2018) Generalizing pooling functions in CNNs: mixed, gated, and tree. IEEE Trans Pattern Anal Mach Intell 40(4):863–875. https://doi.org/10.1109/TPAMI.2017.2703082
80. Wang T, Wu D J, Coates A, et al (2012) End-to-end text recognition with convolutional neural networks. In: Proceedings of the 21st international conference on pattern recognition, Tsukuba, Japan, pp 3304–3308
81. Kaiming H, Xiangyu Z, Shaoqing R et al (2014) Spatial pyramid pooling in deep convolutional networks for visual recognition. In: European conference on computer vision, computer vision—ECCV 2014. Springer International Publishing, Cham, pp 346–361
82. Maas AL, Hannun AY, Ng AY (2013) Rectifier nonlinearities improve neural network acoustic models. In: Proceedings of the 30th international conference on machine learning, Atlanta, 2013. p 3
83. He K, Zhang X, Ren S, et al (2015) Delving deep into rectifiers: surpassing human-level performance on ImageNet classification. In: 2015 IEEE International Conference on Computer Vision (ICCV), 7–13 Dec. 2015, pp 1026–1034. https://doi.org/10.1109/ICCV.2015.123
84. Wang S, Manning C (2013) Fast dropout training. Paper presented at the proceedings of the 30th international conference on machine learning, proceedings of machine learning research
85. Ba J, Frey BJ (2013) Adaptive dropout for training deep neural networks. In: NIPS
86. Tompson J, Goroshin R, Jain A, et al (2015) Efficient object localization using convolutional networks. https://doi.org/10.1109/CVPR.2015.7298664
87. Hutchins J (2006) The history of machine translation in a nutshell. Retrieved December 20 (2009):1–1
88. Johnson M (2009) How the statistical revolution changes (computational) linguistics. Proceedings of the EACL 2009 Workshop on the interaction between linguistics and computational linguistics: virtuous, vicious or vacuous? Association for Computational Linguistics
89. Kamiński B, Jakubczyk M, Szufel P (2018) A framework for sensitivity analysis of decision trees. Cent Eur J Oper Res 26:135–159
90. Rabiner LR (1989) A tutorial on hidden Markov models and selected applications in speech recognition. Proc IEEE 77(2):257–286
91. Cho K, Merriënboer B V, Bahdanau D, et al (2014) On the properties of neural machine translation: encoder-decoder approaches. arXiv preprint arXiv:14091259
92. Gers FA, Schmidhuber J, Cummins F (2000) Learning to forget: continual prediction with LSTM. Neural Comput 12(10):2451–2471. https://doi.org/10.1162/089976600300015015
93. Zhang X, Zhao J, LeCun Y (2015) Character-level convolutional networks for text classification. Adv Neural Inf Proces Syst 28
94. Vaswani A, Shazeer N, Parmar N, et al (2017) Attention is all you need. Paper presented at the Advances in neural information processing systems
95. Devlin J, Chang M-W, Lee K, et al (2018) Bert: pre-training of deep

bidirectional transformers for language understanding. arXiv preprint arXiv:181004805

96. Brown T, Mann B, Ryder N, et al (2020) Language models are few-shot learners. arXiv:200514165. https://doi.org/10.48550/arXiv.2005.14165
97. Lee J, Yoon W, Kim S et al (2020) BioBERT: a pre-trained biomedical language representation model for biomedical text mining. Bioinformatics 36(4):1234–1240
98. Gu Y, Tinn R, Cheng H et al (2021) Domain-specific language model pretraining for biomedical natural language processing. ACM Trans Comput Healthcare (HEALTH) 3(1): 1–23
99. Zeng Z, Shi H, Wu Y et al (2015) Survey of natural language processing techniques in bioinformatics. Comput Math Methods Med 2015. https://doi.org/10.1155/2015/674296
100. Wahab A, Tayara H, Xuan Z et al (2021) DNA sequences performs as natural language processing by exploiting deep learning algorithm for the identification of N4-methylcytosine. Sci Rep 11(1):1–9
101. Li H-L, Pang Y-H, Liu B (2021) BioSeq-BLM: a platform for analyzing DNA, RNA and protein sequences based on biological language models. Nucleic Acids Res 49(22): e129
102. Alzubaidi L, Zhang J, Humaidi AJ et al (2021) Review of deep learning: concepts, CNN architectures, challenges, applications, future directions. J Big Data 8(1):1–74
103. Krizhevsky A, Sutskever I, Hinton GE (2017) ImageNet classification with deep convolutional neural networks. Commun ACM 60(6):84–90. https://doi.org/10.1145/3065386
104. Srivastava R K, Greff K, Schmidhuber J (2015) Highway networks. arXiv preprint arXiv:150500387
105. He K, Zhang X, Ren S, et al (2016) Deep residual learning for image recognition. In: Proceedings of the IEEE conference on computer vision and pattern recognition (CVPR), Las Vegas, NV, USA, pp 770–778
106. Szegedy C, Vanhoucke V, Ioffe S, et al (2016) Rethinking the inception architecture for computer vision. 2016 IEEE conference on computer vision and Pattern Recognition (CVPR):2818–2826
107. Szegedy C, Ioffe S, Vanhoucke V, et al (2017) Inception-v4, inception-resnet and the impact of residual connections on learning. In: Thirty-first AAAI conference on artificial intelligence
108. Szegedy C, Liu W, Jia Y, et al (2015) Going deeper with convolutions. Proceedings of the IEEE conference on computer vision and pattern recognition (CVPR):1–9
109. Chollet F (2017) Xception: deep learning with depthwise separable convolutions. 2017 IEEE conference on computer vision and pattern recognition (CVPR):1800–1807
110. Wang F, Jiang M, Qian C, et al (2017) Residual attention network for image classification. 2017 IEEE conference on computer vision and pattern recognition (CVPR):3156–3164
111. Hu J, Shen L, Sun G (2018) Squeeze-and-excitation networks. 2018 IEEE/CVF conference on computer vision and pattern recognition:7132–7141. https://doi.org/10.1109/CVPR.2018.00745

Chapter 6

Protein Secondary Structure and DNA/RNA Detection for Cryo-EM and Cryo-ET Using Emap2sec and Emap2sec+

Javad Baghirov, Han Zhu, Xiao Wang, and Daisuke Kihara

Abstract

Cryo-electron microscopy (cryo-EM) has become a powerful tool for determining the structures of macromolecules, such as proteins and DNA/RNA complexes. While high-resolution cryo-EM maps are increasingly available, there is still a substantial number of maps determined at intermediate or low resolution. These maps present challenges when it comes to extracting structural information. In response to this, two computational methods, Emap2sec and Emap2sec+, have been developed by our group to address these challenges and benefit the analysis of cryo-EM maps. In this chapter, we describe how to use the web servers of two of our structure analysis software for cryo-EM, Emap2sec and Emapsec+. Both methods identify local structures in medium-resolution EM maps of 5–10 Å to help find and fit protein and DNA/RNA structures in EM maps. Emap2sec identifies the secondary structures of proteins, while Emap2sec+ also identifies DNA/RNA locations in cryo-EM maps. As cryo-electron tomogram (cryo-ET) has started to produce data of this resolution, these methods would be useful for cryo-ET, too. Both methods are available in the form of webservers and source code at https://kiharalab.org/emsuites/.

Key words Cryo-electron microscopy, Cryo-EM, Cryo-electron tomography, Cryo-ET, Emap2sec, Emap2sec+, Low resolution map data, Deep learning, Secondary structure detection

1 Introduction

There has been an increase in the amount of biomolecular structures solved with cryo-EM [1, 2]. The Electron Microscopy Data Bank (EMDB) [3] has seen exponential growth in deposited cryo-EM maps, and the fraction of high-resolution maps, better than 4 Å, has steadily increased. However, despite the remarkable progress, there remains a substantial fraction of maps determined at intermediate (5–10 Å) or low resolutions or even lower. This is because the attainable resolution is influenced by various factors, such as structural flexibility in protein chains, the presence of multiple functional states, and noise in density images [4].

When working with EM maps with intermediate resolutions, some fragments of protein secondary structures may be discernible,

Andrzej Kloczkowski et al. (eds.), *Prediction of Protein Secondary Structure*, Methods in Molecular Biology, vol. 2867,
https://doi.org/10.1007/978-1-0716-4196-5_6,

but tracing the entire main chain becomes exceedingly challenging [5]. To aid in the interpretation of such maps, we developed Emap2sec [6], a novel tool designed specifically for detecting protein secondary structures, α-helix, β-strand, and coil (structures other than α-helix and β-strand) within these maps. Emap2sec leverages the power of deep convolutional neural networks (CNNs) [7, 8] as its fundamental computational framework. CNNs are particularly well-suited for the task of identifying local protein structures within three-dimensional EM maps because they excel at convolving local density features from a small region into a larger context, allowing for the detection of local structures within the broader context of the map.

Building upon Emap2sec, we have taken another step forward by introducing Emap2sec+ [9]. This enhanced method goes beyond the scope of its predecessor by not only detecting protein secondary structures but also extending its capabilities to identify nucleic acids within EM maps of intermediate resolution. Emap2sec+ leverages a more sophisticated convolutional neural network architecture, ResNet [10], to tackle this. The method excels at performing a comprehensive four-class classification for each voxel within an EM map, categorizing them as α-helix, β-strand, coil, or DNA/RNA.

Knowing the position of protein secondary structures in an EM map helps researchers to align known protein structures in the map [11, 12]. For a low-resolution cryo-EM map or subtomogram from cryo-ET for which included proteins are unknown, secondary structure information can also contribute to identifying the proteins. Identifying DNA/RNA in a map is also useful, as it allows researchers to model DNA/RNA and proteins separately in the map [13]. Recently, the resolution of cryo-ET is also improving rapidly, and we are starting to see works that report a resolution better than 10 Å [14, 15]. Emap2sec and Emap2sec+ will also be useful for cryo-ET data of sub-nanometer resolution.

2 Algorithm of Emap2sec

The algorithm of Emap2sec is illustrated in Fig. 1. It first scans an electron microscopy (EM) map using a box with a volume of 11^3 Å^3. This box captures density information and inputs into a deep convolutional neural network (CNN). The pre-trained CNN can then identify the secondary structure of the central component within the box, categorizing it as either an α-helix, β-sheet, or other structural elements.

The convolutional layers in the neural network are responsible for capturing local density patterns within an EM map using convolutional filters. We used different numbers of filters (32, 64, or 128) in each convolutional layer. The output from these

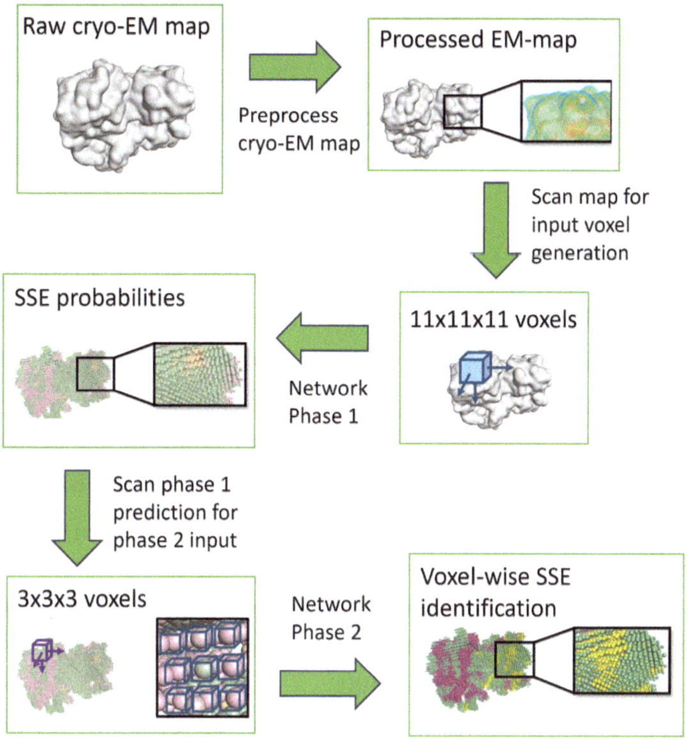

Fig. 1 Algorithm flow of Emap2sec

convolutional layers then goes through a maximum-pooling layer and fully connected layers. Finally, a softmax layer produces a probability indicating whether the structure is an α-helix, β-strand, or some other structure.

In the second phase of the network, the predicted probability values from the first phase are utilized. This phase takes the individual probability values for α-helices, β-strands, and other structures as input. Each input box is with a size of 3^3 Å^3, resulting in 81 values (3 * 3). The input is then processed through a network consisting of five fully connected layers, followed by a softmax layer. The second phase network refine predictions by considering predictions made for neighboring voxels. In experimental maps, the phase 2 network is configured to primarily transition between α-helix and β-strand or from other structures to α-helix or β-strand, reflecting the prevalence of other structures in experimental maps compared to simulated maps. This two-phase network design is inspired by the PSIPRED method [16], which predicts protein secondary structure based on sequence data.

Emap2sec was evaluated on a dataset comprising 43 experimental maps ranging from 5.0 to 9.5 Å, with the number of residues varying from 664 to 11,475, and the number of chains spanning from a single chain to a complex structure with 62 chains. Notably, when examining accuracy, particularly at the residue level, it was observed that the map resolution did not exhibit a strong influence, with consistently high accuracies observed across maps within the resolution range of 5.0–5.7 Å, reaching impressive values ranging from 0.808 to 0.916. Even maps with resolutions around 9.0 Å, such as EMD-1655 (9.0 Å) and EMD-1263 (9.1 Å), achieved substantial accuracy scores of 0.802 and 0.757, respectively.

3 Availability of Emap2sec

Emap2sec is available on four different platforms: Using the Google Colab notebook website, on a web server, using the CodeOcean capsule, or downloading the code and running it locally on a user's computer. Here is the list of URLs for the four releases:

- Google Colab Notebook: https://colab.research.google.com/github/kiharalab/Emap2sec/blob/master/Emap2sec.ipynb
- Webserver: https://em.kiharalab.org/algorithm/emap2sec
- CodeOcean Capsule: https://codeocean.com/capsule/4439990
- Source code at GitHub: https://github.com/kiharalab/Emap2sec

4 How to Use Emap2sec Google Colab

Here we explain how to use Emap2sec on Google Colab using Fig. 2.

1. On the notebook, first we select and connect GPUs. To connect to a GPU machine in the Google Colab notebook, click the "Connect" button on the top right corner.
2. Install dependencies. This is to install libraries (components running Emap2sec code).
3. Input cryo-EM map. Upload the input cryo-EM map file (.map/.mrc format).
4. Input contour level. The contour level is used on the map to specify which part of the map should be considered for the computation. For our example of EMD-1733, we use a contour level of 2.5. This can be found in the EMDB database (https://www.ebi.ac.uk/emdb/EMD-1733).
5. Run Emap2Sec. This will run the algorithms and the resulting .pdb file will be generated.
6. Download predicted output. You have the option to download the .pdb file here for visualization locally.

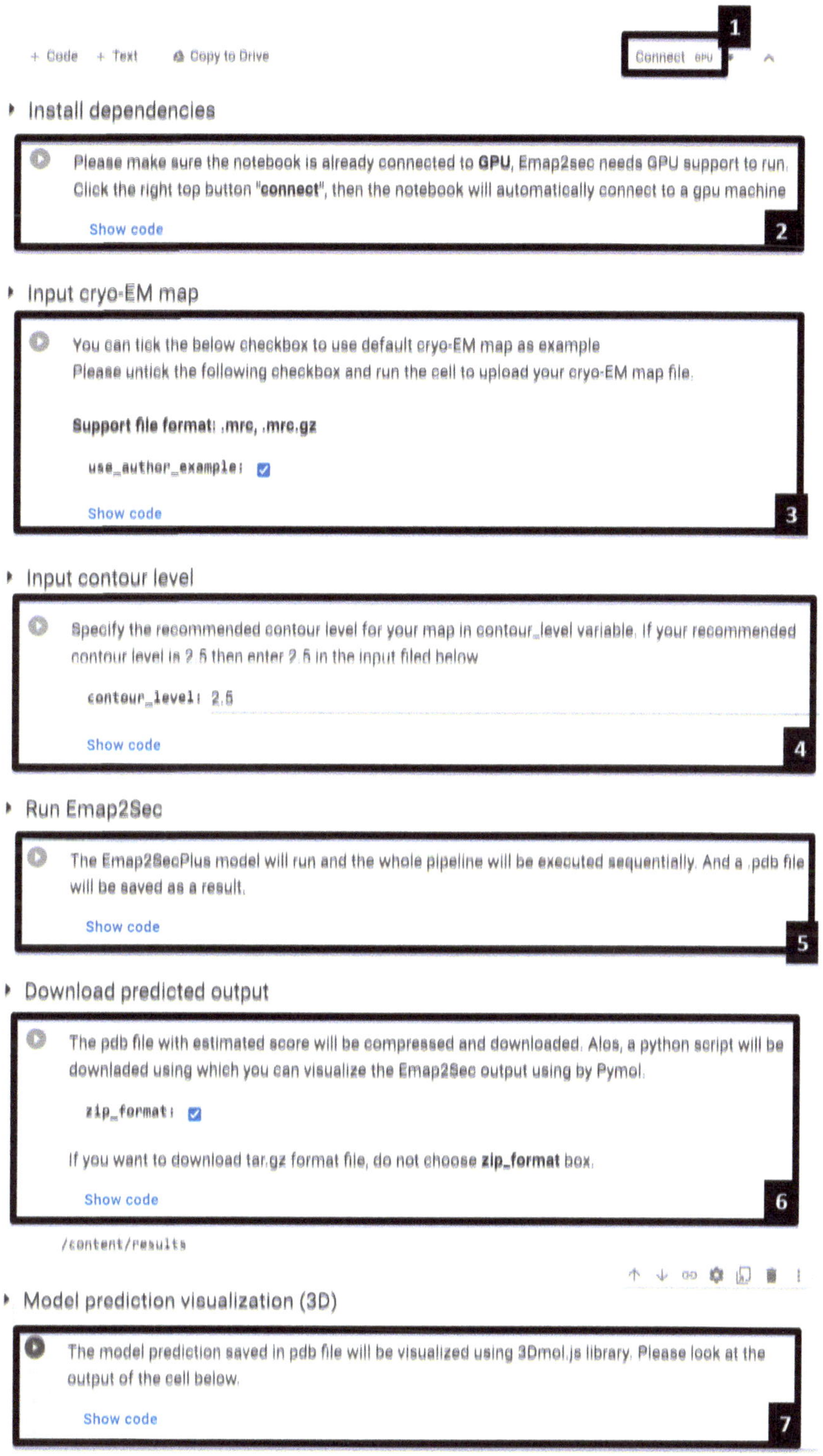

Fig. 2 Google Colab notebook for Emap2sec

7. Model prediction visualization. The generated pdb file will be visualized directly in the notebook.

5 How to Use Emap2sec Web Server

The web server provides another user-friendly use of Emap2sec. It would be easier to use than the Google Colab because users do not need to download or set up anything. Before submitting any job on the web server, the user must sign up and make an account (Fig. 3).

1. Visit the website at https://em.kiharalab.org.
2. Click on the "Register" button located in the top-right corner of the website. Alternatively, you can use https://em.kiharalab.

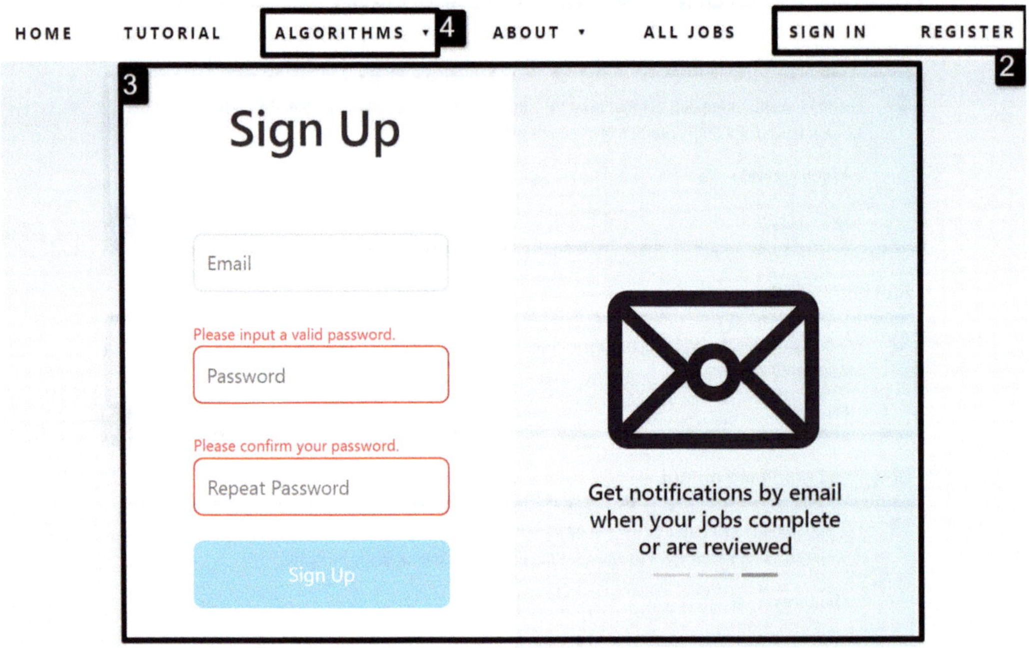

Fig. 3 Sign Up page of the EM tool webserver

Fig. 4 Job submission form of the Emap2sec webserver

org/users/registerpage to directly access the sign-up page (Fig. 3).

3. Complete the registration process by providing the required information and activating your account. You will need to verify your email address to activate the account. The email will usually arrive instantly, but it might also take up to 1–2 h.
4. After logging in, on the navigation bar, locate and click on the "ALGORITHMS" option, from the drop down please click Emap2sec.
5. You will be directed to a page where you need to input specific parameters for your job (Fig. 4).
   - (i) Contour Level: You can specify your desired contour level for the map. For our example, we use a contour level of 2.5. Read the Google Colab section above on how to find the contour level for a map.
   - (ii) Cryo-EM Map: Upload the map file (.mrc or .map files only).
   - (iii) Once you have filled in all the required information and uploaded the necessary files, submit the job.
   - (iv) If you are uploading a large file, you can monitor the progress bar at the bottom of the page to track the upload progress.
6. After submission, you will be redirected to the jobs table. This table will display all the jobs you have submitted, with the most recent one appearing at the top. The status of each job will be noted next to it, and you will be notified by email once the job you submitted has been completed.

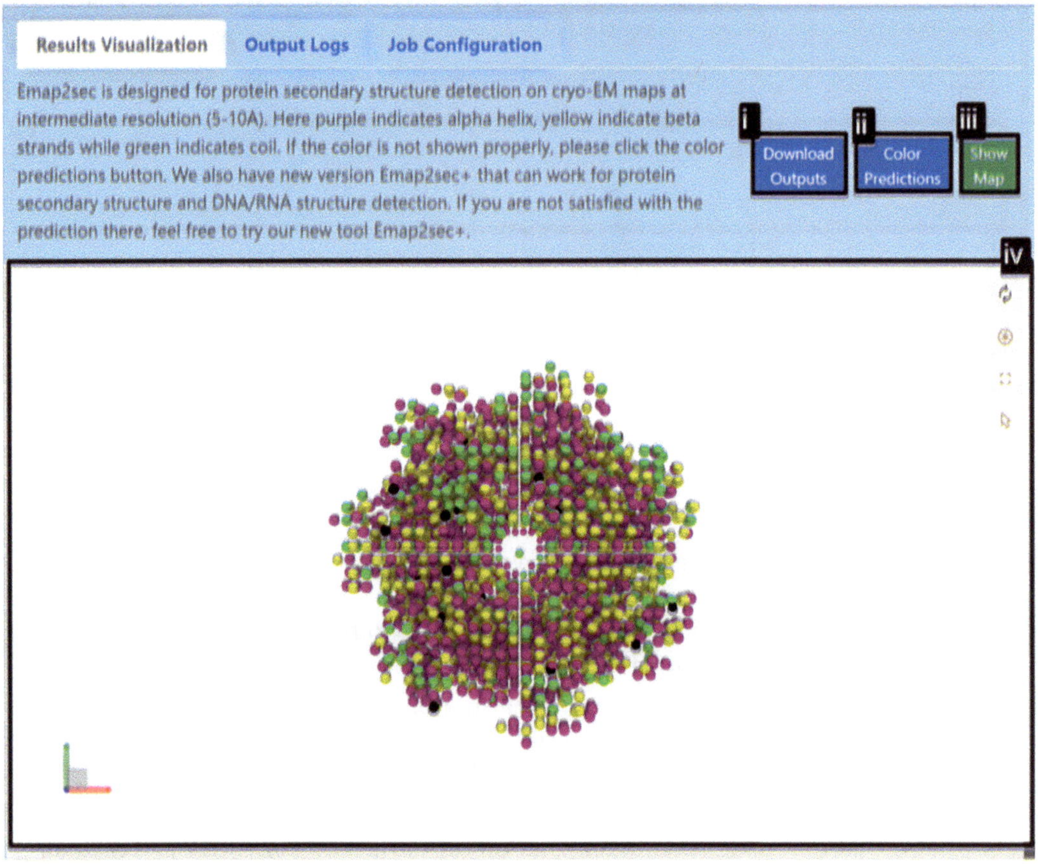

Fig. 5 Example of Emap2sec output on the webserver. This shows the output of Emap2sec with the input map EMD-1733 with the contour level of 2.5. EMD-1733 is the cryo-EM structure of the Archaeal 20S proteasome with an open gate determined at 6.8 Å

7. Locate the Emap2sec job you just scheduled and click on the "View" button to access the 3D outputs generated by the algorithm (Fig. 5). If there is no output, then the job is most likely still running or waiting for server resources.
8. The output is stored in a .pdb file. The .pdb file can be seen in the PDBe Molstar plugin (Fig. 5). (i) The raw .pdb file can also be downloaded by clicking the "Download Outputs" button found on the top right of the "Results Visualization" tab. (ii) If certain colors are not showing or showing incorrectly then click the "Color Predictions" button to refresh the colors. (iii) Additionally, the input map can be viewed by clicking the "Show Map" button. If you wish to share the job with others, you can use the social media sharing options. Alternatively, you can also just copy the link to the job and share it manually.

6 Code Ocean Capsule for Emap2sec

Code Ocean is a capsule of codes, which includes the code itself and the environment for running the code, such as libraries that the code depends on. It is used to easily test and run your code online and download the entire capsule and run it locally on a user's computer. To run the code, here is a short instruction:

1. First, you need to create an account on code ocean using academic credentials and login into your account.
2. Then, navigate to the Code Ocean link. (https://codeocean.com/capsule/4439990).
3. To make a reproducible run, i.e., run the code on an example input provided by us, click on the "Reproducible Run" button in the top right corner. This will start running the code on our example, and the results will be generated in the results folder at the bottom right after the execution is complete.
4. To run the code on an input of your choice, first go to our capsule and click on the "Edit Capsule" button in the top right corner. This will make a copy of our capsule, which you can edit. Follow the instructions in README about how to upload and run your input in this copied capsule by reading the readme file present in the respective code ocean capsules.

7 Example Output from Emap2sec

Figure 6 is an example of Emap2sec output on *E. coli* replicative DNA polymerase complex. The density map and its associated protein structures are shown on the left and the secondary structure detection result from the algorithm is shown on right. Within this visual depiction, red, yellow, and green spheres meticulously pinpoint detected α-helices, β-strands, and other distinctive structural components, respectively. It is worth noting that the reported map resolution for this example is low, 8.34 Å, and despite of the low resolution the Emap2sec algorithm managed to achieve a noteworthy F1 score of 0.406 and an accuracy score of 0.447.

Regarding the inherent structural composition, it reveals a complex mix of α-helices, β-strands, and other intriguing structural facets. More specifically, as we focus on the left side of the structure, the protein chains predominantly manifest themselves as an intricate network of exposed β-strands on the surface region and α-helices inside. This secondary structure pattern is well captured by Emap2sec. When examining the domain on the right, it predominantly aligns with α-helical structures, interleaves with a delicate interweaving of β-strands. For this region, Emap2sec's

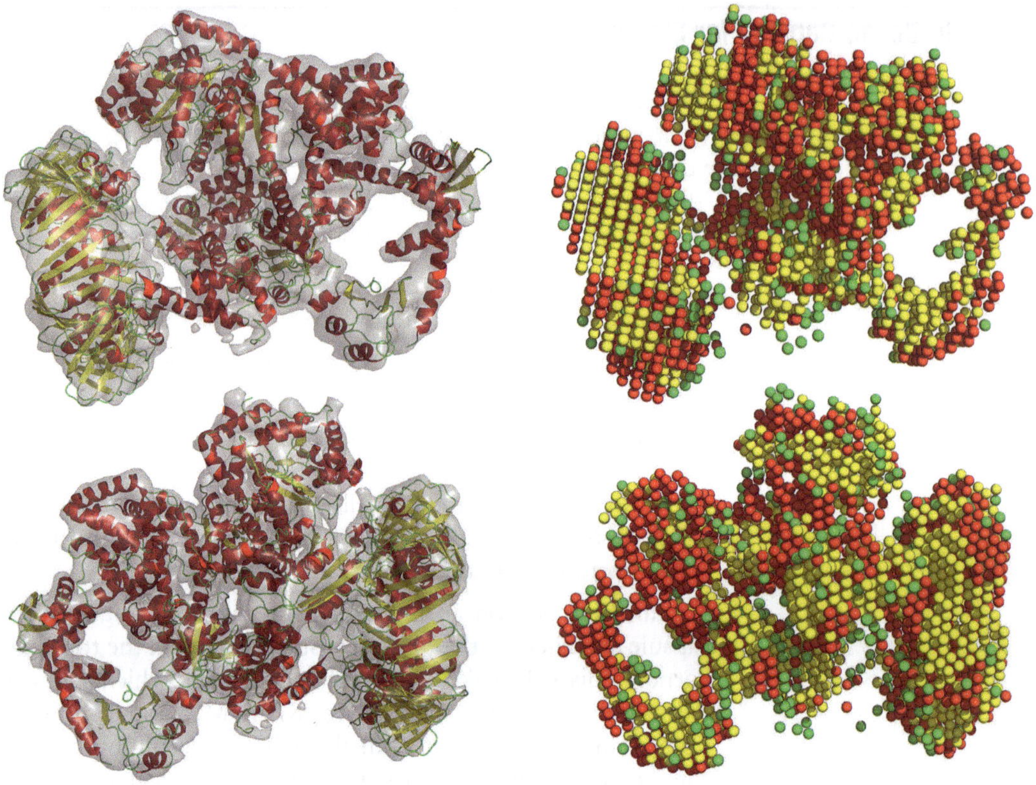

Fig. 6 Example of Emap2sec application to cryo-EM map. Iti is *E. coli* replicative DNA polymerase complex (EMD ID: EMD-3201; PDB ID: 5FKU). The map resolution was 8.34 Å. This protein complex has 5 chains with 2219 residues. The map and structures are shown on the left, and the corresponding predictions are shown on the right. The bottom shows the same structure rotated around the vertical axis by 180°

detection also shows a good correspondence with the protein structure.

8 Algorithm of Emap2sec+

Now we move on to the second method, Emap2sec+ [9]. It first adjusts the grid size of the input map to 1.0 Å. Similar to Emap2-sec, Emap2sec+ scans the input EM map with a 11^3 Å^3 box and identifies the central structure within each box, which can be DNA/RNA or various protein secondary structures like α-helices, β-sheets, or other structures. Essentially, Emap2sec+ classifies each box into one of these four structural categories. The box is moved through the EM map with a 2 Å stride in three different directions, and the structure assignment is made for every other grid in the input map through iterative scanning. Except that Emap2sec+ uses ResNet architecture, the algorithm is similar to Emap2sec. For details, please see the original paper [9].

Emap2sec+ was tested on 43 experimental maps with resolutions ranging from 5.0 to 9.5 Å. It demonstrated consistent high accuracy, especially within the 5.0–5.7 Å resolution range, where accuracy values reached a range of 0.808–0.916. Even when applied to maps with resolutions around 9.0 Å, Emap2sec+ achieved substantial accuracy scores, averaging approximately around 0.78.

9 Availability of Emap2sec+

Emap2sec+ is available on four different platforms: Using the Google Colab notebook website, on a web server, using the CodeOcean capsule, or downloading the code and running it locally on a user's computer. Here is the list of URLs for the four releases:

- Google Colab Notebook: https://colab.research.google.com/github/kiharalab/Emap2secPlus/blob/master/Emap2sec%2B.ipynb
- Webserver: https://em.kiharalab.org/algorithm/emap2sec+
- CodeOcean Capsule: https://codeocean.com/capsule/9499750
- Source code at GitHub: https://github.com/kiharalab/Emap2secPlus

10 How to Use Emap2sec+ Google Colab Notebook

Visit the Google Colab notebook. The steps are indicated in Fig. 7.

1. Install Conda Package Manager. This will be used to install the components running Emap2sec+ code.
2. Install dependencies. This is to install libraries (components running Emap2sec+ code).
3. Input cryo-EM map. Upload the input cryo-EM map file (.map and or .mrc),
4. Input contour level. For our example of EMD- 7103, we use a contour level of 2.5. This can be found in the EMDB database. (https://www.ebi.ac.uk/emdb/EMD-7103). For more information about contour levels check the Google Colab section for Emap2sec.
5. Run Emap2SecPlus. This will run the algorithms and the resulting .pdb file will be generated.
6. Download predicted output. You have the option to download the .pdb file in.

Fig. 7 The Google Colab Notebook form for Emap2sec+

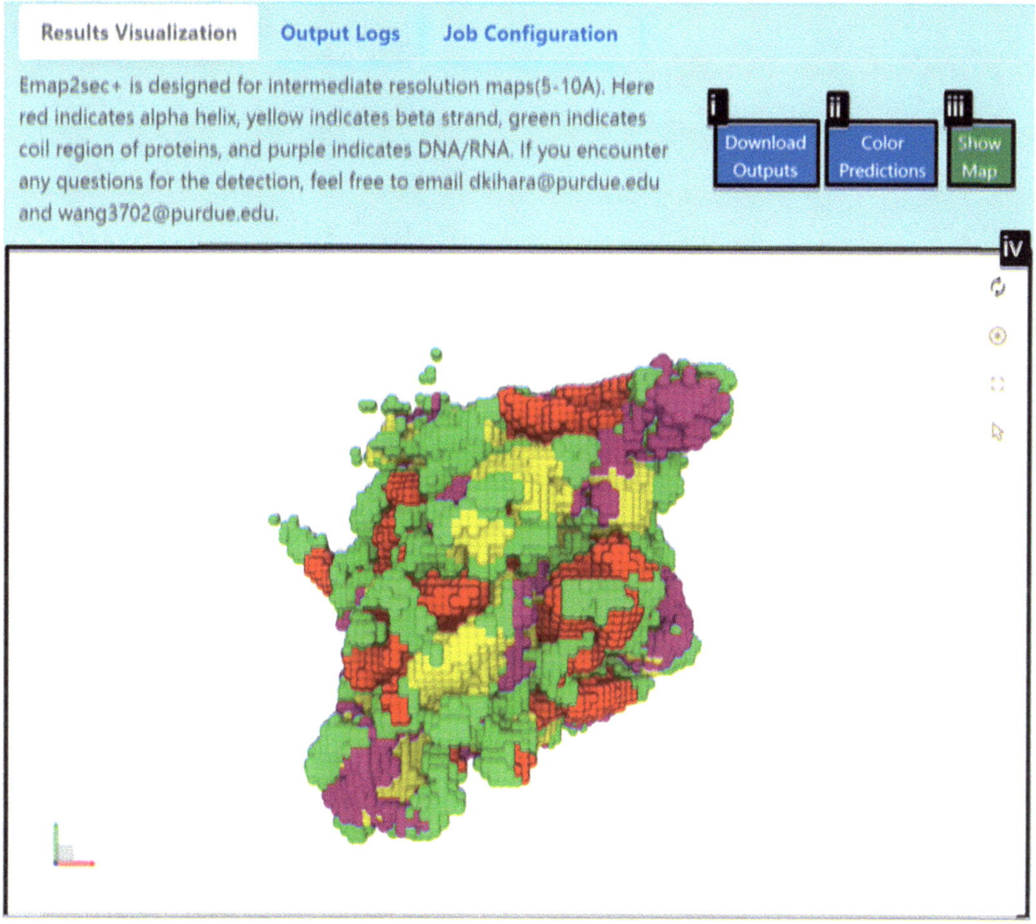

Fig. 8 The output of Emap2sec+ with the input map EMD-7103 with the contour level of 0.006. EMD-7103 is the cryo-EM structure of *E. coli's* pause elongation complex without pause hairpin

7. Model prediction visualization. The generated pdb file will be visualized directly in the notebook (Fig. 8).

11 How to Use Emap2sec+ Web Server and Code Ocean Capsule

Emap2sec+ has the same procedure and input fields as Emap2sec for the webserver and Code Ocean. For information check the Emap2sec section.

12 Example Output by Emap2sec+

Figure 9 shows the results of Emap2sec+ on a cryo-EM map of human maedi-visna virus (MVV) strand transfer complex, which is determined at an 8.2 Å resolution. The density maps (EMD-4139)

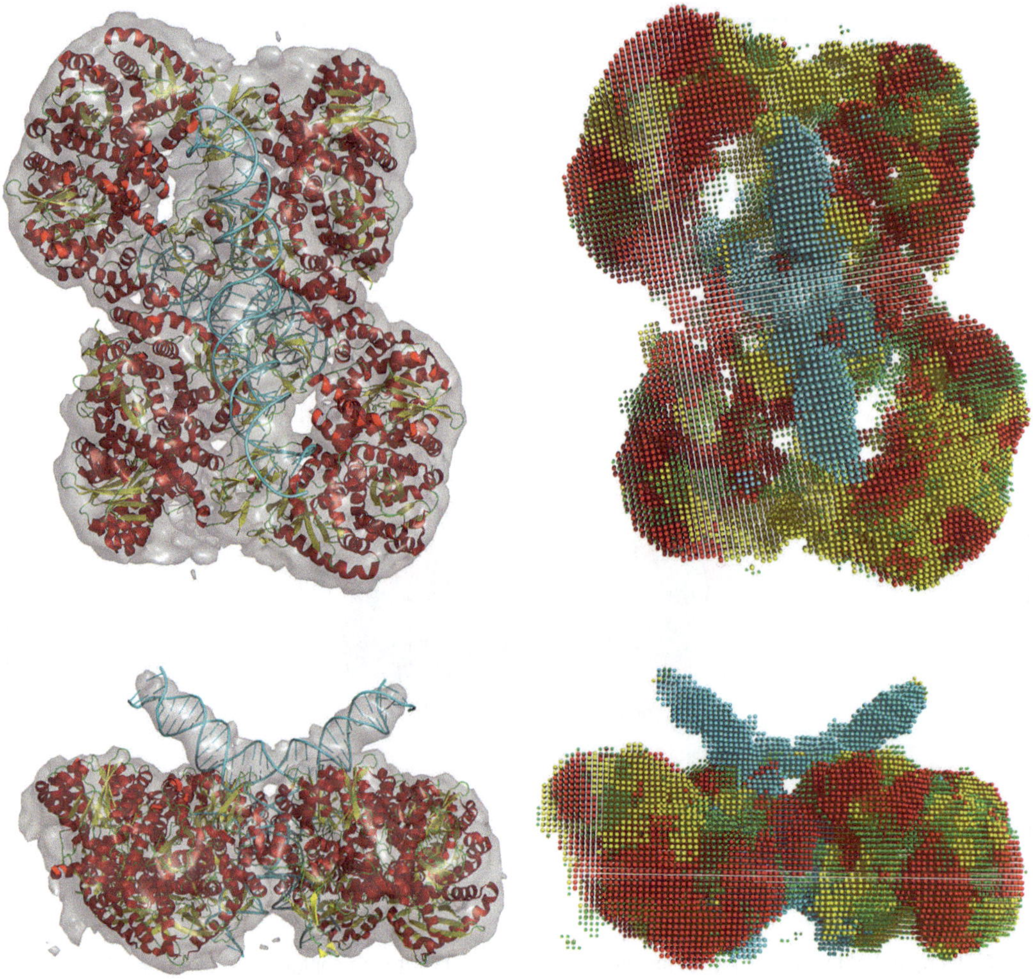

Fig. 9 An example of Emap2sec+ application to the cryo-EM map of the maedi-visna virus (MVV) strand transfer complex (EMD-4139; PDB-5M0R). The map resolution was 8.2 Å. This protein complex has 4064 amino acid residues and 188 nucleotides. The map and structures are shown on the left and the corresponding predictions are shown on the right. The bottom shows the side view of the same structure

and associated structures (5M0R) are presented on the left, and the detection results of Emap2sec+ are shown on the right. Spheres in red represent detected helices; yellow, strands; green, other structures; and cyan, RNA/DNA. Despite its low resolution, the algorithm recognizes different types of secondary structures in protein-DNA complexes with mostly correct predictions and achieved a voxel-based F1 score of 0.480 and an accuracy of 0.482 in 4-class classification. The F1 score for binary classification between protein and DNA/RNA is 0.921 and the accuracy is 0.915. Overall, the DNA structure highlighted in cyan is clearly distinguished from the others, while the secondary structure is successfully modeled within the surrounding regions, and the prediction corresponds well with the PDB entry.

13 Summary

Interpreting cryo-EM maps need various software. Deep learning can make significant contribution for cry-EM because it is suitable for image processing. We described how to use Emap2sec and Emapsec+ on different platforms. Our group has developed various other software for structure analysis and modeling for cryo-EM maps. The methods include VESPER for structure fitting [12, 17], EM-GAN for enhancing maps for facilitating structure modeling [18], structure model assessment [19, 20], and protein [21] and DNA/RNA structure modeling [22]. To use the software, please visit https://kiharalab.org/emsuites/.

Acknowledgments

This work was partly supported by the National Institutes of Health (R01GM133840 and 3R01GM133840-02S1) and the National Science Foundation (CMMI1825941, MCB1925643, IIS2211598, DMS2151678, DBI2146026, and DBI2003635).

References

1. Cheng Y (2018) Single-particle cryo-EM-how did it get here and where will it go. Science 361:876–880. https://doi.org/10.1126/science.aat4346
2. Kühlbrandt W (2014) Cryo-EM enters a new era. eLife 3:e03678. https://doi.org/10.7554/eLife.03678
3. Patwardhan A (2017) Trends in the Electron Microscopy Data Bank (EMDB). Acta Crystallogr D Struct Biol 73:503–508. https://doi.org/10.1107/S2059798317004181
4. Nogales E (2016) The development of cryo-EM into a mainstream structural biology technique. Nat Methods 13:24–27
5. Esquivel-Rodríguez J, Kihara D (2013) Computational methods for constructing protein structure models from 3D electron microscopy maps. J Struct Biol 184:93–102. https://doi.org/10.1016/j.jsb.2013.06.008
6. Maddhuri Venkata Subramaniya SR, Terashi G, Kihara D (2019) Protein secondary structure detection in intermediate-resolution cryo-EM maps using deep learning. Nat Methods 16: 911–917
7. Maturana D, Scherer S (2015) Voxnet: a 3d convolutional neural network for real-time object recognition. In: 2015 IEEE/RSJ international conference on intelligent robots and systems (IROS). IEEE, pp 922–928
8. Russakovsky O, Deng J, Su H et al (2015) Imagenet large scale visual recognition challenge. Int J Comput Vis 115:211–252
9. Wang X, Alnabati E, Aderinwale TW et al (2021) Detecting protein and DNA/RNA structures in cryo-EM maps of intermediate resolution using deep learning. Nat Commun 12:2302. https://doi.org/10.1038/s41467-021-22577-3
10. He K, Zhang X, Ren S, Sun J (2016) Deep residual learning for image recognition. In: Proceedings of the IEEE conference on computer vision and pattern recognition, pp 770–778
11. Alnabati E, Esquivel-Rodriguez J, Terashi G, Kihara D (2022) MarkovFit: structure fitting for protein complexes in electron microscopy maps using Markov random field. Front Mol Biosci 9:935411. https://doi.org/10.3389/fmolb.2022.935411

12. Han X, Terashi G, Christoffer C et al (2021) VESPER: global and local cryo-EM map alignment using local density vectors. Nat Commun 12:2090. https://doi.org/10.1038/s41467-021-22401-y
13. Wang Y, Guan Z, Wang C et al (2022) Cryo-EM structures of Escherichia coli Ec86 retron complexes reveal architecture and defence mechanism. Nat Microbiol 7:1480–1489. https://doi.org/10.1038/s41564-022-01197-7
14. Tegunov D, Xue L, Dienemann C et al (2021) Multi-particle cryo-EM refinement with M visualizes ribosome-antibiotic complex at 3.5 Å in cells. Nat Methods 18:186–193. https://doi.org/10.1038/s41592-020-01054-7
15. von Kügelgen A, Tang H, Hardy GG et al (2020) In situ structure of an intact lipopolysaccharide-bound bacterial surface layer. Cell 180:348–358.e15. https://doi.org/10.1016/j.cell.2019.12.006
16. Jones DT (1999) Protein secondary structure prediction based on position-specific scoring matrices. J Mol Biol 292:195–202. https://doi.org/10.1006/jmbi.1999.3091
17. Alnabati E, Terashi G, Kihara D (2022) Protein structural modeling for electron microscopy maps using VESPER and MAINMAST. Curr Protoc 2:e494. https://doi.org/10.1002/cpz1.494
18. Maddhuri Venkata Subramaniya SR, Terashi G, Kihara D (2023) Enhancing cryo-EM maps with 3D deep generative networks for assisting protein structure modeling. Bioinformatics 39: btad494. https://doi.org/10.1093/bioinformatics/btad494
19. Nakamura T, Wang X, Terashi G, Kihara D (2023) DAQ-score database: assessment of map–model compatibility for protein structure models from cryo-EM maps. Nat Methods 20: 775–776. https://doi.org/10.1038/s41592-023-01876-1
20. Terashi G, Wang X, Maddhuri Venkata Subramaniya SR et al (2022) Residue-wise local quality estimation for protein models from cryo-EM maps. Nat Methods 19:1116–1125. https://doi.org/10.1038/s41592-022-01574-4
21. Terashi G, Wang X, Prasad D, Kihara D (2023) Integrated protocol of protein structure modeling for Cryo-EM with deep learning and structure prediction. Nat Methods
22. Wang X, Terashi G, Kihara D (2023) De novo structure modeling for nucleic acids in cryo-EM maps using deep learning. Nat Methods 20:1739

Chapter 7

Beyond AlphaFold2: The Impact of AI for the Further Improvement of Protein Structure Prediction

Ahmet Gurkan Genc and Liam J. McGuffin

Abstract

Protein structure prediction is fundamental to molecular biology and has numerous applications in areas such as drug discovery and protein engineering. Machine learning techniques have greatly advanced protein 3D modeling in recent years, particularly with the development of AlphaFold2 (AF2), which can analyze sequences of amino acids and predict 3D structures with near experimental accuracy. Since the release of AF2, numerous studies have been conducted, either using AF2 directly for large-scale modeling or building upon the software for other use cases. Many reviews have been published discussing the impact of AF2 in the field of protein bioinformatics, particularly in relation to neural networks, which have highlighted what AF2 can and cannot do. It is evident that AF2 and similar approaches are open to further development and several new approaches have emerged, in addition to older refinement approaches, for improving the quality of predictions. Here we provide a brief overview, aimed at the general biologist, of how machine learning techniques have been used for improvement of 3D models of proteins following AF2, and we highlight the impacts of these approaches. In the most recent experiment on the Critical Assessment of Techniques for Protein Structure Prediction (CASP15), the most successful groups all developed their own tools for protein structure modeling that were based at least in some part on AF2. This improvement involved employing techniques such as generative modeling, changing parameters such as dropout to generate more AF2 structures, and data-driven approaches including using alternative templates and MSAs.

Key words Protein refinement, Machine learning, Data-driven approaches, AlphaFold, CASP

1 Introduction

Proteins are complex molecules that play a vital role in the functioning of living organisms. They are composed of long chains of amino acids, which fold into a specific 3D structure that is determined by the sequence of these amino acids. Determining the 3D structure of a protein is crucial for understanding its function and how it interacts with other molecules [1]. In addition to the ordered structures in protein structure, such as alpha helices, beta sheets, and coils [2], there are also intrinsically disordered regions within a protein structure that play a functional role by inducing

Andrzej Kloczkowski et al. (eds.), *Prediction of Protein Secondary Structure*, Methods in Molecular Biology, vol. 2867,
https://doi.org/10.1007/978-1-0716-4196-5_7,

conformational changes in the 3D structure of a protein. These regions lack a well-defined structure and exhibit flexibility, allowing proteins to adopt different forms to perform alternative functions not based on sequence-structure-function paradigm [3, 4]. Furthermore, the presence of intrinsically unstructured or disordered proteins [5], which are associated with diseases like Alzheimer's and Parkinson's [6], along with the existence of approximately up to 30% disorder-promoting regions in eukaryotic proteomes [7], is widely acknowledged by bioinformaticians in protein modeling. Protein modeling is the process of predicting the 3D structure of a protein based on its amino acid sequence. This can be done using various computational methods, such as homology modelling, which involves using the 3D structure of a similar protein as a template, or de novo modelling, which involves predicting the structure without the use of a template. Once a protein model has been generated, it is often necessary to refine it to correct any errors and improve its accuracy. This process, known as refinement, can involve using experimental data or other predicted information to optimize the model and make it more consistent with known biophysical constraints. Refinement is an important step in protein modelling as it can improve the accuracy and reliability of the model, which is essential for its use in downstream applications such as drug design or protein engineering [1].

2 The History of Protein Modelling

With the advent of modern biochemistry, a significant problem for biochemists in the 1950s was how to deduce the structural forms and functions of protein molecules. It was known that the protein amino acid sequences, also known as primary structures, govern the unique chemical characteristics of distinct proteins, and most biochemists have guided their work based on this understanding [8]. In light of this, Frederick Sanger was the first person to design an experimental method in 1955 to obtain the protein sequence of insulin [9]. Christian Anfinsen also made significant contributions to our understanding of proteins, developing his theory of protein folding and summarizing in 1972, that the native conformation, or the unique 3D structure of a given protein [10], is determined by the sum of interatomic interactions generated from the information stored in the primary structure of the protein [8].

The goal of protein structure prediction is to determine the Euclidean coordinates of every atom in a protein using computational methods, starting from the protein's amino acid sequence. Traditionally, there have been two main approaches to protein structure prediction: template-based modeling (TBM) and template-free modeling (FM). In TBM, models are constructed using the structures of similar proteins as templates, which are stored in the Protein Data Bank (PDB). FM, on the other hand, seeks to predict protein structures without using templates of known structures. However, recent advances in deep learning have significantly affected the field of protein structure prediction, leading to the development of tools that can accurately predict high-quality geometric properties of proteins, such as inter-residue contacts, distances, and torsion angles. These tools, which combine structural features generated by deep learning with classical folding methods, have greatly improved the quality of FM protein structure prediction, where there are no homologous templates available [11].

Modern prediction systems of protein structure typically consist of four main modules: input, a computational "trunk," output, and refinement. The input module converts the protein sequence into various forms of information, such as a protein-specific scoring matrix, pairwise interactions (co-evolutionary information), Markov random fields, and a raw multiple sequence alignment. The trunk module, usually a neural network, converts this information into spatial information about the protein's structure. The output module produces representations of the protein's structure, such as a contact map, distogram, or orientogram. Finally, the refinement module improves the quality of the 3D structure and generates atomic coordinates if needed. Traditionally, these operations have relied on a combination of physics-based energy functions, knowledge-based statistical reasoning, and heuristic algorithms. However, more recently, protein structure prediction tools have started using more sophisticated neural networks and machine learning algorithms in an end-to-end approach, rather than individually modeling each component [12].

The extension of a residual neural network to predict distance information between two residues within a certain range greatly improved residue–residue contact prediction [13]. Using this distance map, generated by 220 residual blocks from a protein sequence, AlphaFold (AF) demonstrated its capabilities in the CASP13 experiment [14]. These predicted distances were then used in structural fragment assembly and folding simulation through gradient descent. AF2, the second version of AF developed by the DeepMind team, made an exciting breakthrough in the field of protein structure prediction in CASP14 [11]. Its modeled structures achieved unprecedented accuracy, making the longstanding problem of predicting protein folding seem trivial [15].

One of the main differences between the first and second versions of AF was in their neural network architectures. While AF used a convolutional neural network and employed distance map prediction, AF2 used an attention-based neural network and took richer input information from the full multiple sequence alignment (MSA). Instead of using gradient descent optimization, as in AF, AF2 evolved a fully end-to-end deep learning algorithm that predicts the 3D structure solely from the sequence, with an iterative refinement-like process, termed recycling, based on its local error estimation score [11]. While there were many AF2 models that reached near experimental accuracy in CASP14, for many targets, the AF2 models still include significant errors, and so there is still room for further improvement. To ensure the more active utilization of structure prediction beyond AF2, the importance of refinement will be further emphasized.

3 Methods for the Refinement of 3D Models of Protein Structures Before AF2

The objective of computational protein structure refinement is to optimize a starting structure to its native state, such that it approaches the accuracy of experimental measurements. To achieve this, refinement methods of protein structure utilize conformational sampling to optimize an atomic force field to bring the starting structure closer to its native conformation [16]. Two basic concepts have been so far used to obtain and improve the three-dimensional structures of proteins from protein sequences: 1) physical interactions and 2) evolutionary information from bioinformatics analysis. Evolutionary approaches have become more useful because physical interactions, such as thermodynamic or kinetic information from protein physics or statistical approaches, can create a computational burden for molecular simulations [17]. Therefore, tools based on physical laws could only be used for refinement of predicted structures following initial modeling using evolutionary approaches.

Before the development of AF2, refinement methods were generally divided into three categories: molecular dynamics (MD) as shown in Fig. 1 and Monte Carlo simulation, energy minimization, and fragment assembly. Molecular simulations and physics-based approaches sample multiple MD trajectories according to the physical principles governing atom-atom interactions. Energy minimization methods aim to find the lowest energy structure by repackaging both backbone and side-chain atoms based on physical and knowledge-based force fields, while fragment-based methods use template fragment information from the PDB in conjunction with statistical potentials. However, these refinement methods require extensive conformational sampling, which can be time-consuming and computationally demanding [18]. A common approach in refinement protocols without the use of machine learning is the application of restraints. Many protocols have adopted this method in different ways. For example, the latest version of the ReFOLD tool—ReFOLD3 [19]—uses restraints based on quality estimation scores, while approaches developed by Feig use restraints on backbone atoms [20].

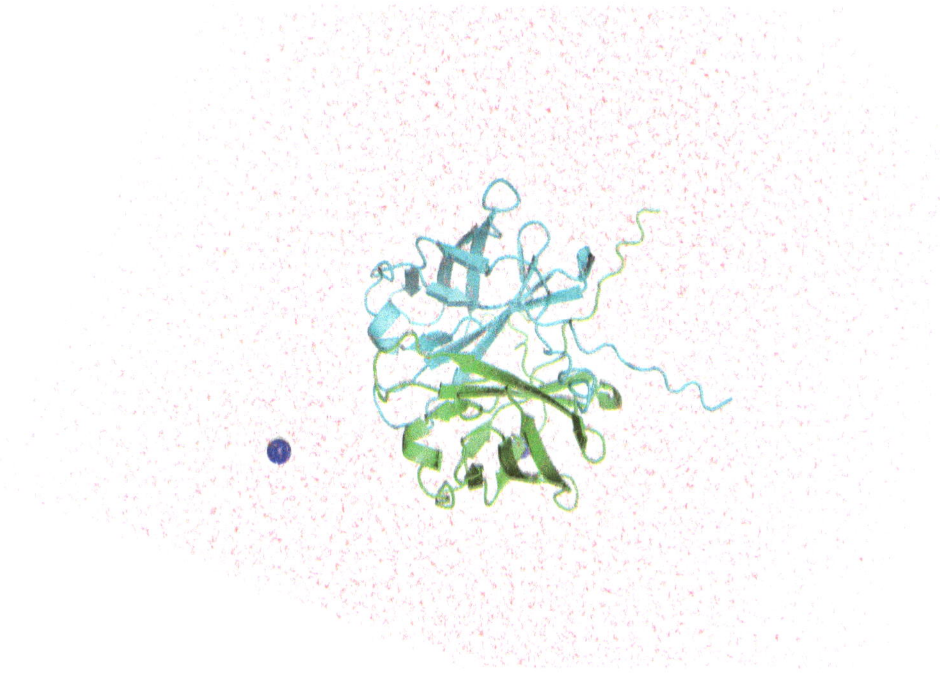

Fig. 1 View of molecular dynamic simulation, which is the traditional and most popular refinement method. It depicts the multimeric protein structure (T1078) and its environment, resulting from molecular dynamics simulations. It is a visual representation of the simulation output from Pymol. In this simulation, the 3D protein structure was generated within a cubic framework, surrounded by water molecules (red color) and ions (blue spherical structures) according to chosen parameters, which correspond to a single MD trajectory for the given protein

Even prior to the release of AF2, some machine learning approaches had been used for refinement. RefineD [16] applied a machine learning-based approach for generating restraints to effectively guide conformational sampling. In machine learning, feature extraction is a useful technique for preparing training datasets, which is beneficial in terms of constraining sampling space. However, extracting features from raw data requires manual effort. End-to-end approaches, which do not require feature extraction, outperformed traditional methods such as principal component analysis (PCA). One of the first end-to-end deep learning methods, with feature neutralization, for protein structure prediction was SASNet [21], which was developed for protein interface prediction. rawMSA [22] also used an end-to-end method based on MSA embedding as input to a convolutional neural network (CNN). However, the protein refinement area had ignored the concept of feature neuralization until recently. NEMO [23] was the first tool to use neuralization employing an explicit 3D simulator to fold proteins. NEMO's principle is to build an end-to-end model from sequence by updating to protein position based on Langevin dynamics, which proves to learn itself without co-evolutionary information, leading to competitive results [12]. It is clear that by using end-to-end methods like AF2, it is possible to capture more distant evolutionary relationships from MSAs and to explore the protein space more effectively on both a sequence and structural basis, than slower MD-based methods, based on physical laws.

4 Machine Learning Approaches for the Refinement of 3D Models of Protein Structure Via AF2

Machine learning–based approaches have had a more significant impact on various aspects of protein research [12], while the deep learning approaches, using different kinds of neural networks, have been focused on 3D protein structure refinement in recent years. In the first machine learning–based refinement methods after AF2, such as DeepRefiner [24], models were generated by using deep neural networks to calculate the error rate at the residue level from the initial structures, followed by energy minimization [24]. However, these methods do not always predict with the same accuracy for all residues in the protein because of existing disorder regions in protein structures and the problem of using of limited-size MSA [25]. With the development of deep learning methods, different data other than the sequence or the 3D structure of the protein have been used in training to refine protein models. For example, cryo-EM density data have been used to ensure that high-quality atom-level protein structures are further refined. In this regard, AlphaFold2-Phenix made use of both Cryo-EM data and deep learning via AF2 [26], which is one of the basic methods for refining AF2 structures.

Thanks to AF2, the significance of backbone refinement has decreased as the backbone accuracies have increased up to 0.96 Å [27]. Rather than backbone refinement, the improvement of protein structure predictions based on the side chains has been a long-standing goal, and the methods commonly used for this purpose are based on energy minimization techniques, rotamer libraries, or energy functions. However, these approaches tend to be weak in detecting long-range interactions. One of the successes of AF2 is that it correctly predicts such interactions by combining residue covariations with deep learning. OPUS-rota4 [28], on the other hand, has been developed to better predict the FM-type models predicted by AF2, which is the first side-chain repacking method using machine learning, by taking MSA as an input. This approach utilizes a sequential deep network architecture to predict side-chain coordinates. The method first derives an initial model using side-chain orientation information and subsequently refines it through a submodule, with gradient descent optimization based on predicted distance constraints, resulting in a final protein structure prediction [29].

Now that tertiary protein structures can be predicted at the near experimental level, the focus has shifted more toward the prediction of protein complexes. Traditional machine learning and deep learning approaches, particularly after the release of AF2, have gained attention for use in modeling protein complexes. Deep learning models have been utilized to aid in the refinement of residue positions within tertiary protein structures or predict refined residue positions using indirect target values, such as inter-residue distances. However, these methods, after refinement, necessitate all-atom restoration procedures as a subsequent step in order to recover the positions of backbone and side-chain atoms. Deep learning methods were initially demonstrated to be successful for tertiary structures rather than complex structures due to their high computational memory complexity, which requires large datasets for training [30]. Currently, there is no standalone machine learning–based tool for refining protein-protein interaction (PPI) models after AF2, and existing standalone PPI refinement methods use a combination of different protocols. While the AF2-Multimer version has been developed to successfully model complexes, there is still a gap between the predicted structures and experimental structures. Therefore, refinement approaches are still required for both AF2- Multimer and non-AF2-based models of complexes. One limitation of AF2-Multimer is that it did not predict structures with more than two chains in a protein complex very well. This interaction gap was addressed using the Monte Carlo Tree algorithm, but it was noted that the MolPC [31] method still had a problem with estimating stoichiometry [31].

5 Data-Driven Approaches for the Refinement of 3D Models of Protein Structure

Knowledge-based methods are widely used for the improvement methods for protein modeling as a guide to sampling based on existing data [32]. This information can be derived from experimental methods such as X-ray crystallography and nuclear magnetic resonance [33] or the use of other structural information, for example, fragments [34]. In recent times, the most widely used experimental data in knowledge-based approaches is cryo-EM data (as mentioned previously), which is integrated with AF2 to improve predictions made by deep learning methods [25]. Luckily, with the increase in data availability, there has been a rapid increase in the use of data-driven methods, and deep generative modeling has gained great interest in providing new sequence production. One advantage of these methods is that they overcome the limitations of traditional statistical sequence methods used in protein analysis, which are not effective in capturing the relationships between

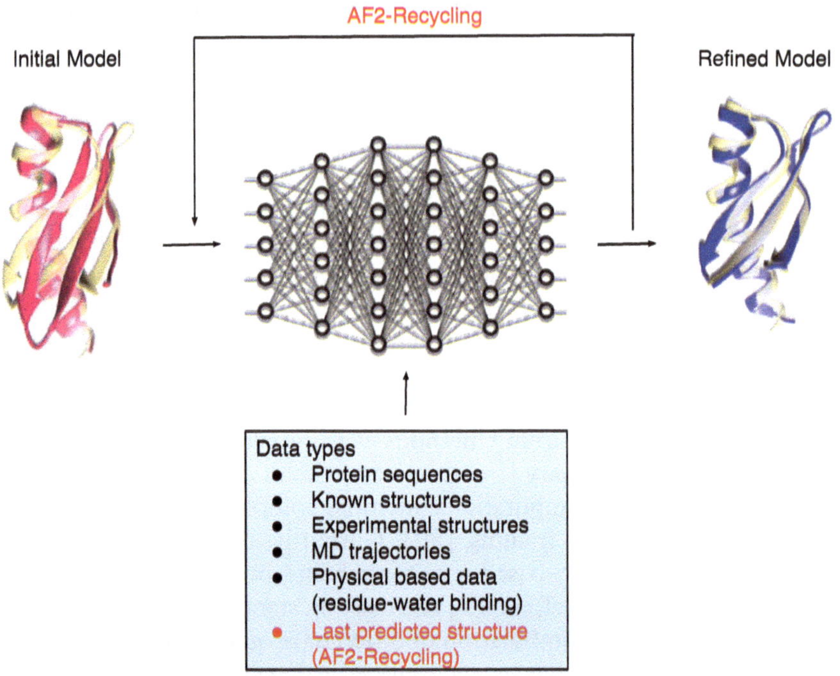

Fig. 2 The figure illustrates the general workflow of methods for improving a modeled structure using deep learning techniques. The aim is to obtain a protein structure of higher quality using deep neural networks designed by the researcher. The use of deep neural networks for various types of data, where their efficient utilization for training has been researched, has become widespread after AF2. The red-highlighted component represents the iterative improvement process referred to as recycling, which is included in the AF2 algorithm

different protein families. This enables the extraction of better evolutionary information from sequences and better definition of intrinsic restraints [35]. Figure 2 demonstrates the workflow of general refinement based on data and the different approaches are detailed in the following subsections.

5.1 Coevolution Data-Based Approaches

Multiple sequence alignments (MSAs) provide the inputs to extract coevolution information to machine learning approaches, based on the principle that a residue that is in close contact with another residue may co-mutate with that residue in order to preserve protein structure and function. The more homologous the sequences that there are in an MSA, the more accurately the protein structure can be predicted [36]. A general approach for enhancing the quality of predicted structures from methods such as AF2 can be based on improving the input MSAs. One of the key features that sets AF2 apart from other state-of-the-art methods is the use and inference of MSAs. The success achieved here has inspired the community by creating higher quality (or deeper) MSAs using various deep alignment methods. One of the tools that has gained popularity is MMseq2 [37], which is used by ColabFold [38] and is computationally efficient. Hence, ColabFold often produces deeper MSAs using MMseq2 and, as a result, can be used to generate better structures than DeepMind"s AF2 [38].

One alternative method for directly extracting evolutionary information is to use deep learning methods thanks to deep latent representation, which encode information from billions of known protein sequences by adapting language models from natural language processing (NLP). Using generative models of protein sequences has been a successful approach for obtaining deeper evolutionary information and functional protein sequences [39]. In addition to protein sequences, metagenomic data contain a vast amount of information for discovering new proteins with functional structures. As a result, metagenomic data have received particular attention for uncovering evolutionary information on proteins of interest by incorporating it into MSAs, which can be used as input to deep learning tools to improve protein structure quality. While there are many types of metagenomic databases, the combination of genomic and metagenomic databases is becoming increasingly important, as there is limited information on them in UniProtKB [40]. What is significant here is the increase in the Neff (number of effective sequences) value with metagenomic data. However, using more metagenomic data may not always result in a more accurate MSA [41]. Yang et al. [42] discovered that using one or a few specific microbes linked to the target protein family can be more beneficial in constructing MSAs. The other significant issue is the technique that is used to search the metagenomic databases, particularly when used for modeling protein interactions.

Co-evolutionary approaches, such as direct coupling analysis (DCA), can be utilized to predict the interactions between protein chains. Originally, these methods were applied to predict the interactions of bacterial two-component signaling proteins. However, the accuracy of these predictions has since been improved through the incorporation of machine learning techniques. Bryant' et al. [43] emphasized the importance of producing optimal MSAs for generating complex proteins using the Fold and Dock approach, which utilizes distance and angle information as constraints. In this approach, the sequence for each chain is first classified according to their taxonomy, and matching was achieved between the sequences with the highest number of hits for the same organism in each chain. This customized MSA approach led to improvements in the Dock-Q score [44], indicating improved model quality [43].

Due to its reliance on coevolutionary information derived from multiple sequence alignments, AF2 demonstrates suboptimal performance when applied to proteins that lack homologous sequences. This is projected to include 20% of all metagenomic protein sequences and 11% of eukaryotic and viral proteins. In addition, the integration of MSAs with deep learning has been shown to be weak in terms of the antibody-antigen model of the complex protein class, which often has highly variable paratopes and lacks MSA data [45]. It has been demonstrated that by utilizing deep learning techniques on individual sequences, it is possible to circumvent the dependency on multiple sequence alignment-based methods. For instance, RGN2 [46] was developed as a single sequence approach which was trained using a language-based method and highlighted as a potential alternative to MSA. Furthermore, the results of this method were emphasized to be superior to those of AlphaFold2 and RoseTTAFold [47]. In addition, ESM-Fold [48] and OmegaFold [49] were designed subsequently using a similar approach. The Meta group created the Metagenomic Atlas, which provides predicted structures for new metagenomic sequences. However, the use of such single sequence methods for protein prediction is often limited to smaller proteins (<400 amino acids) and often inadequate for orphan proteins, and Lin et al. stated that language models may have memorized multiple sequence alignments (MSA) to some extent [48].

5.2 Molecular Dynamics Trajectory-Based Approaches

Molecular dynamics trajectories are generated from molecular dynamics simulations. Trajectories depict atomic coordinates of a simulated molecular system at specific time intervals, represented as a series of sequential snapshots [50]. The trajectory information utilized encompasses the changes in the conformations of a protein, including unfolding or folding events, as the simulation progresses over time. The focus of the analysis is to determine if, and when, the protein reaches improved conformations from a range of starting model qualities [51]. If the issue of computationally expensive simulations can be resolved, then molecular dynamics trajectories will become practical for refinement purposes. However, solving these problems typically requires the integration of high performance computing systems with many graphical processing units (GPUs). Given that AF2 creates a single structure for a given protein sequence, using data sets for training that contain various dynamic conformational changes will be highly effective [52]. However, such time series data is not suitable for use in deep learning architectures, and some efforts are being made to provide them as inputs to deep learning methods [51].

5.3 Cryo-EM-Based Approaches

With the advancements in cryo-EM density mapping and deep learning methods, there has been rapid progress in the field of structural biology, particularly in producing and refining protein structures at an atomic level with high accuracy [26, 53], as mentioned earlier. While AF2 often generates highly precise models using information from protein sequences, the main challenge with using AF2 is that the predicted structure is a single snapshot and may not match the conformation seen in a specific experimental context or conditions, such as those obtained through cryo-EM.

To this end, improved structure reconstruction can be achieved by combining AlphaFold's predicted structure models with Cryo-EM data. A main example of this is the AlphaFold-Phenix collaboration, which has successfully improved the final predicted structure by using cryo-EM structures as templates [54]. However, determining atomic structures from cryo-EM maps with a resolution of 4 to 6 Å is a major challenge in the field of structural biology [55]. Therefore, the utilization of Cryo-EM data with these resolutions may pose difficulties. The Phenix version has been found to produce suitable results for density maps up to 4 Å, and the situation for structures beyond 4.5 Å is being investigated in the next Phenix version [25]. Another challenge in obtaining high-resolution 3D structures is the diversity of dynamic macromolecules. However, using single-particle cryo-EM data can reveal this heterogeneity [56].

5.4 Structure-Based Approaches

The realization of experimental-level structure predictions in deep learning is achieved through not only protein language learning but also the means of geometric learning [57]. Indeed, CNNs, have significantly improved the performance of methods addressing a range of problems, including sequence embedding for feature extraction and protein structure and function prediction. These networks overcome the limitations of traditional feature-based machine-learning methods by extracting task-specific features from protein sequences or structures. While various neural network applications, such as 3D CNNs, facilitate feature extraction, geometric-based neural networks can be a solution to this problem where they require a large amount of memory to explicitly process and store high-quality protein structures [58].

Despite the impressive success of CNNs in protein structure prediction, CNNs are not rotation-invariant and require data augmentation for the network to function properly. In contrast, graph neural networks (GNNs) perform convolution operations without relying on Cartesian coordinates by iteratively updating the properties of nodes and their neighboring nodes. As a result, GNNs can handle graphs of any size and more naturally represent protein structures [59, 60]. This can be a refinement of AF2 or non-AF2 models. AF2 structures are used as training examples for GNNs and are expanding the training sets for additional problems such as protein function prediction [61].

AF2 employed a neutralized refinement principle by utilizing a transformer that considers the translational and rotational symmetries of space. In contrast to NEMO, AF2 appears to refine proteins using only a limited number of iterations, which may result in the displacement of protein coordinates in a nonrealistic manner [12]. Most machine learning-based tools that improve current protein structures focus on improving the backbone structure without fully utilizing the structure of all atoms, including side chains. There is still a gap in terms of improving the structure of all atoms. Refinement methods have recently been developed using equivariant networks in GNNs. These networks have been shown to improve AF2 structures according to various evaluation metrics, including all-atom contacts, bond length, atom clashes, torsion angles, and side-chain rotamers [18]. AF2 already implemented a new attention mechanism called "Invariant Point Attention (IPA)" for three-dimensional structures. This mechanism is similar to the SE(3) equivariant used in RoseTTAFold, but it is a simplified version [36]. When considering the potential for improved results in protein structure modeling and refinement, it may be possible that GNN together with structure-based information will be more widely utilized in the field of deep learning. Table 1 presents a list of tools that employ GNN architectures, specifically limited to the area of protein modellng.

6 CASP Past and Future

The Critical Assessment of Techniques for Protein Structure Prediction (CASP) evaluates the current state of protein structure modeling from amino acid sequences. The first CASP (CASP-1) competition was held in 1994, and the most recent CASP15 competition at the time of writing was held in 2022. It would be useful to emphasize once again the innovations that affected the field in the history of CASP. During the CASP12 period, the most significant advancements have been the utilization of residue contacts, for example, via RaptorX [62]. Subsequently, in CASP13, the updated version of RaptorX and the first version of AF used predicted distance measurements for protein structure modeling [63], and AlQuraishi [12] also introduced end-to-end methods for protein modeling without using coevolutionary information. In CASP14, the DeepMind team introduced their new AF2 architecture, which was also used in CASP-covid [64] and differs significantly from the original AF. This structure consists of two main modules, the Evoformer and structure modules, as well as performing recycling. In CASP14, AF2 achieved outstanding performance [17], such that similar deep learning architectures were definitively established for protein structure prediction. After the DeepMind group shared the code source of AF2 with the community, almost all groups integrated the AF2 method into their own pipelines and were able to both benefit from and improve upon the models obtained. Following the success of AF2 in CASP14, the model refinement category was excluded from CASP15; nevertheless, many groups continued to seek ways to further improve upon the AF2 models. There was still room for improvement to achieve experimental accuracy as deviations between predicted and experimental

Table 1
A list of tools that employ graph neural networks for the prediction of specific protein modeling, protein structure refinement, or related intermediate processes

GNN-based tool	Brief method	Reference	Link
RoseTTAfold	SE(3)-equivariant graph transformer to improve backbone structure without leveraging side chain.	[47]	https://github.com/RosettaCommons/RoseTTAFold
GNNRefine	Its main aim is to provide per residue accuracy score and distance error between pair residues in order via Convolutional Neural Network to guide the refinement process of Rosetta.	[69]	http://raptorx.uchicago.edu
ATOMRefine	It employed the combination of a transformer with 3D-equivariant (not dependent on translation and rotation) by learning the main structural information in order to output refined coordinates.	[18]	https://github.com/BioinfoMachineLearning/ATOM Refine
DeepRank-GNN	The main point is to convert interface protein–protein complexes from 3D coordinates PDB file into graph via GNN.	[59]	https://github.com/DeepRank/deeprank
InterPepScore	It generates DockQ score via GNN whose training set was trajectories from the initial coarse by perturbing the native peptide (the improvement of FlexPepDock refinement tool by adding additional scoring term).	[70]	http://wallnerlab.org/InterPepScore
Equivariant graph refiner (EGR)	The integration of SE(3)-equivariance into message-passing neural networks and using refinement loss between the model's coordinates and ground truth coordinates for the refinement of 3D structures.	[30]	https://github.com/BioinfoMachineLearning/DeepR efine
Paragraph	The main aim is to predict paratope. It uses equivariant graph neural network to predict the possible relevant residues of paratope when the antibody structure is taken as input.	[71]	www.github.com/oxpig/Paragraph
VoroCNN	It predicts the local-CAD score for all residues in the original structure using a graph neural network, even if the main part consists of Convolutional Neural network.	[72]	https://team.inria.fr/nano-d/software/vorocnn/
DeepBindPPI	It includes the inclusion of an attention mechanism to GCN to predict the binding sites to proteins.	[73]	https://github.com/Sharon1989Sunny/DBP

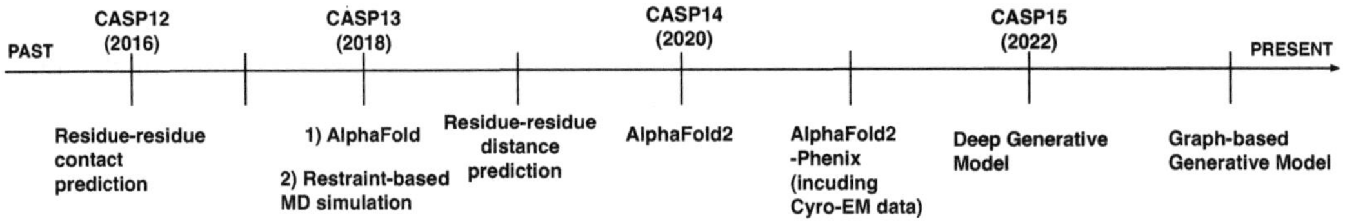

Fig. 3 This graph encompasses the significant milestones and methods, either directly or indirectly impacting structure improvement, in the field of protein bioinformatics over time for structure refinement. The critical turning points have been selected as the CASP competitions and the methods and tools that have been commonly effective in these competitions, as the competition has gained substantial importance in guiding the community

structures had been observed, in terms of backbone and/or side chain positions [24, 65].

Despite DeepMind not participating in CASP15, AF2 has continued to have an impact and the groups with pipelines that integrated AF2 were among the most successful participants. Two primary methods were used by these groups in order to improve upon AF2: the utilization of more effective templates and/or MSA techniques, and the enhancement of AF2 through the implementation of dropout methods [66]. The option of using dropout during inference allows for activation of the stochastic component of the model, resulting in varying predictions. By iterating through different seeds, it is possible to sample different structural predictions from the uncertainty of the model or imprecision of the coevolutionary constraints inferred from the input MSA [38]. Figure 3 shows some of the milestones that have influenced improvements in model quality in subsequent CASP competitions.

The use of the AF2 recycling process for improvement of models was another successful approach in CASP15 (Fig. 2). The term "recycling" refers to the process of tidying up some degree of disorganization before extracting valuable information. The initial recycling outputs, as well as the quality of templates/MSAs, play a crucial role in determining the amount of recycling required to successfully model the protein. High-quality templates, a large number of MSAs, or a well-known sequence may only require minor adjustments from an initial recycling output. However, missing templates or synthetic sequences may require extensive cleaning during recycling. By repeatedly running the model through the network, recycling improves the model quality. In CASP15, our groups (McGuffin & MultiFOLD) used the AF2 recycling process to improve the quality of both tertiary and multimeric models beyond the quality of the baseline AF2 models [67, 68]. Our MultiFOLD server [67] ranked as the ninth-best server group on monomeric targets according to the GDT_TS scores and the eighth-best server group for multimers by the assessors' formula, outperforming both the baseline NBIS-AF2-standard (AlphaFold2) group and NBIS-AF2-multimer (AlphaFold2-multimer) group across the board. Therefore, the MultiFOLD server has been independently verified to produce higher quality tertiary and quaternary structure models than AlphaFold2 according to all measures [67].

7 Conclusion

The development of methods for protein structure prediction has been ongoing for several decades, with continual improvements that have recently resulted in the availability of models for tertiary structures with near experimental quality for many targets. The efficacy of deep learning methods in this field was demonstrated with the success of AF2 in CASP14, and after the source code was made publicly available by DeepMind, the application of AF2 and other deep learning-based methods has become widespread. Subsequently, a process of improving upon AF2 began, with various types of data utilized to obtain more accurate 3D models, with the most noteworthy example being the use of Cryo-EM data. In CASP15, more studies were conducted on coevolution information from MSA and geometric representation of structures. The methods used included generative models for obtaining more homologous sequences, protein language modeling, and graph neural network variations, as well as recycling and dropout methods for improving or refining structures. Although the validity of using more traditional refinement methods, such as MD, has not been fully proven for deep learning-based models, they are now only commonly used as minimization methods, due to the inadequacy of force fields and the high computational costs required. The recent improvements made to AF2 and other deep learning-based methods will be crucial in fields such as drug design and will continue to be relevant as we move towards experimental quality prediction of quaternary structures.

Acknowledgments

This work was supported by the Republic of Turkey Ministry of National Education (to A.G.G.) and the Biotechnology and Biological Sciences Research Council (BBSRC) [BB/T018496/1 to L.J.M.].

References

1. Kuhlman B, Bradley P (2019) Advances in protein structure prediction and design. Nat Rev Mol Cell Biol 20(11):681–697. https://doi.org/10.1038/s41580-019-0163-x
2. Chou P, Fasman GD (2009) Amino acid sequence. Adv Enzymol Relat Areas Mol Biol 47:45

3. Dunker AK, Lawson JD, Brown CJ et al (2001) Intrinsically disordered protein. J Mol Graph Model 19(1):26–59. https://doi.org/10.1016/s1093-3263(00)00138-8
4. Dunker AK, Oldfield CJ (2015) Back to the future: nuclear magnetic resonance and bioinformatics studies on intrinsically disordered proteins. Adv Exp Med Biol 870:1–34. https://doi.org/10.1007/978-3-319-20164-1_1
5. Bondos SE, Dunker AK, Uversky VN (2021) On the roles of intrinsically disordered proteins and regions in cell communication and signaling. Cell Commun Signal 19(1):88. https://doi.org/10.1186/s12964-021-00774-3
6. Coskuner O, Uversky VN (2019) Intrinsically disordered proteins in various hypotheses on the pathogenesis of Alzheimer's and Parkinson's diseases. Prog Mol Biol Transl Sci 166: 145–223. https://doi.org/10.1016/bs.pmbts.2019.05.007
7. Basile W, Salvatore M, Bassot C et al (2019) Why do eukaryotic proteins contain more intrinsically disordered regions? PLoS Comput Biol 15(7):e1007186. https://doi.org/10.1371/journal.pcbi.1007186
8. Wooley JC, Ye Y (2007) A historical perspective and overview of protein structure prediction. In: Xu Y, Xu D, Liang J (eds) Computational methods for protein structure prediction and modeling, vol 1. Basic Characterization. Springer, New York, pp 1–43. https://doi.org/10.1007/978-0-387-68372-0_1
9. Sanger F, Thompson EO, Kitai R (1955) The amide groups of insulin. Biochem J 59(3): 509–518. https://doi.org/10.1042/bj0590509
10. Anfinsen CB, Redfield RR, Choate WL et al (1954) Studies on the gross structure, cross-linkages, and terminal sequences in ribonuclease. J Biol Chem 207(1):201–210
11. Pearce R, Zhang Y (2021) Deep learning techniques have significantly impacted protein structure prediction and protein design. Curr Opin Struct Biol 68:194–207. https://doi.org/10.1016/j.sbi.2021.01.007
12. AlQuraishi M (2019) End-to-end differentiable learning of protein structure. Cell Systems 8(4):292–301.e293. https://doi.org/10.1016/j.cels.2019.03.006
13. Xu J (2019) Distance-based protein folding powered by deep learning. Proc Natl Acad Sci 116(34):16856–16865. https://doi.org/10.1073/pnas.1821309116
14. Senior AW, Evans R, Jumper J et al (2020) Improved protein structure prediction using potentials from deep learning. Nature 577(7792):706–710. https://doi.org/10.1038/s41586-019-1923-7
15. Gomes P, Gomes DEB, Bernardi RC (2022) Protein structure prediction in the era of AI: challenges and limitations when applying to in silico force spectroscopy. Front Bioinform 2: 983306. https://doi.org/10.3389/fbinf.2022.983306
16. Bhattacharya D (2019) refineD: improved protein structure refinement using machine learning based restrained relaxation. Bioinformatics 35(18):3320–3328. https://doi.org/10.1093/bioinformatics/btz101
17. Jumper J, Evans R, Pritzel A et al (2021) Applying and improving AlphaFold at CASP14. Proteins 89(12):1711–1721. https://doi.org/10.1002/prot.26257
18. Wu T, Guo Z, Cheng J (2023) Atomic protein structure refinement using all-atom graph representations and SE(3)-equivariant graph transformer. Bioinformatics 39(5). https://doi.org/10.1093/bioinformatics/btad298
19. Adiyaman R, McGuffin LJ (2021) ReFOLD3: refinement of 3D protein models with gradual restraints based on predicted local quality and residue contacts. Nucleic Acids Res 49(W1): W589–W596. https://doi.org/10.1093/nar/gkab300
20. Feig M, Mirjalili V (2016) Protein structure refinement via molecular-dynamics simulations: what works and what does not? Proteins 84 Suppl 1 (Suppl 1):282-292:282. https://doi.org/10.1002/prot.24871
21. Townshend RJL, Bedi R, Suriana PA et al (2019) End-to-end learning on 3D protein structure for interface prediction. In: Proceedings of the 33rd International Conference on Neural Information Processing Systems. Curran Associates Inc., p Article 1401
22. Mirabello C, Wallner B (2019) rawMSA: end-to-end deep learning using raw multiple sequence alignments. PLoS One 14(8): e0220182. https://doi.org/10.1371/journal.pone.0220182
23. Ingraham J, Riesselman AJ, Sander C et al (2018) Learning protein structure with a differentiable simulator. In: International Conference on Learning Representations
24. Shuvo MH, Gulfam M, Bhattacharya D (2021) DeepRefiner: high-accuracy protein structure refinement by deep network calibration. Nucleic Acids Res 49(W1):W147–W152. https://doi.org/10.1093/nar/gkab361
25. Terwilliger TC, Poon BK, Afonine PV et al (2022) Improved AlphaFold modeling with implicit experimental information. Nat

Methods 19(11):1376–1382. https://doi.org/10.1038/s41592-022-01645-6

26. Zhang B, Liu D, Zhang Y et al (2022) Accurate flexible refinement for atomic-level protein structure using cryo-EM density maps and deep learning. Brief Bioinform 23(2). https://doi.org/10.1093/bib/bbac026
27. Jumper J, Evans R, Pritzel A et al (2021) Highly accurate protein structure prediction with AlphaFold. Nature 596(7873):583–589. https://doi.org/10.1038/s41586-021-03819-2
28. Xu G, Wang Q, Ma J (2021) OPUS-Rota4: a gradient-based protein side-chain modeling framework assisted by deep learning-based predictors. Brief Bioinform 23(1). https://doi.org/10.1093/bib/bbab529
29. McPartlon M, Xu J (2023) An end-to-end deep learning method for protein side-chain packing and inverse folding. Proc Natl Acad Sci USA 120(23):e2216438120. https://doi.org/10.1073/pnas.2216438120
30. Morehead A, Chen X, Wu T et al. (2022) EGR: Equivariant graph refinement and assessment of 3D protein complex structures. https://doi.org/10.48550/arXiv.2205.10390
31. Bryant P, Pozzati G, Zhu W et al (2022) Predicting the structure of large protein complexes using AlphaFold and Monte Carlo tree search. Nat Commun 13(1):6028. https://doi.org/10.1038/s41467-022-33729-4
32. Adiyaman R, McGuffin LJ (2019) Methods for the refinement of protein structure 3D models. Int J Mol Sci 20(9). https://doi.org/10.3390/ijms20092301
33. Cárdenas R, Martínez-Seoane J, Amero C (2020) Combining experimental data and computational methods for the non-computer specialist. Molecules 25(20). https://doi.org/10.3390/molecules25204783
34. Vetrivel I, Mahajan S, Tyagi M et al (2017) Knowledge-based prediction of protein backbone conformation using a structural alphabet. PLoS One 12(11):e0186215. https://doi.org/10.1371/journal.pone.0186215
35. Bepler T, Berger B (2021) Learning the protein language: evolution, structure, and function. Cell Systems 12(6):654–669.e653. https://doi.org/10.1016/j.cels.2021.05.017
36. Rubiera CO (2021) AI3SD video: how good are protein structure prediction methods at predicting folding pathways? Paper presented at the AI 4 Proteins Seminar Series 2021, 14/04/21 – 17/06/21
37. Mirdita M, Steinegger M, Söding J (2019) MMseqs2 desktop and local web server app for fast, interactive sequence searches. Bioinformatics 35(16):2856–2858. https://doi.org/10.1093/bioinformatics/bty1057
38. Mirdita M, Schütze K, Moriwaki Y et al (2022) ColabFold: making protein folding accessible to all. Nat Methods 19(6):679–682. https://doi.org/10.1038/s41592-022-01488-1
39. Wu Z, Johnston KE, Arnold FH et al (2021) Protein sequence design with deep generative models. Curr Opin Chem Biol 65:18–27. https://doi.org/10.1016/j.cbpa.2021.04.004
40. Consortium TU (2022) UniProt: the universal protein knowledgebase in 2023. Nucleic Acids Res 51(D1):D523–D531. https://doi.org/10.1093/nar/gkac1052
41. Hou Q, Pucci F, Pan F et al (2022) Using metagenomic data to boost protein structure prediction and discovery. Comput Struct Biotechnol J 20:434–442. https://doi.org/10.1016/j.csbj.2021.12.030
42. Yang P, Zheng W, Ning K et al (2021) Decoding the link of microbiome niches with homologous sequences enables accurately targeted protein structure prediction. Proc Natl Acad Sci 118(49):e2110828118. https://doi.org/10.1073/pnas.2110828118
43. Bryant P, Pozzati G, Elofsson A (2022) Improved prediction of protein-protein interactions using AlphaFold2. Nat Commun 13(1):1265. https://doi.org/10.1038/s41467-022-28865-w
44. Basu S, Wallner B (2016) DockQ: a quality measure for protein-protein docking models. PLoS One 11(8):e0161879. https://doi.org/10.1371/journal.pone.0161879
45. Jin W, Barzilay DR, Jaakkola T (2022) Antibody-antigen docking and design via hierarchical structure refinement. Paper presented at the Proceedings of the 39th International Conference on Machine Learning, Proceedings of Machine Learning Research
46. Chowdhury R, Bouatta N, Biswas S et al (2022) Single-sequence protein structure prediction using a language model and deep learning. Nat Biotechnol 40(11):1617–1623. https://doi.org/10.1038/s41587-022-01432-w
47. Baek M, DiMaio F, Anishchenko I et al (2021) Accurate prediction of protein structures and interactions using a three-track neural network. Science 373(6557):871–876. https://doi.org/10.1126/science.abj8754
48. Lin Z, Akin H, Rao R et al (2023) Evolutionary-scale prediction of atomic-level protein structure with a language model. Science 379(6637):1123–1130. https://doi.org/10.1126/science.ade2574

49. Wu R, Ding F, Wang R et al. (2022) High-resolution *evol* structure prediction from primary sequence. bioRxiv:2022.2007.2021.500999. https://doi.org/10.1101/2022.07.21.500999
50. Likhachev IV, Balabaev NK, Galzitskaya OV (2016) Available instruments for analyzing molecular dynamics trajectories. Open Biochem J 10:1–11. https://doi.org/10.2174/1874091x01610010001
51. Pfeiffenberger E, Bates PA (2018) Predicting improved protein conformations with a temporal deep recurrent neural network. PLoS One 13(9):e0202652. https://doi.org/10.1371/journal.pone.0202652
52. Sathvik Kolli AL, Geng X, Kumar A, Levine S (2022) Data-driven optimization for protein design: workflows, algorithms and metrics. Paper presented at the ICLR Workshop on Machine Learning for Drug Discovery
53. Glaeser RM (2016) How good can cryo-EM become? Nat Methods 13(1):28–32. https://doi.org/10.1038/nmeth.3695
54. Giri N, Roy RS, Cheng J (2023) Deep learning for reconstructing protein structures from cryo-EM density maps: recent advances and future directions. Curr Opin Struct Biol 79:102536. https://doi.org/10.1016/j.sbi.2023.102536
55. Alshammari M, He J, Wriggers W (2022) Refinement of ALPHAFold2 models against experimental cryo-EM density maps at 4-6Å resolution. In: 2022 IEEE International Conference on Bioinformatics and Biomedicine (BIBM), 6–8 Dec. 2022, pp 3423–3430. https://doi.org/10.1109/BIBM55620.2022.9995676
56. Doerr A (2016) Single-particle cryo-electron microscopy. Nat Methods 13(1):23–23. https://doi.org/10.1038/nmeth.3700
57. Laine E, Eismann S, Elofsson A et al (2021) Protein sequence-to-structure learning: is this the end(-to-end revolution)? Proteins: structure. Function, and Bioinformatics 89(12):1770–1786. https://doi.org/10.1002/prot.26235
58. Gligorijević V, Renfrew PD, Kosciolek T et al (2021) Structure-based protein function prediction using graph convolutional networks. Nat Commun 12(1):3168. https://doi.org/10.1038/s41467-021-23303-9
59. Réau M, Renaud N, Xue LC et al (2022) DeepRank-GNN: a graph neural network framework to learn patterns in protein–protein interfaces. Bioinformatics 39(1). https://doi.org/10.1093/bioinformatics/btac759
60. Wang X, Flannery ST, Kihara D (2021) Protein docking model evaluation by graph neural networks. Front Mol Biosci 8. https://doi.org/10.3389/fmolb.2021.647915
61. Ma W, Zhang S, Li Z et al (2022) Enhancing protein function prediction performance by utilizing AlphaFold-predicted protein structures. J Chem Inf Model 62(17):4008–4017. https://doi.org/10.1021/acs.jcim.2c00885
62. Källberg M, Wang H, Wang S et al (2012) Template-based protein structure modeling using the RaptorX web server. Nat Protoc 7(8):1511–1522. https://doi.org/10.1038/nprot.2012.085
63. Pakhrin SC, Shrestha B, Adhikari B et al (2021) Deep learning-based advances in protein structure prediction. Int J Mol Sci 22(11). https://doi.org/10.3390/ijms22115553
64. Kryshtafovych A, Moult J, Billings WM et al (2021) Modeling SARS-CoV-2 proteins in the CASP-commons experiment. Proteins 89(12):1987–1996. https://doi.org/10.1002/prot.26231
65. Schreiner M (2022) CASP15: AlphaFold's success spurs new challenges in protein-structure prediction. https://the-decoder.com/casp15-alphafolds-success-brings-new-challenges/, vol 2023
66. Elofsson A (2022) Protein structure prediction until CASP15. arXiv:221207702. https://doi.org/10.48550/arXiv.2212.07702
67. McGuffin LJ, Edmunds NS, Genc AG et al (2023) Prediction of protein structures, functions and interactions using the IntFOLD7, MultiFOLD and ModFOLDdock servers. Nucleic Acids Res 51(W1):W274–W280. https://doi.org/10.1093/nar/gkad297
68. Adiyaman R, Edmunds NS, Genc AG et al (2023) Improvement of protein tertiary and quaternary structure predictions using the ReFOLD refinement method and the AlphaFold2 recycling process. Bioinformatics. Advances 3(1). https://doi.org/10.1093/bioadv/vbad078
69. Jing X, Xu J (2021) Fast and effective protein model refinement using deep graph neural networks. Nat Comput Sci 1(7):462–469. https://doi.org/10.1038/s43588-021-00098-9
70. Johansson-Åkhe I, Wallner B (2022) InterPepScore: a deep learning score for improving the FlexPepDock refinement protocol. Bioinformatics 38(12):3209–3215. https://doi.org/10.1093/bioinformatics/btac325
71. Chinery L, Wahome N, Moal I et al (2022) Paragraph—antibody paratope prediction using graph neural networks with minimal

feature vectors. Bioinformatics 39(1). https://doi.org/10.1093/bioinformatics/btac732

72. Igashov I, Olechnovič K, Kadukova M et al (2021) VoroCNN: deep convolutional neural network built on 3D Voronoi tessellation of protein structures. Bioinformatics 37(16): 2332–2339. https://doi.org/10.1093/bioinformatics/btab118

73. Sunny S, Prakash PB, Gopakumar G et al. (2023) DeepBindPPI: protein–protein binding site prediction using attention based graph convolutional network. Protein J https://doi.org/10.1007/s10930-023-10121-9, 42, 276

Chapter 8

AlphaFold2, SPINE-X, and Seder on Four Hard CASP Targets

Eshel Faraggi, Robert L. Jernigan, and Andrzej Kloczkowski

Abstract

We analyzed four cases from the CASP15 experiment with low prediction accuracy and compared AlphaFold2, SPINE-X, and Seder on these cases. We find that overall, AlphaFold2 performs better than SPINE-X in predicting secondary structure (SS) and solvent accessible surface area (ASA). For some cases, SPINE-X better predicts sheet and coil regions. We also find that AlphaFold2 is better than Seder in selecting the best matching tertiary structure model for one case and is worse in another case. For two cases Alphafold2 and Seder selected the same models. From the cases presented here, it appears that AlphaFold2 predicts more compact structures than the native one. We find that while, as widely reported, AlphaFold2 significantly improved protein tertiary structure prediction, there are cases, such as the four presented here, for which the tertiary structure prediction could still be significantly enhanced. The source code, license, and documentation for SPINE-X and Seder are available from Research and Information Systems, LLC at http://mamiris.com.

Key words Secondary structure, Accessible surface area, Protein structure, AlphaFold2, Seder, SPINE-X

1 Introduction

Since the first publication of our book *Prediction of Protein Secondary Structure* in October 2016 [1], there has been enormous progress in the field due to application of deep learning–based methods such as AlphaFold or Protein Language Models that use Natural Language Processing (NLP) techniques. Summarizing secondary structure (SS) in the post-AlphaFold (AF) [2, 3] era is challenging. In what follows, we examine some shortfalls still in existence for SS prediction as well as briefly discuss the use of the SPINE-X (SPX) [4–6] and Seder [7–9] servers and compare their prediction to AF models submitted to CASP15 on a few hard targets. The Seder server is one of the early instances of using deep learning for protein structure prediction.

Proteins are the biological machinery that manifest the information of the genetic code. As such, proteins are responsible on

Andrzej Kloczkowski et al. (eds.), *Prediction of Protein Secondary Structure*, Methods in Molecular Biology, vol. 2867, https://doi.org/10.1007/978-1-0716-4196-5_8,

some level for much of biological form and function. Proteins provide cells the building blocks, signaling mechanisms, etc., for their different functions. These various effects are associated with the three-dimensional (3D) structure of a protein. Their 3D structure is determined by their genetic amino-acid sequence and the conditions they are in [10]. Based on the 3D structure of a protein, one can assign a structure class to individual residues. An example is the three-state classification with Helix (H), Sheet (E), and Coil (C) states. Other classifications, such as with eight states, are also widely used. These assignments create a sequence parallel to the amino-acid sequence, containing information about the structure of the protein. This sequence is known as the SS of the protein while the 3D structure is known as tertiary. The assembly of protein chains to biological units is known as quaternary structure. Historically, SS prediction is used as a major feature and building block in tertiary structure prediction. This hierarchy enables proteins to bridge the size gap between the atomic scale and the biological scale.

Besides secondary structure [5, 11–18] other invariant one-dimensional structural properties that can be associated with individual residues exist. Most notably maybe is the Accessible Surface Area (ASA) [19–25], which measures how much of a given residue can interact with the solvent. Typically, one calculates the Relative ASA (RASA) by normalizing the ASA by the maximum ASA for a given residue type in a peptide chain. Other one-dimensional properties are the dihedral angles along the peptide backbone [26–32]. It should be noted that the most useful, one-dimensional structural properties should be invariant under a coordinate transformation.

Prediction of protein structure usually employs the structural hierarchy mentioned above as well. Secondary structure predictions are used to set initial conditions and act as constraints in three-dimensional prediction schemes [31, 33–38]. We have shown that substitution of dihedral angles for secondary structure constraints in free (non-homology) modeling of three-dimensional structures results in a 100% improvement in the prediction accuracy [31]. Part of that approach of predicting dihedral angles uses secondary structure predictions as input features. More recently it was shown that predicted contact maps, a two-dimensional structural property of a protein that accounts for the set of distances between pairs of residues, is instrumental in predicting protein structure [39–41].

Accurate prediction of protein secondary structure and other one-dimensional structural features is essential for accurate sequence alignment [42, 43], three-dimensional structure modeling [44–46], and function prediction [47, 48]. In recent years, deep learning coupled with big data has significantly improved the accuracy of tertiary and in some cases even quaternary structure prediction. Most notably the AF server has produced results that far exceed previous ones. The industrial application of deep learning-based methods developed for facial recognition and language processing, combined with decades of research by numerous groups involved in protein structure initiative aimed at accelerating discovery in structural genomics for better understanding of biological function, has given rise to tools that are far better at predicting structural models from protein sequences than any earlier methods [49, 50]. In many cases, these tools predict structural models that deviate from experimental ones within the experimental error. This is true from the macro point of view. Looking deeper, these tools still have limitations.

Most important is the fact that the protein sequence alone does not determine protein structure. The sequence plus the external (boundary) conditions, determine its structure. For many cases, these conditions result in similar structures, and thus, predictions solely from sequence are meaningful and useful. AlphaFold2 (AF2) [2, 3, 49] has various other limitations [50], such as difficulty with intrinsically disordered regions [51], loops [52], membrane proteins [53], apo and holo forms [54], docking [55, 56], and mutations [57, 58].

2 Material and Methods

The Critical Assessment of Protein Structural Prediction (CASP) [59] is a longstanding experiment to test the quality of protein structure prediction methods. We have selected four targets for which the top predictions in the most recent CASP15 were not accurate. These cases are T1114s1, T1122, T1125, and T1137s4. For these cases, we compared the top AF2 structure to the native structure. We compared the prediction of SS, RASA, and tertiary structure. We also compared SPX and Seder.

Here SPINE-X (SPX) is a protein secondary structure prediction method developed by Faraggi et al. [4–6], which uses a multistep neural network approach and integrates predictions of secondary structure, solvent accessibility, and backbone torsion angles. This improves the accuracy of protein secondary structure prediction, which is essential for sequence alignment and three-dimensional structure modeling. Seder denotes our newest methodology based on a knowledge-based protein scoring function termed Seder [7], combined with our GEneral Neural Network (GENN) [8] for learning trends from existing data and making predictions of unknown information.

To compare tertiary structure, we use the TM-score [60, 61]. The TM-score is calculated from the distances between the residues of two aligned protein sequences:

$$TM - \text{score} = \max\left[\frac{1}{L_t}\sum_{i=1}^{L_a}\frac{1}{1+\left(\frac{d_i}{d_0}\right)^2}\right], \quad (1)$$

where the maximum is taken over all possible alignments. L_t is the length of the target protein, L_a is the length of the protein that is aligned to it, d_i is the distance.

between the residues at alignment location i, and d_0 is a scaling distance, optimized to $d_0 = 1.24\sqrt[3]{L_t - 15} - 1.8$. The TM-score does not depend on the protein length, a value of 1, indicates a perfect match, while a TM-score below 0.2 indicates structurally unrelated proteins [60–62]. To estimate the success of SS prediction we use the accuracy: the number of correct predictions divided by the total number of prediction attempts. For estimating RSA prediction, we use the root mean square deviation and Pearson's correlation coefficient.

The SPX server [4–6] predicts the SS, ASA, and dihedral angles of a given protein sequence. It consists of six steps of iterative prediction. The first five steps predict real value dihedral angles (both φ and ψ) [31], using the Position Specific Scoring Matrix (PSSM) [63] and other features describing the sequence [31, 64]. In the first step, a neural network is set up to predict secondary structure defined according to SKSP+, the consensus of STRIDE, KAKSI, SECSTR, and P-SEA secondary structure assignment methods [65]. The first two steps correspond to Real-SPINE 3.0 [18]. Then, predicted RASA and SS are used to predict the torsion angles. The fourth step predicts secondary structure with all previous information. Then a new round of dihedral angle prediction is performed using previous predictions as an input. The sixth and final step is a neural network that is trained to predict DSSP [66] assigned secondary structure using the results from the first five rounds. This step is useful when compared with other methods that use the DSSP assignment. The eight state DSSP assignments were grouped as follows: the 3-helix (G), alpha-helix (H), and pi-helix (I) into state H; beta-bridge (B) and extended-strand (E) into state E; and hydrogen bonded-turn (T), bend (S), and other () into state C. To calculate the ASA, we also use the DSSP program.

In each step, the general form of the neural networks is the same. It consists of two hidden layers with 101 hidden nodes. All weights were guided based on sequence separation [18]. A 21-residue window is employed. For a given 21-residue input window the target output is the one-dimensional property of the central residue in the window. Vacant locations in the windows around residues near the terminals of a protein were explicitly excluded from the training by limiting the range of the input window. We employed a bipolar activation function given by

$f(x) = \tanh(\alpha x)$, with $\alpha = 0.2$, momentum of 0.4, and the back-propagation method with a learning rate of 0.001. These parameters were optimized in previous studies of torsion angles and solvent accessibility [5, 18, 30, 31].

The overall accuracy estimates for SPX are as follows. For secondary structure, the accuracy is between 81% and 84% depending on the dataset and choice of tests. The Pearson correlation coefficient for accessible surface area prediction is 0.75 and the mean absolute error from the φ and ψ dihedral angles are 20° and 33°, respectively. We refer to the original publications [5, 18, 31] of these methods for a more detailed analysis of the accuracy of SPX.

The current version of Seder [9] uses the same methodology as in the original version [7]. Protein residue sequences are scored based on their similarity to the native protein structure of the sequence as in the PDB. We used a recurrent two-hidden-layer feed-forward neural network with momentum and associative memory described in detail elsewhere [8, 9]. For optimization, we use a hybrid of the Levenberg-Marquardt and back-propagation algorithms. The input features we use are the partial energy sums between pairs of residues, the distance profile of individual atoms to the solvent, and residue-residue contact maps with a cutoff distance of 5 and 7 Å. A final z-score normalization was applied to all features. The output° layer is trained to predict the TM-score to the native structure for the given model.

3 Results and Discussion

We wish to analyze the prediction accuracy of AF2 models for which there is room for improvement in their fitness and compare this to the predictions of Seder and SPX. In Table 1, we give the results of this analysis. In general, we see TM-score values at or below 0.5, indicating low structural similarity between the AF2 top one model and its native counterpart. It is interesting that even though the overall structural similarity is low, the AF2 prediction accuracy for the H SS type is high for all four cases. Given the lower accuracy for the C and E states, this points towards a bias of AF2 for the H state. The prediction accuracy of the E and C SS types vary. For T1122 the entire beta sheet portion of the structure is missed. For the C type we get values that are standard for SS prediction [5, 67, 68]. For RASA we find two cases (T1114s1, T1137s4) with relatively well-predicted values and two cases (T1122, T1125) with relatively poor prediction. The Seder server trained on hard targets selected the second AF2 model for T1114s1 which improved the TM-score by 26%. For the case of T1122 Seder top-ranked the third AF2 model with the TM-score reduced by 36%. Overall, the predictions of SPX for 1D properties are not as good as those obtained from top AF2 models. The only exception is predicted E

Table 1
Prediction fitness for sample of CASP targets

CASP-ID[a]	PDB-ID[b]	TMS[c]	SSA[d]	SSA-H[e]	SSA-E[f]	SSA-C[g]	RASA-C[h]	RASA-R[k]
T1114s1	7UUSQ	0.35	0.92	0.95	N/A	0.80	0.66	0.13
		[0.44][m]	(0.90)[n]	(0.97)	(N/A)	(0.57)	(0.45)	(0.32)
T1122	8BBTA	0.50	0.75	0.83	0.00	0.68	0.47	0.28
		[0.32]	(0.62)	(0.54)	(0.89)	(0.74)	(0.31)	(0.31)
T1125	8H2NA	0.32	0.71	0.85	0.36	0.80	0.43	0.31
		[0.32]	(0.70)	(0.93)	(0.45)	(0.59)	(0.65)	(0.17)
T1137s4	8FEFD	0.34	0.90	0.96	1.00	0.82	0.84	0.12
		[0.34]	(0.79)	(0.75)	(0.66)	(0.88)	(0.32)	(0.30)

[a] CASP target ID. [b] PDB ID for native structure, fifth letter denotes the chain [69]. [c] TM-score between native and AF2 top pick model. [d] AF2 three states SS prediction accuracy (number of correct predictions divided by total). [e] AF2 three states SS prediction accuracy for H state. [f] AF2 three states SS prediction accuracy for E state. [g] AF2 three states SS prediction accuracy for C state. [h] AF2 Pearson's correlation for RASA. [k] AF2 root mean square deviation for RASA. [m] TM-score between native and Seder top pick among AF2 models given in square brackets. [n] SPX prediction fitness given in parenthesis

state, with average accuracies of 45% and 67% for AF2 and SPX, respectively. The overall SS accuracy is 82% and 75% for AF2 and SPX, respectively. The RASA predicted by SPX is reduced in quality by about a third.

We further examine the quality of RASA prediction. Figure 1 shows the histogram of the directional error: the predicted RASA minus the native RASA at a given residue site. The y-axis is normalized by the number of residues in the native model for the given protein.

The directional error on the x-axis is binned into twenty bins. We compare the AF2 prediction as obtained from the tertiary structure model using the DSSP program (red) to the SPX RASA prediction (green). We see that in general, AF2 produces errors that are better distributed around the original, and in two targets (T1114s1 and T1137s4), SPX produced predictions that were significantly skewed towards a more bulk structure with lower ASA. For the target T1125, SPX produced a better prediction with a more peaked directional error distribution and a higher correlation, lower error, as displayed also in Table 1.

We also give cartoon representation of these four cases in Fig. 2. In each case, the native structure is given in red. The predicted AF2 top one model is given in green and blue. Since the native structures contain missing regions while the predicted models cover the entire sequence, we plot in green those parts of the sequence for which there are native coordinates. In blue we plot the entire predicted model. From the figures we can confirm the

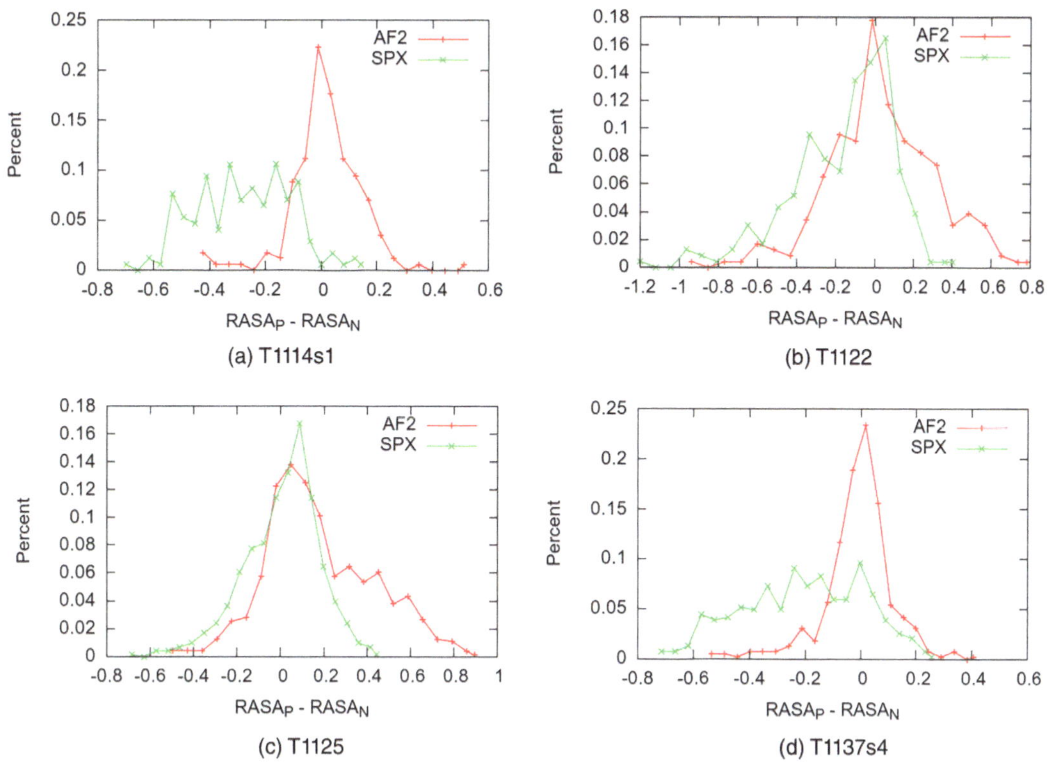

Fig. 1 Distribution of RASA predictions. The x-axis of the histogram is the directional error: the predicted RASA minus the native RASA at a given residue site. The directional error on the x-axis is binned into 20 bins. The y-axis is normalized by the number of residues in the native model for the given protein. We compare the AF2 prediction (red) with the SPX RASA prediction (green)

results of the TM-score, low similarity between native and model structures. For the case of T1122, which has a marginal TM-score of 0.5 we see that indeed part of the native structure is covered by the model, especially in the interior of the structure. The periphery, especially the beta sheet form on the top right of the figure is missed by AF2 prediction.

In the combination of Table 1 and Fig. 2, we can observe the critical effect of coil prediction. The overall AF2 accuracy for predicting the C state is at or below 0.8. Misidentification of the C state can result in significant error in the tertiary structure. This is evident in Fig. 2a. In this case, the native structure has a coil region between residues 79LEU and 90ALA. This region acts as a flexible linker between two helical regions. In the AF2 predicted model this coil region is only between residues 81ALA and 86ILE. This smaller coil region is interpreted by AF2 as connecting two parts of a compact structure. In Fig. 2d, we see misidentification of helices as coils, and again AF2 predicts a significantly more compact structure. One should note that less compact native structures of protein chains, such as the ones depicted in Fig. 2a and d, are stabilized by

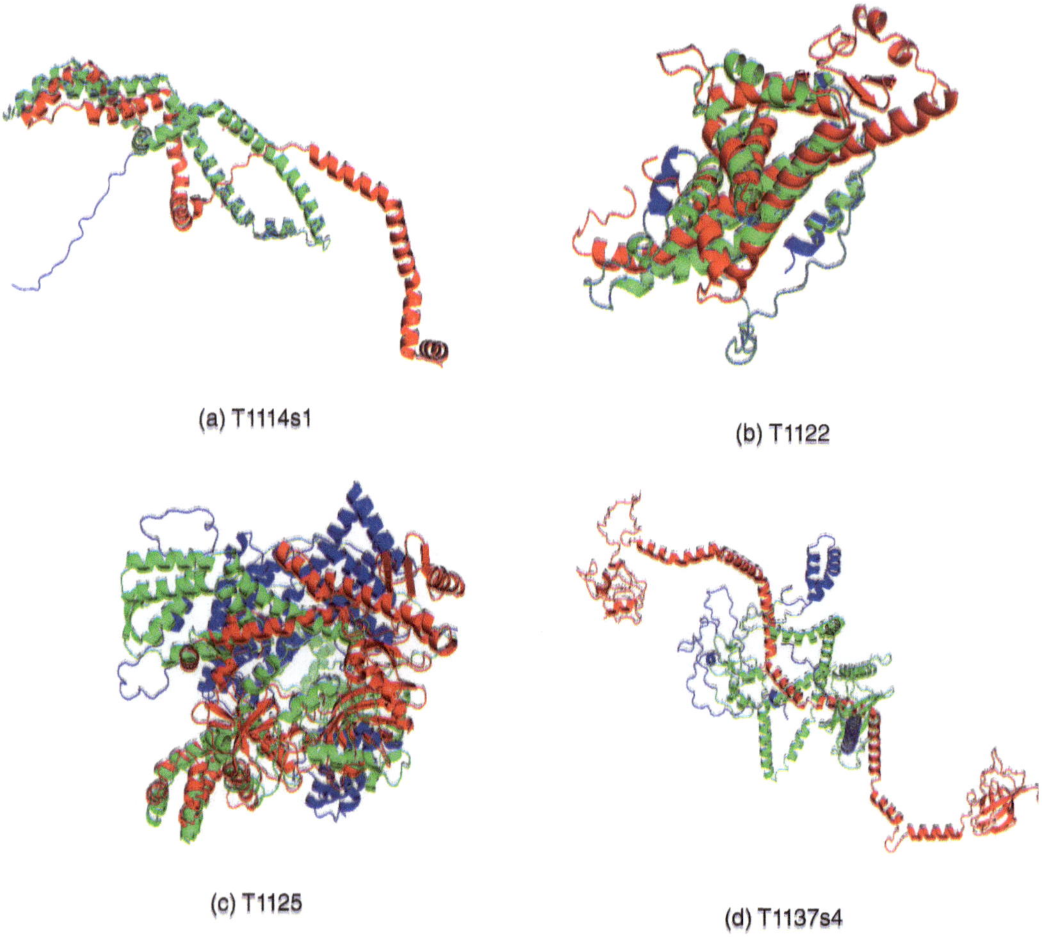

Fig. 2 Cartoon representations for four example targets from CASP15. We selected the first four available cases with low prediction accuracy. In each case, the native structure is given in red, and the predicted AF2 top one model is given in green and blue. In green we plot those parts of the sequence for which there are native coordinates. In blue we plot the entire predicted model (overlapping with the green)

interaction with other chains that are part of the gene. This information may be unavailable or harder to interpret and learn for AF2.

4 Notes for SPX and Seder

1. *Source and Availability:* SPX and Seder were implemented in FORTRAN90 and are supplied in source code format. Running SPX requires supplying a sequence file in the FASTA format, or if available, a PSSM matrix. The use of a PSSM matrix will significantly reduce the running time. SPX returns the predicted SS and ASA and other parameters as described in its documentation. Seder takes a PDB file for a model as input and produces a predicted TM-score. Both servers also give an

error estimate obtained by taking the standard deviation on the ensemble of trained weights. Helpful documentation, executables, and sources are available from Research and Information Systems, LLC at http://mamiris.com.

2. *License:* The license for SPX is available from a file called "LICENSE" in its distribution directory. It allows for academic users to use and modify it freely with proper citations while retaining some rights for commercial use.
3. *Installation:* The original programs were compiled on a Linux system using the Intel Fortran compiler (ifort), freely available for academic use at Intel's website. The compile script given in the code directory assumes that users have ifort or gfortran installed and that its location is in a proper path. Note: If you are using a Linux system, there is some chance that the compiled binaries supplied with this distribution will work on your system. Test cautiously at your own risk while backing up anything important. To compile the source code, go to the code directory and issue the command. */compile.* Consult the script "compile" in the directory "code/" if errors come up. Further information is available in the "README" files that are packed with the distribution.

5 Conclusions

In conclusion, we analyzed four cases from the CASP experiment for which the prediction accuracy was low with a TM-score below 0.5 for the top one AF2 model.

We find that some aspects of the predictions are still useful, albeit the low tertiary match. For example, helical regions are identified well. On the other hand, we find that identification of the coil and sheet SS states is not very successful. We also find that overall AF2 performs better than SPX in predicting SS and ASA. However, for some cases such as T1122, SPX better predicts sheet and coil regions. For T1125 SPX predicts RASA better. We also find that AF2 is better than Seder in selecting the best matching tertiary structure model for one case (T1122) and is worse in another case (T1114s1). For two other cases, AF2 and Seder selected the same models. We note that from the cases presented here, it appears that AF2 predicts more compact structures than the native one. This may be due to unavailability of the predictor of interactions outside the chain or otherwise difficulty accounting for them. In response, the predictor creates false interactions to stabilize the single chain as is evident for the cases of T1114s1 and T1137s4. In summary, we find that while AF2 has significantly improved protein tertiary structure prediction in general, there are cases, such as the four presented here, for which the tertiary structure prediction could still be significantly upgraded. The source code, license, and documentation for SPX and Seder are available from Research and Information Systems, LLC at http://mamiris.com.

Acknowledgments

R.L.J. and A.K. acknowledge financial support from NIH Grants R01GM127701 and R01HG012117.

References

1. Zhou Y, Kloczkowski A, Faraggi E, Yang Y et al (2017) Prediction of protein secondary structure. Springer
2. Jumper J, Evans R, Pritzel A, Green T, Figurnov M, Ronneberger O, Tunyasuvunakool K, Bates R, Zidek A, Potapenko A et al (2021) Highly accurate protein structure prediction with alphafold. Nature 596(7873):583–589
3. Jumper J, Evans R, Pritzel A, Green T, Figurnov M, Ronneberger O, Tunyasuvunakool K, Bates R, Zidek A, Potapenko A et al (2021) Applying and improving alphafold at casp14. Proteins: Structure, Function, and Bioinformatics 89(12):1711–1721
4. Faraggi E, Yang Y, Zhang S, Zhou Y (2009) Predicting continuous local structure and the effect of its substitution for secondary structure in fragment-free protein structure prediction. Structure 17(11):1515–1527
5. Faraggi E, Zhang T, Yang Y, Kurgan L, Zhou Y (2012) Spine x: improving protein secondary structure prediction by multistep learning coupled with prediction of solvent accessible surface area and backbone torsion angles. J Comput Chem 33(3):259–267
6. Faraggi E, Kloczkowski A (2017) Accurate prediction of one-dimensional protein structure features using spine-x. Prediction of Protein Secondary Structure pages:45–53
7. Faraggi E, Kloczkowski A (2014) A global machine learning based scoring function for protein structure prediction. Proteins: Structure, Function, and Bioinformatics 82 (5): 752–759
8. Faraggi E, Kloczkowski A (2015) Genn: a general neural network for learning tabulated data with examples from protein structure prediction. Artificial. Neural Netw:165–178
9. Faraggi E, Jernigan RL, Kloczkowski A (2021) A hybrid levenberg–marquardt algorithm on a recursive neural network for scoring protein models. Artificial. Neural Netw:307–316
10. Creighton TE (1993) Proteins: structures and molecular properties. Macmillan
11. Garnier J, Osguthorpe D, Robson B (1978) Analysis of the accuracy and implications of simple methods for predicting the secondary structure of globular proteins. J Mol Biol 120(1):97–120
12. Gibrat JF, Garnier J, Robson B (1987) Further developments of protein secondary structure prediction using information theory: new parameters and consideration of residue pairs. J Mol Biol 198(3):425–443
13. Holley LH, Karplus M (1989) Protein secondary structure prediction with a neural network. Proc Natl Acad Sci 86(1):152–156
14. Kneller D, Cohen F, Langridge R (1990) Improvements in protein secondary structure prediction by an enhanced neural network. J Mol Biol 214(1):171–182
15. Rost B, Sander C, Schneider R (1994) Phd-an automatic mail server for protein secondary structure prediction. Computer applications in the biosciences: CABIOS 10(1):53–60
16. Garnier J, Gibrat JF, Robson B et al (1996) Gor method for predicting protein secondary structure from amino acid sequence. Methods Enzymol 266:540
17. Dor O, Zhou Y (2007) Achieving 80% ten-fold cross-validated accuracy for secondary structure prediction by large-scale training. Proteins: structure. Function, and Bioinformatics 66:838–845
18. Faraggi E, Xue B, Zhou Y (2009) Improving the prediction accuracy of residue solvent accessibility and real-value backbone torsion angles of proteins by guided-learning through a two-layer neural network. Proteins: Structure, Function, and Bioinformatics 74(4): 847–856
19. Lee B, Richards FM (1971) The interpretation of protein structures: estimation of static accessibility. J Mol Biol 55(3):379-IN4

20. Chothia C (1974) Hydrophobic bonding and accessible surface area in proteins. Nature 248(5446):338–339
21. Chothia C (1976) The nature of the accessible and buried surfaces in proteins. J Mol Biol 105(1):1–12
22. Holbrook SR, Muskal SM, Kim SH (1990) Predicting surface exposure of amino acids from protein sequence. Protein Eng 3(8): 659–665
23. Rost B, Sander C (1994) Conservation and prediction of solvent accessibility in protein families. Proteins Struct Funct Genet 20 (3): 216–226
24. Faraggi E, Zhang T, Yang Y, Kurgan L, Zhou Y (2012) Spine x: improving protein secondary structure prediction by multistep learning coupled with prediction of solvent accessible surface area and backbone torsion angles. J Comput Chem 33:259–267
25. Faraggi E, Zhou Y, Kloczkowski A (2014) Accurate single-sequence prediction of solvent accessible surface area using local and global features. Proteins: Structure, Function, and Bioinformatics 82 (11):3170–3176
26. Ramachandran GN, Sasisekharan V (1968) Conformation of polypeptides and proteins. Adv Protein Chem 23:283–437
27. Wood MJ, Hirst JD (2005) Protein secondary structure prediction with dihedral angles. Proteins: structure. Function, and Bioinformatics 59:476–481
28. Zimmermann O, Hansmann UHE (2006) Support vector machines for prediction of dihedral angle regions. Bioinformatics 22: 3009–3015
29. Dor O, Zhou Y (2007) Real-spine: an integrated system of neural networks for real-value prediction of protein structural properties. PROTEINS: Structure, Function, and Bioinformatics 68 (1):76–81
30. Xue B, Dor O, Faraggi E, Zhou Y (2008) Real value prediction of backbone torsion angles. Proteins: structure. Function, and Bioinformatics 72:427–433
31. Faraggi E, Yang Y, Zhang S, Zhou Y (2009) Predicting continuous local structure and the effect of its substitution for secondary structure in fragment-free protein structure prediction. Structure 17:1515–1527
32. Kountouris P, Hirst JD (2009) Prediction of backbone dihedral angles and protein secondary structure using support vector machines. BMC bioinformatics 10(1):437
33. Rost B (1995) TOPITS: Threading one-dimensional predictions into three-dimensional structures. Third International Conference on Intelligent Systems for Molecular Biology AAAI Press, pp 314–321
34. Rost B, Sander C (1997) Protein fold recognition by prediction-based threading. J Mol Biol 270:471–480
35. Przybylski D, Rost B (2004) Improving fold recognition without folds. J Mol Biol 341: 255–269
36. Qiu J, Elber R (2006) SSALN: an alignment algorithm using structure-dependent substitution matrices and gap penalties learned from structurally aligned protein pairs. Proteins: Structure, Function, and Bioinformatics 62: 881–891
37. Cheng J, Baldi P (2006) A machine learning information retrieval approach to protein fold recognition. Bioinformatics 22:1456–1463
38. Liu S, Zhang C, Liang S, Zhou Y (2007) Fold recognition by concurrent use of solvent accessibility and residue depth. Proteins: structure. Function, and Bioinformatics 68:636–645
39. Marks DS, Hopf TA, Sander C (2012) Protein structure prediction from sequence variation. Nat Biotechnol 30(11):1072–1080
40. Rollins NJ, Brock KP, Poelwijk FJ, Stiffler MA, Gauthier NP, Sander C, Marks DS (2019) Inferring protein 3d structure from deep mutation scans. Nat Genet 51(7):1170–1176
41. Zheng W, Li Y, Zhang C, Pearce R, Mortuza S, Zhang Y (2019) Deep-learning contact map guided protein structure prediction in casp13. Proteins: Structure, Function, and Bioinformatics 87(12):1149–1164
42. Huang YM, Bystroff C (2006) Improved pairwise alignments of proteins in the twilight zone using local structure predictions. Bioinformatics 22:413–422
43. Simossis V, Heringa J (2004) Integrating protein secondary structure prediction and multiple sequence alignment. Curr Protein Pept Sci 5(4):249–266
44. Zhang W, Liu S, Zhou Y (2008) SP^5: improving protein fold recognition by using predicted torsion angles and profile-based gap penalty. PLoS One 3:e2325
45. Kihara D, Lu H, Kolinski A, Skolnick J (2001) Touchstone: an ab initio protein structure prediction method that uses threading-based tertiary restraints. Proc Natl Acad Sci 98(18): 10125–10130
46. Kolinski A et al (2004) Protein modeling and structure prediction with a reduced representation. Acta Biochim Pol 51:349–372
47. Liang S, Zhang C, Liu S, Zhou Y (2006) Protein binding site prediction using an empirical scoring function. Nucleic Acids Res 34(13): 3698–3707

48. Radivojac P, Clark WT, Oron TR, Schnoes AM, Wittkop T, Sokolov A, Graim K, Funk C, Verspoor K, Ben-Hur A et al (2013) A large-scale evaluation of computational protein function prediction. Nat Methods 10(3): 221–227
49. Bryant P, Pozzati G, Elofsson A (2022) Improved prediction of protein-protein interactions using alphafold2. Nat Commun 13(1): 1265
50. Bertoline LM, Lima AN, Krieger JE, Teixeira SK (2023) Before and after alphafold2: an overview of protein structure prediction. Front Bioinform 3:1120370
51. Ruff KM, Pappu RV (2021) Alphafold and implications for intrinsically disordered proteins. J Mol Biol 433(20):167208
52. Stevens AO, He Y (2022) Benchmarking the accuracy of alphafold 2 in loop structure prediction. Biomol Ther 12(7):985
53. Azzaz F, Yahi N, Chahinian H, Fantini J (2022) The epigenetic dimension of protein structure is an intrinsic weakness of the alphafold program. Biomol Ther 12(10):1527
54. Saldano T, Escobedo N, Marchetti J, Zea DJ, Mac Donagh J, Velez Rueda AJ, Gonik E, Garcia Melani A, Novomisky Nechcoff J, Salas MN et al (2022) Impact of protein conformational diversity on alphafold predictions. Bioinformatics 38(10):2742–2748
55. Wong F, Krishnan A, Zheng EJ, Stark H, Manson AL, Earl AM, Jaakkola T, and Collins JJ¨ (2022) Benchmarking alphafold-enabled molecular docking predictions for antibiotic discovery. Mol Syst Biol 18 (9):e11081
56. Scardino V, Di Filippo JI, Cavasotto CN (2023) How good are alphafold models for docking-based virtual screening? Iscience 26(1)
57. Buel GR, Walters KJ (2022) Can alphafold2 predict the impact of missense mutations on structure? Nat Struct Mol Biol 29(1):1–2
58. Pak MA, Markhieva KA, Novikova MS, Petrov DS, Vorobyev IS, Maksimova ES, Kondrashov FA, Ivankov DN (2023) Using alphafold to predict the impact of single mutations on protein stability and function. PLoS One 18(3): e0282689
59. Kryshtafovych A, Schwede T, Topf M, Fidelis K, Moult J (2021) Critical assessment of methods of protein structure prediction (casp)2014ˇ round xiv. Proteins Struct Funct Bioinform 89(12):1607–1617
60. Zhang Y, Skolnick J (2004) Scoring function for automated assessment of protein structure template quality. Proteins: Structure, Function, and Bioinformatics 57 (4):702–710
61. Zhang Y, Skolnick J (2005) Tm-align: a protein structure alignment algorithm based on the tm-score. Nucleic Acids Res 33(7): 2302–2309
62. Xu J, Zhang Y (2010) How significant is a protein structure similarity with tm-score= 0.5? Bioinformatics 26 (7):889–895
63. Altschul SF, Madden TL, Schaffer AA, Zhang J, Zhang Z, Miller W, Lipman DJ (1997) Gapped BLAST and PSI-BLAST: a new generation of protein database search programs. Nucl Aci Res 25:3389–3402
64. Meiler J, Muller M, Zeidler A, Schmaschke F (2001) Generation and evaluation of dimension reduced amino acid parameter representations by artificial neural networks. J Mol Model 7:360–369
65. Zhang W, Dunker AK, Zhou Y (2008) Assessing secondary-structure assignment of protein structures by using pairwise sequence-alignment benchmarks. Proteins: structure. Function, and Bioinformatics 71:61–67
66. Kabsch W, Sander C (1983) Dictionary of protein secondary structure: pattern recognition of hydrogen-bonded and geometrical features. Biopolymers 22:2577–2637
67. Dor O, Zhou Y (2007) Achieving 80% ten-fold cross-validated accuracy for secondary structure prediction by large-scale training. Proteins: Structure, Function, and Bioinformatics 66 (4):838–845
68. Zhou Y, Faraggi E (2010) Prediction of one-dimensional structural properties of proteins by integrated neural networks. Methods and Algorithms, Introduction to Protein Structure Prediction, pp 45–74
69. Berman HM, Westbrook J, Feng Z, Gilliland G, Bhat T, Weissig H, Shindyalov IN, Bourne PE (2000) The protein data bank. Nucleic Acids Res 28(1):235–242

Chapter 9

Improving the Annotations of JCVI-Syn3a Proteins

Mesih Kilinc, Kejue Jia, and Robert L. Jernigan

Abstract

The JCVI-Syn3 organism is a minimal organism derived from *Mycoplasma mycoides capri*, which is capable of self-replication. While the ancestor has 863 genes, the synthetic progeny has only 473, with 434 of these coding for proteins. Despite initial efforts to understand all functions of the organism, a significant number of these protein-coding genes still have unknown functions, and subsequent studies have been only partially successful in elucidating their roles. In this study, we employ our innovative method PROST to identify homologs and better understand these previously unidentified genes. PROST employs protein language embeddings and enables the identification of remote homologs with as low as 16% sequence identity. PROST successfully finds functionally annotated homologs for 93% of the minimal genome with a high level of accuracy, both confirming previously identified functions, as well as proposing new functions for others. The results of our study can be accessed at https://bit.ly/prost-syn3a.

Key words Minimal organism, JCVI-Syn3 organism, PROST, Alphafold2, FATCAT, Functional annotation

1 Introduction

Unraveling the complexities of biological systems is a formidable challenge. Despite advances in state-of-the-art technologies, fully comprehending the mechanisms of life remains a daunting task due to an incomplete comprehension of the underpinning atomistic machinery. To address this challenge, researchers have reduced the genome of the small *Mycoplasma mycoides capri* to the bare minimum that can still reproduce. The starting point had only 863 genes. Through rigorous cycles of mutagenesis experiments and verification, JCVI-Syn3 was created, having only 473 genes and is the smallest known organism capable of self-replication [1]. Despite the small size of the JCVI-Syn3 genome, a significant number of genes still have no known function. Initial annotation

Supplementary Information The online version contains supplementary material available at https://doi.org/10.1007/978-1-0716-4196-5_9.

Andrzej Kloczkowski et al. (eds.), *Prediction of Protein Secondary Structure*, Methods in Molecular Biology, vol. 2867,
https://doi.org/10.1007/978-1-0716-4196-5_9,

efforts classified the genome into five parts: Equavolog, probable, putative, generic, and unknown. The first three categories provide sufficient information to infer functionality whereas the latter two lack sufficient homology information for reliable functional classification.

Several efforts have pursued the annotations of the unknown genes in the minimal organism JCVI-Syn3. The first effort by [2] was able to annotate 76 out of 134 unknown genes. Later [3] used secondary structure alignments to identify similar fold homologs, while [4] utilized a consensus of multiple methods to annotate additional proteins, reducing the unknown cases to 15, most of which were transmembrane proteins. The new version of the minimal organism, JCVI-Syn3a, was also published, with those 16 additional proteins these cells had improved doubling time and more uniform shapes [5]. Further annotation efforts employed various techniques to improve the annotation of the minimal genome such as [6] utilized deep learning-assisted contact predictions to improve threading performance, resulting in better structure predictions. They were able to confidently annotate 88% of the genome with specific biological process Gene Ontology (GO) annotations, including the newly added Syn3a proteins, although 6 out of those 16 proteins were annotated with less specific, generic annotations. Recently, [7] focused specifically on the essential proteins involved in phenotypical properties such as cell growth and morphology. They utilized Alphafold 2 [8] to predict structures for the entire genome and conducted pairwise structural alignments using the FATCAT tool [9]. FATCAT is a flexible alignment tool that importantly applies twists and rotations to the structures to improve alignment accuracies. Moreover, [7] also used gene localization information in a closely related species to support the homologs identified. This is appropriate because the arrangement of genes in the genome, as well as gene localization, tend to be conserved among related organisms.

The method PROST used here takes a completely different approach for homolog detection, as detailed in our publication [10]. It selects layers from ESM1b large protein language model [11] from Meta that encodes protein sequences into a context-aware numerical representation capturing physical, biochemical, and evolutionary information. The method also applies a cosine transformation to capture global aspects. Subsequently, PROST reduces the size of this representation to optimize the accuracy and efficacy of homolog detection. Notably, PROST has been optimized to be particularly adept at identifying globally related homologs, thereby enabling the detection of remote homologs that are notoriously challenging to identify using traditional sequence similarity-based methods.

In this study, we investigate the proteome of the JCVI-Syn3 organism, a self-replicating minimal organism. Despite significant efforts to understand its functions, a considerable number of the protein-coding genes still have unknown functions. To address this, we used PROST, which utilizes protein language embeddings to identify homologs and shed light on the functions of the remaining proteins/genes. We also use BLAST [12] and Foldseek [13], along with our own PROST [10] method to support findings. Notably, PROST can detect remote homologs even for cases having as low as 16% sequence identity. These three different methods have different strengths and so using them all improves the final results. In the end, the structures from AlphaFold2 [8] are considered with FATCAT [9] structure alignments to confirm the relationships by using a flexible structural alignment. Our results succeed in functionally annotating 93% of the minimal genome, offering a high level of accuracy. This approach not only confirms the functions of previously identified genes but also proposes new functions for others.

2 Methods

PROST can identify functions for proteins that previously had no known function. In order to corroborate the homolog-based predictions from PROST, we employ some additional computational tools for homolog identification. Specifically, we utilize both BLAST and Foldseek along with PROST, to identify potential homologs. Foldseek is a novel method that leverages structure matches to identify homologs. We have utilized it by using predicted structures generated by Alphafold2 [8] for the SwissProt database [14]. The Foldseek method first converts the 3D structure into a 1D sequence, based on each amino acid atom and its surroundings, and subsequently uses the MMSeq2 algorithm to search the SwissProt database for 1D sequence matches. The method then reports TM scores [15]. BLAST is used to identify homologs as a means of creating a comparison baseline for the results obtained from both PROST and Foldseek; it is used here because its strength is accounting for local considerations, whereas PROST is better with global considerations. To validate our findings, we align each homolog structure from all three methods with its corresponding query of the Syn3a structure (all structures taken from Alphafold 2 [8] predictions) using FATCAT [9], which provides a statistical significance p-value to assess the alignment's reliability. Ultimately, the homolog with the lowest p-value and having an annotation is taken as the most informative putative homolog for the query protein and is taken as the best homolog. Homologs that are uncharacterized or have unknown functions are excluded from the best homolog selection. Figure 1 depicts the workflow of the best homolog identification process.

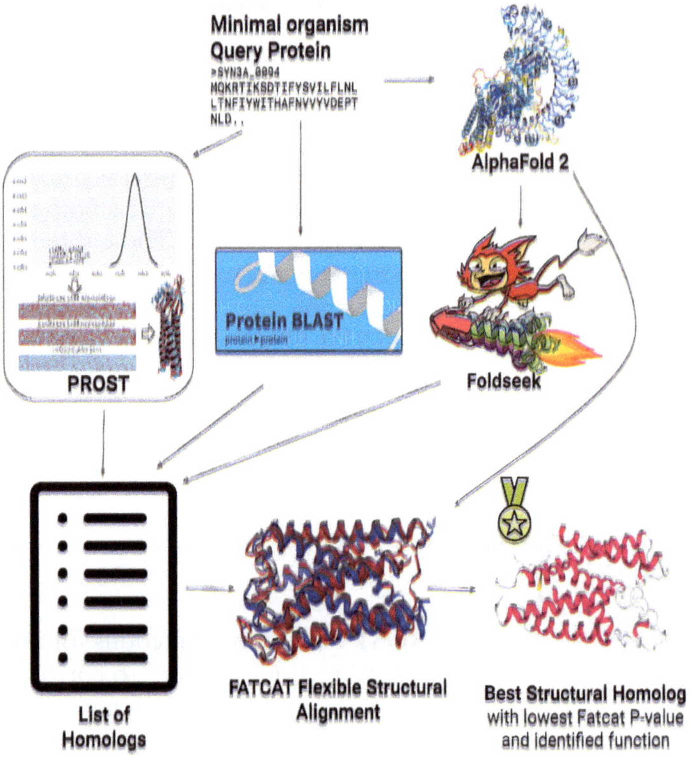

Fig. 1 The workflow for identifying best homologs for minimal organism query proteins. This schematic illustrates the process of identifying the best homolog and from it annotating the functions with various methods. Homolog identification is carried out with three methods: PROST [10], Foldseek [13], and BLAST [12]. Arrows indicate the flow of information and the successive steps involved in the process. Alphafold 2 [8] is used to predict the query and homolog protein structures. All homologs predicted by all methods are collected and their structures are predicted with Alphafold 2. FATCAT [9] is used to align these structures in the last step to identify the homolog having the best structural alignment with the query protein. Finally, the best structural homolog with a known function is found and used to infer the query protein's predicted function

3 Results

Table 1 provides the number of best structural homologs found with each individual method. Figure 2 shows that among the three methods, PROST is more effective by independently identifying the best homologs for 61 proteins, with Foldseek uniquely identifying homologs for only 6 proteins, and BLAST only 1 such case. Furthermore, PROST and Foldseek together discovered the 22 best homologs. Similarly, BLAST and Foldseek yield homologs for 9 proteins. Notably, PROST and BLAST together identify homologs for 5 proteins. However, it is important to note that there are still 15 protein sequences that have no homologs with

Table 1
Total Number of Best Homolog Proteins from Each Individual Method

Method	Number
PROST	424
Foldseek	373
BLAST	351
No homologs	15

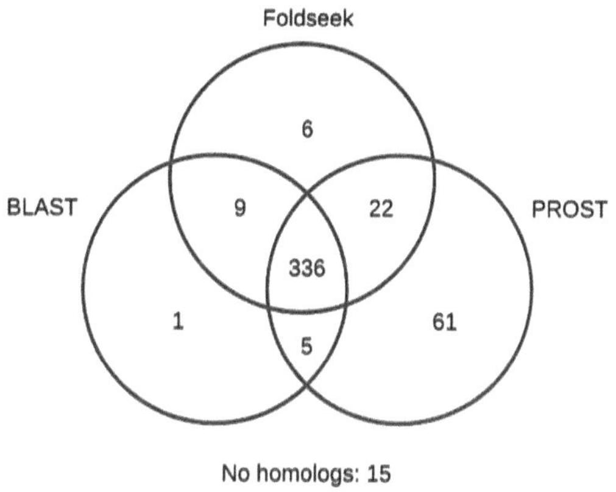

Fig. 2 Diagram identifying the overlapping functions found by the three different methods. The outside parts show the GO terms that are found uniquely by each of the three methods: PROST [10], BLAST [12], and Foldseek [13], with PROST showing the largest number of 61 cases. The largest number 336 are found with all three methods. PROST and Foldseek have the largest overlap of 22 cases. BLAST shows the weakest performance overall. The totals for each individual method are shown in Table 1

significant structural similarity to any protein present in the SwissProt database, for any of the methods used. The largest number of proteins have the same best structural homolog identified by all three methods. PROST demonstrates its efficacy by independently identifying homologs for a significant number of additional proteins, while the other methods contributed to the identification of homologs for a smaller number of proteins.

All results are available on the website as described next.

4 Website

To disseminate the PROST results, we have developed an interactive website using JSONWP tool [16], with the landing page shown in Fig. 3. The website's main page displays a results table that lists the entire JCVI-Syn3a proteome, including its NCBI id,

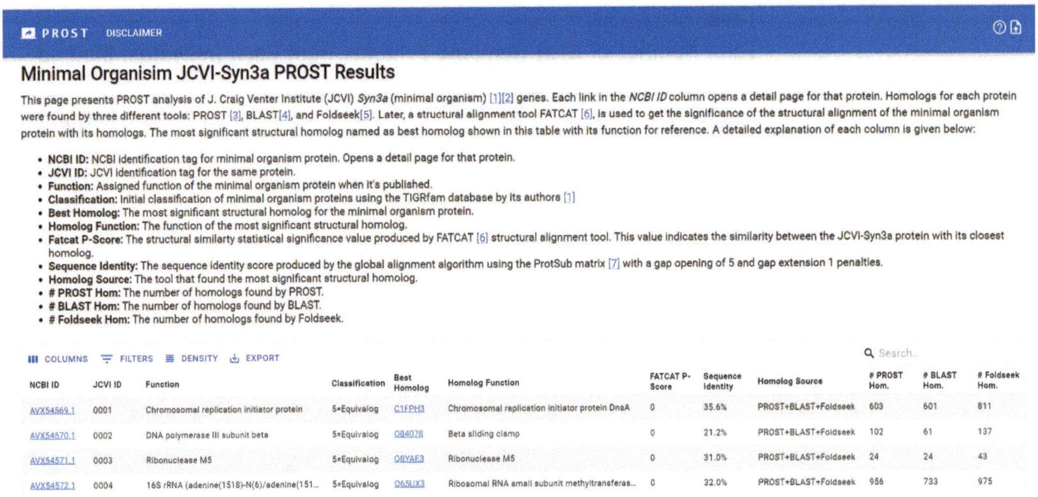

Fig. 3 Landing page of the interactive website for PROST results for JCVI-Syn3a proteome. The main page of the website shows a results table that lists the entire JCVI-Syn3a proteome with various information such as NCBI id, JCVI id, function, TIGRfam classification, essentiality, best homolog, structural alignment score, sequence identity score, and number of homologs from the different methods. The table can be searched, sorted, paginated, and customized by the user. The links in the first and the fifth columns lead to the detailed result pages for each protein and Uniprot pages for each homolog, respectively

JCVI id, function, the original TIGRfam classification from [1], and essentiality. Additionally, we provide information on the best homolog identified by FATCAT (flexible structural alignment tool) using the homologs supplied by PROST, FATCAT, and BLAST. This information includes the homolog's Uniprot ID, structural alignment score (p-value), sequence identity score when aligned using the ProtSub matrix [17], as well as the number of homologs identified by PROST, FoldSeek, and BLAST.

The table is interactive and provides users with a search box in the top right corner that performs searches across all columns. The table is paginated, and the bottom right buttons enable users to navigate. Users can also sort each column by clicking the up-down arrow located next to the column label. The top left dropdown menu allows users to specify the number of results displayed per page. The bottom left label displays the total number of results found and the number of results currently shown. The link in the first column directs users to a detailed results page designed for each protein. Similarly, the link provided in the PROST homolog column directs users to the Uniprot website for the specific homolog found.

The homolog with the most significant structural alignment and a known function is reported in the "Best Homolog" column. We also indicate the tool that identified the best homolog in the "Hom. Source" column. Furthermore, we provide the sequence alignment using the ProtSub matrix [17]. To see the unique

Fig. 4 Example of the Summary section of the result page for each JCVI-Syn3a protein. This section displays categorical information such as NCBI id, JCVI id, functional annotation, and TIGRfam classification. It also shows statistics about homologs and GO annotations from different sources: PROST [10], FoldSeek [13], BLAST [12], and Uniprot [14]. Additionally, it provides a literature summary of the protein based on the previous annotations by five different studies. This section helps the reader to assess each protein comprehensively

contributions of PROST to illuminate functions of the Syn3a proteome, one can filter the results by searching with "Only PROST" keyword. Each JCVI-Syn3a protein has a unique results page that consists of a summary, sequence and structural alignment, GO terms, and homologs sections. We display the aligned structures and the individual structures of the query and the best homolog. The function of the best homolog comes from its name, which does not necessarily conform to a GO annotation. We also present the sequence alignment colored by type of the secondary structures. The summary section, shown in Fig. 4, provides categorical information such as NCBI id, JCVI id, functional annotation, and TIGRfam classification. We also provide statistics about homologs and GO annotations. We use three different sources of homologs in this study: PROST, FoldSeek, and BLAST. GO annotations have four different sources: the homolog search methods and Uniprot GO annotations as of July 24, 2022, which we downloaded and extracted for each JCVI-Syn3a protein. For each homology method, we generate a list of homologs and use it to find enriched GO terms compared to the natural frequency of occurrence in the SwissProt database. We obtain all the associated GO terms for a given list of homologs and compare them with the frequency of each term in the SwissProt database. We then report the statistically significant terms. We use the same method to obtain GO terms for PROST, BLAST, and Foldseek. We report the size of the union of all three different sets of GO terms as "Total GO annotation." We also give the total number of unique GO term predictions with all three different methods, such as "Unique PROST Go." Next, we show the total number of homologs found by PROST, BLAST, and Foldseek and then the number of unique homologs with each

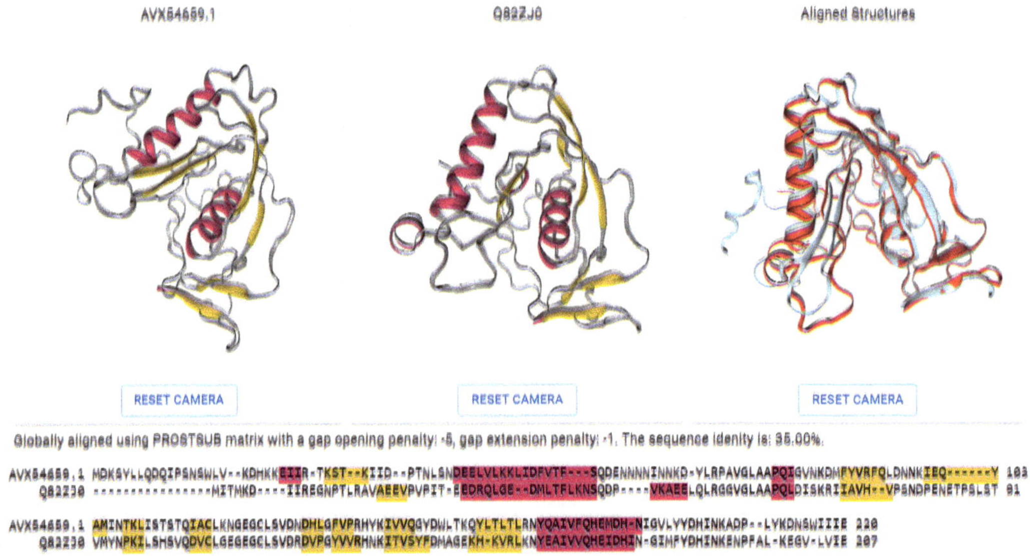

Fig. 5 Example of the structure and sequence alignments of the query JCVI-Syn3a protein and its homolog found by PROST [10]. The 3D protein structures are taken from Alphafold2 [8] and are colored based on secondary structures. Beta sheets are yellow and alpha helices are red. The aligned 3D protein structures on the right display the query protein in red and the homolog in blue. The structures can be inspected interactively with a mouse. The sequence alignment is performed with ProtSub matrix [17] and shows the secondary structures as well. The secondary structure assignments done using TM-align [15] method using Alphafold2 [8] predicted structures

method. For simplicity, Uniprot GO terms are shown in a separate table. This section also provides a literature summary of the JCVI-Syn3a protein under investigation. It gives the annotations made by [2–4, 6, 7]. This aids the user in evaluating the protein in a relatively comprehensive way.

The next section shows the structure and sequence alignments of query JCVI-Syn3a proteins and their homologs found by PROST, as illustrated in Fig. 5. The homolog displayed is the most significant structurally aligned protein to the query Syn3a protein. The 3D protein structures are from Alphafold2 [8] and are colored based on secondary structures. This section also presents the aligned 3D protein structures on the right. Structures are visualized with NGL.js [18] and can be moved interactively with the mouse. In addition, this section provides the corresponding sequence alignment. The sequence alignment is done with ProtSub matrix [17], gap opening penalty 5, and gap extension penalty 1. The secondary structure assignments are predicted using the TM align method [15] and applied to the Alphafold2 [8] predicted structures.

The next section on the results page provides the GO annotations. Figure 6 shows this section. Here, we listed all the GO terms found by three different methods. Each method generates a list of

Go Annotations

1. **PBF** indicates the go terms that are found by both PROST and BLAST and Foldseek.
2. **PF** indicates the go terms that are found by only PROST and Foldseek.
3. **BF** indicates the go terms that are found by only BLAST and Foldseek.
4. **PB** indicates the go terms that are found by both PROST and BLAST.
5. **P** indicates the go terms that are found by only PROST.
6. **F** indicates the go terms that are found by only Foldseek.
7. **B** indicates the go terms that are found by only BLAST.

COLUMNS FILTERS DENSITY EXPORT Search...

Source	GO	Description
1. PBF	GO:0042586	peptide deformylase activity
1. PBF	GO:0006412	translation
1. PBF	GO:0043686	co-translational protein modification
1. PBF	GO:0018206	peptidyl-methionine modification
1. PBF	GO:0046872	metal ion binding
1. PBF	GO:0008463	formylmethionine deformylase activity
1. PBF	GO:0031365	N-terminal protein amino acid modification
3. BF	GO:0006464	cellular protein modification process
4. PB	GO:0005829	cytosol
4. PB	GO:0009570	chloroplast stroma

Rows per page: 10 1–10 of 17

Fig. 6 GO annotations section of the result page for each JCVI-Syn3a protein. This section lists all the GO terms found by three different methods: PROST [10], BLAST [12], and Foldseek [13]. The GO terms are divided into seven sections according to the key at the top of Fig. 6

homologs and finds the GO terms that are enriched in this list compared to the natural distribution of GO terms in the SwissProt database. The three different sets of GO terms are divided into seven sections. The first section (1. PBF) shows the GO terms that are found by all three methods. The second section (2. PF) displays the GO terms annotated with PROST and Foldseek. All GO term sections are listed in Fig. 6. The.

GO annotations result table is paginated and searchable. The user can enter their search query in the top right input box. The search will be performed on all columns. Moreover, each column can be sorted by clicking the up-down arrow next to the column label.

The Uniprot database has GO annotations for the parent organism *Mycoplasma mycoides capri*. We obtain these annotations by finding one-to-one homolog gene matches between

Homologs

1. **PBF** indicates the homologs that are found by both PROST and BLAST and Foldseek.
2. **PF** indicates the homologs that are found by only PROST and Foldseek.
3. **BF** indicates the homologs that are found by only BLAST and Foldseek.
4. **PB** indicates the homologs that are found by both PROST and BLAST.
5. **P** indicates the homologs that are found by only PROST.
6. **F** indicates the homologs that are found by only Foldseek.
7. **B** indicates the homologs that are found by only BLAST.

COLUMNS FILTERS DENSITY EXPORT Search

Source	Homolog	Description	Fatcat P-value	PROST E-value	BLAST E-value	Foldseek TM-Score
1. PBF	Q0ASK2	Peptide deformylase	3.44e-14	4.71e-16	1.53e-04	0.785
1. PBF	C0MFA6	Peptide deformylase	0.00e+00	1.27e-51	2.85e-28	0.8564
1. PBF	B2A2K1	Peptide deformylase	1.11e-15	3.12e-31	2.05e-08	0.8061
1. PBF	Q7NQ75	Peptide deformylase	5.27e-13	4.30e-22	0.011	0.7391
1. PBF	Q9K9I9	Peptide deformylase	0.00e+00	1.25e-43	6.89e-34	0.8647
1. PBF	Q7VBG6	Peptide deformylase 1	2.38e-13	7.71e-15	9.36e-05	0.7199
1. PBF	B8FHH0	Peptide deformylase	8.63e-14	2.02e-24	7.12e-11	0.7965
1. PBF	Q74JW2	Peptide deformylase	0.00e+00	1.84e-41	3.52e-25	0.9369
1. PBF	Q8DDE3	Peptide deformylase 1	6.78e-13	1.52e-20	6.60e-07	0.7653
1. PBF	Q1GAR4	Peptide deformylase	0.00e+00	3.02e-42	8.46e-25	0.9356

Rows per page: 10 1–10 of 672

Fig. 7 Homolog list on the result page for each JCVI-Syn3a protein. This section lists all homologs found by BLAST, PROST, and Foldseek. The homologs are marked according to the key at the top. The table shows the e-values from PROST [10] and BLAST [12] methods, the TM-Scores [15] from Foldseek [13], and the p-values from the FATCAT [9] structural alignment. The links lead to the Uniprot [14] website for details. A short description is also provided. The table is searchable, sortable, and paginated

JCVI-Syn3a and *Mycoplasma mycoides capri* and displayed them in a separate table on the website.

The last section on the results page is the homology list, as shown in Fig. 7. Each protein in the JCVI-Syn3a proteome is searched using BLAST, PROST, and Foldseek. Common homologs are marked as "1. PBF." Unique homologs found by PROST are marked with "5. P" label, according to the key at the top of the page. The table shows the e-values from PROST and BLAST methods in individual columns. Foldseek reports TM-Scores, shown in a separate column. All the homologs are aligned to the query structure with FATCAT and the alignment significance as a p-value is reported in another column. The links lead to the Uniprot website for further details. A short description is also provided for quick inspection. This table, like the GO annotations table, is searchable, sortable, and paginated.

In summary, the website is a convenient and powerful way to explore the JCVI-Syn3a genome with the all homolog results for each protein. It provides structural information as well as sequence alignments. It collects all the literature. Moreover, it displays Uniprot annotations and the annotations from the three different methods: PROST, BLAST, and Foldseek. The results are presented in a searchable, paginated, and sortable table.

The aim of this study is to identify gene/protein functions in the minimal organism that have not been assigned. We use PROST, a tool for finding remote homologs with better accuracy, and verified the homolog relationships using structural alignment with Alphafold2 predicted structures. We also use BLAST and Foldseek, a method for searching homologs based on predicted structures rather than sequences. Interestingly, PROST identifies more homologs and more functions than the other two methods.

5 Identification of Missing Genes for Essential Functions

We hypothesize that some of the unknown or uncharacterized genes in the minimal organism are responsible for 6% of the metabolic reactions that have no assigned gene. To test this hypothesis, we use PROST and Foldseek to detect remote homologs of these genes based on sequence and structure similarity, respectively. We also verify the homolog relationships using structural alignments using Alphafold2 predicted structures. We report three cases where we found evidence for missing genes encoding enzymes or transporters involved in essential functions.

The first case is a Na+ /H+ antiporter, required for ion transport. We find gene JCVISYN3A_0870 has the function GO:0015385 (sodium:proton antiporter activity) annotation in PROST, which found 133 homologs with the same activity. Among these homologs, 29 were also detected by Foldseek, providing further support for this annotation.

The second case is a transaldolase (TALA), which is essential for the pentose phosphate pathway. We find gene JCVISYN3A_0262 to have the function ribulose-phosphate 3-epimerase, with three homologs with TALA activity in PROST and six in Foldseek. This suggests that JCVISYN3A_0262 may function as both an epimerase and a transaldolase.

The third case is a lactate transporter, which is needed for lactate metabolism. We found that gene JCVISYN3A_0325 has monocarboxylate transporter type activity in PROST and L-lactate transporter activity in Foldseek. This indicates that JCVISYN3A_0325 may be involved in lactate transport. In addition, we found that JCVISYN3A_0325 also has homologs with nucleoside permease NupG and purine ribonucleoside efflux pump NepI activities in both PROST and Foldseek. This implies that JCVISYN3A_0325 may also participate in nucleoside uptake, which has not been fully characterized in mycoplasmas [5].

Another essential function is acetate transport, which is required for its experimental two-hour doubling time [5]. We found that the gene JCVISYN3A_0878 has homologs with cation/acetate symporter ActP activity in both PROST and Foldseek.

We also investigate the substrate specificity of two Ecf transporter subunits S, JCVISYN3A_0345 and JCVISYN3A_0314, which were not known previously. We find that JCVISYN3A_0345 has a homolog with riboflavin transporter activity in PROST, with a FATCAT p-value of 6.99e-13. Similarly, we found that JCVI-SYN3A_0314 has a homolog with folate transporter activity in both PROST and BLAST, with a FATCAT p-value of 5.46e-14.

Another missing enzyme in lipid metabolism was phosphatidate phosphatase. The Syn3a paper [5] suggested that JCVI-SYN3A_0214 may have this activity as a secondary function. We confirm this by finding a putative phosphatidate phosphatase homolog for this protein with PROST.

5.1 PROST Helps Explain the Improved Growth Phenotype

We identify two Syn3a special genes, JCVISYN3A_0527 and JCVI-SYN3A_0592, that have structural homologs with known functions in other organisms. We use FATCAT to compare the predicted structures of these genes and find the best matches. We also used PROST, BLAST, and Foldseek to find sequence and structure homologs for these genes.

JCVISYN3A_0527 has a structural homolog with the large ribosomal RNA subunit accumulation protein (YceD) from *E. coli*, with a FATCAT p-value of 1.72e-05 and a sequence identity of 23.1%. This homolog was also predicted by [7]. We find six YceD homologs in PROST, five of which were also detected by Foldseek. YceD is involved in the maturation and stability of the 23S rRNA, which catalyzes the peptidyl transferase reaction in ribosomal synthesis. A knockout of YceD in *E. coli* reduces the amount of 23S rRNA [19]. We hypothesize that JCVISYN3A_0527 may have a similar role in Syn3a and contribute to protein synthesis and hence its improved doubling time.

JCVISYN3A_0592 has a homolog with the DNA recombination protein RmuC from E. coli, with a FATCAT p-value of 2.06e-11 and a sequence identity of 25.1%. This gene has no previous annotation in the literature. We found 10 RmuC homologs in PROST, but none in BLAST or Foldseek. Figure 8 shows the sequence and structure alignment of the JCVISYN3A_0592 and RmuC. RmuC is a DNA endonuclease that cleaves DNA at specific secondary structures and protects DNA from nuclease damage. A mutation of RmuC in *E. coli* increases the frequency of DNA inversions [20]. We speculate that JCVISYN3A_0592 may play a role in DNA repair/protection in Syn3a and act to enhance its growth rate.

5.2 PROST Improves the Annotation of JCVI-Syn3a Proteins

We have used PROST to find homologs for five JCVI-Syn3a proteins that had no previous annotations in the literature. We verify the homology relationships using FATCAT to compare the AlphaFold2 predicted structures of these proteins with their homologs. We report the following:

JCVISYN3A_0520 has a structural homolog with monoacylglycerol lipase (O07427) from *Mycobacterium tuberculosis*, with a FATCAT p-value of 1.44e-13. O07427 and is a virulence factor

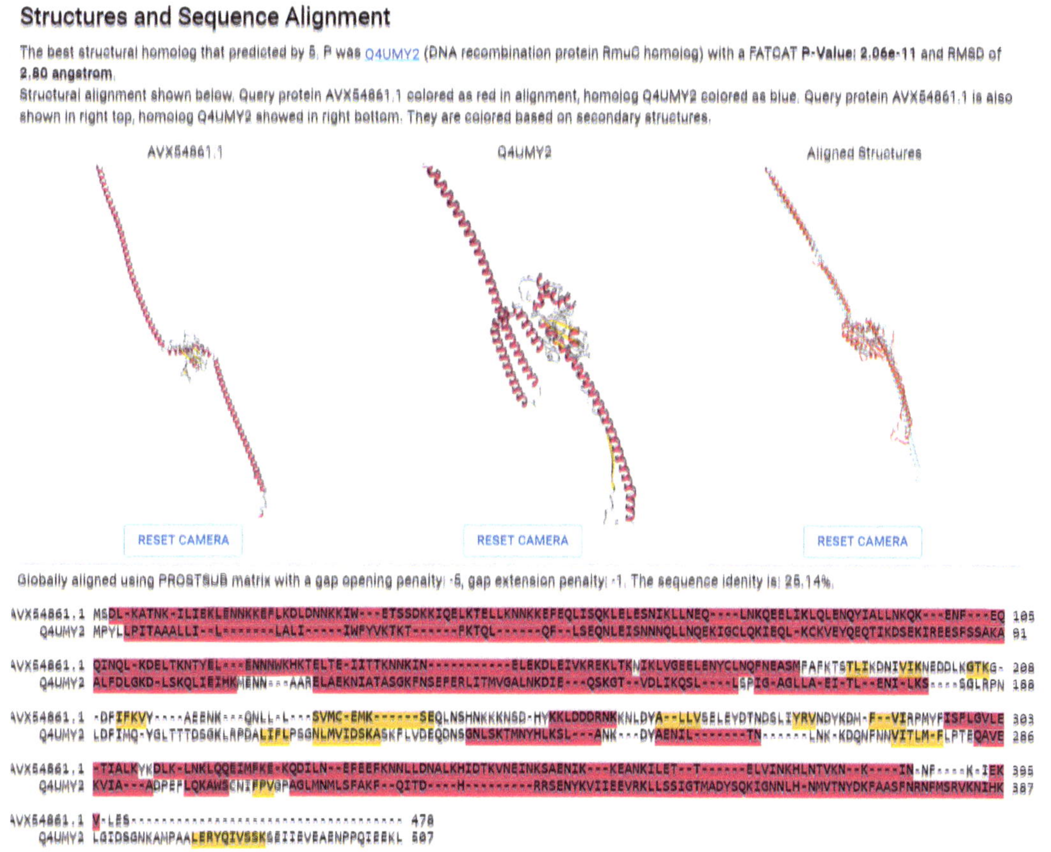

Fig. 8 Sequence and Structural alignment of JCVISYN3A_0592 and Q4UMY2, DNA recombination protein RmuC homolog. This structural alignment has a FATCAT [9] P-Value of 2.06e-11 and RMSD of 2.80 Å. The sequence alignment identity is 25.1%

that induces cell death and inflammatory response in the host organism.

JCVISYN3A_0778 is located near the ATP operon. We find that this protein has a structural homolog with ATP synthase subunit C (Q83G86) from *Methanosarcina mazei*, with a FATCAT p-value of 1.54e-07 and a sequence identity of 23.2%. This protein is part of the F0 complex that translocates protons across a membrane.

JCVISYN3A_0827 may be involved in Na+/H+ transport. We find this protein has a homolog with proton extrusion protein PcxA (B0CG09) from *Prochlorococcus marinus*, with a FATCAT p-value of 0.00238 and a sequence identity of 18.4%. This protein mediates Na + dependent H+ extrusion in response to light.

JCVISYN3A_0830 may be involved in DNA import. We find this protein has a structural homolog with DNA import protein CedA1 (Q4JB68) from *Sulfolobus tokodaii*, with a FATCAT p-value

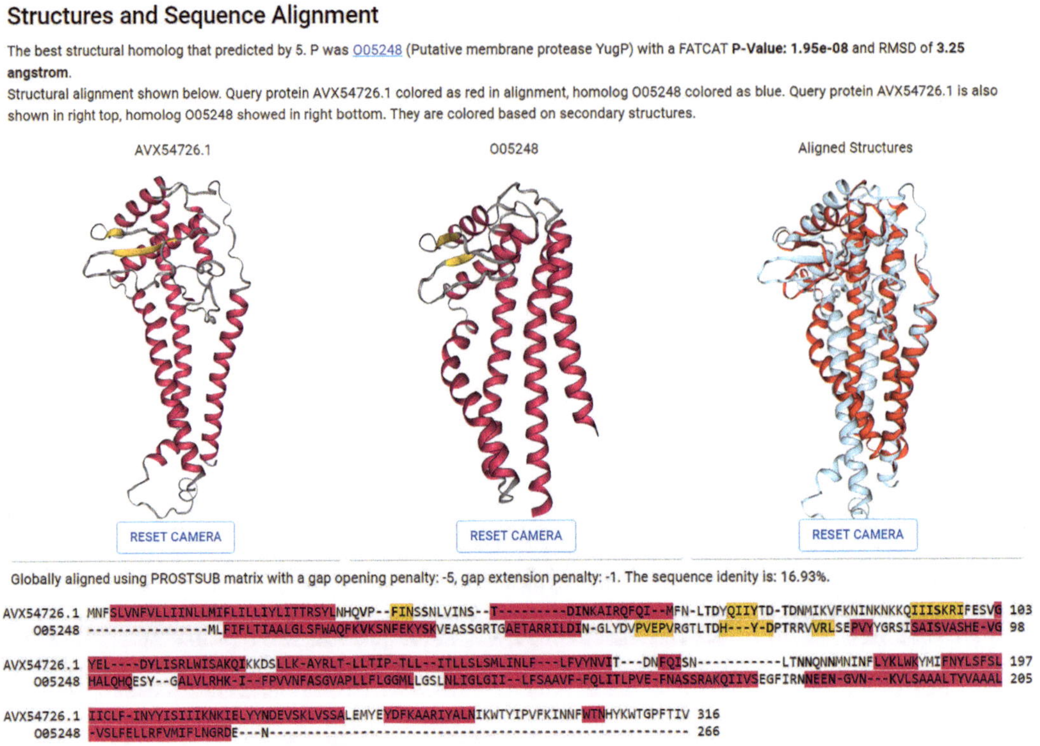

Fig. 9 Sequence and structural alignment of JCVISYN3A_0332 and O05248, the putative membrane protease YugP homolog. This structural alignment has a FATCAT [9] *P*-Value of 1.95e-08 and RMSD of 3.25 Å. The sequence alignment identity is only 16.9%

of 7.21e-11 and a sequence identity of 26.3%. This protein participates in the crenarchaeal system for exchange of DNA (Ced).

JCVISYN3A_0873 may be involved in NAD(P)H pathway. We find this protein has a structural homolog with NAD(P)H-quinone oxidoreductase subunit O (B7KGC8) from *Rhodopseudomonas palustris*, with a FATCAT p-value of 0.000233 and a sequence identity of 10.9%. This protein is part of the NDH-1 complex that transfers electrons from NAD(P)H to quinone.

Another interesting case is JCVISYN3A_0332, where PROST identified 421 Protease HtpX homologs out of the total of 440 homologs. The best structurally aligned homolog was identified as a putative membrane protease, YugP homolog (O05248). The sequence and structural alignment are shown in Fig. 9. These two proteins exhibit only 16.9% sequence identity, which suggests that the low sequence identity may be attributable to variability in the transmembrane regions. It is well-established that transmembrane regions typically have lower sequence conservation, in contrast to globular proteins [21]. This highlights PROST's efficacy in identifying structural similarities for cases having only low sequence identity, which is particularly important for membrane proteins.

6 Discussion and Conclusion

In this study, we demonstrate the utility of PROST for protein function annotation by using the minimal organism JCVI-Syn3a as a case study. We have identified homologous proteins in the Syn3a genome using three different methods: Foldseek, BLAST, and PROST. Out of the 455 Syn3a proteins analyzed, PROST uniquely and successfully identified the best homolog for 61 proteins. Notably, five of these proteins were newly added to the Syn3a [5]. Among the identified homologs, 56 proteins were from Syn3 and had TIGRFAM classification, indicating their known functional roles. Additionally, nine proteins were classified as generic, one as putative, two as probable, and one as Equivalog. However, the remaining 44 proteins were classified as unknown, indicating their functions having yet to be determined. Importantly, PROST was the only method that yielded significant structural homologs for these 61 proteins, suggesting its high level of efficacy in predicting homology within the Syn3a genome. These proteins are all listed in Supplemental Table 1. We identify several missing genes for metabolic reactions that occur in Syn3a but have no assigned function. These are JCVISYN3A_0262, which may have transaldolase activity and JCVISYN3A_0214, which may be the missing phosphatidate phosphatase enzyme. We also characterize two novel Syn3a proteins that are likely related to its enhanced growth rate. These are JCVISYN3A_0527 RNA subunit accumulation protein and JCVISYN3A_0592 DNA recombination protein. Moreover, we identify numerous transporters such as Na+/H+ antiporters, lactate, acetate, riboflavin, folate transporters. We have assembled all results into an interactive and accessible web page to display and facilitate further exploration. PROST is a powerful tool that can help to rapidly and accurately annotate whole genomes by finding homologs with lower sequence identities than other methods and validate these by demonstrating their high structural similarities. This improves our understanding of protein functions and enables us to study organisms presently having limited annotations.

Acknowledgments

We thank Dr. John Glass and his team at J. Craig Venter Institute and Dr. Zaida (Zan) Luthey-Schulten and her team at the University of Illinois at Urbana-Champaign for their valuable discussions. We gratefully acknowledge the support of NIH grants R01GM127701 and R01HG012117 and DOE grant DE-SC0022090.

References

1. Hutchison CA, Chuang R-Y, Noskov VN et al (2016) Design and synthesis of a minimal bacterial genome. Science 351:aad6253. https://doi.org/10.1126/science.aad6253
2. Danchin A, Fang G (2016) Unknown unknowns: essential genes in quest for function. Microb Biotechnol 9:530–540. https://doi.org/10.1111/1751-7915.12384
3. Yang Z, Tsui SK-W (2018) Functional annotation of proteins encoded by the minimal bacterial genome based on secondary structure element alignment. J Proteome Res 17:2511–2520. https://doi.org/10.1021/acs.jproteome.8b00262
4. Antczak M, Michaelis M, Wass MN (2019) Environmental conditions shape the nature of a minimal bacterial genome. Nat Commun 10:3100. https://doi.org/10.1038/s41467-019-10837-2
5. Breuer M, Earnest TM, Merryman C et al (2019) Essential metabolism for a minimal cell. eLife 8:e36842. https://doi.org/10.7554/eLife.36842
6. Zhang C, Zheng W, Cheng M et al (2021) Functions of essential genes and a scale-free protein interaction network revealed by structure-based function and interaction prediction for a minimal genome. J Proteome Res 20:1178–1189. https://doi.org/10.1021/acs.jproteome.0c00359
7. Bianchi DM, Pelletier JF, Hutchison CA et al (2022) Toward the complete functional characterization of a minimal bacterial proteome. J Phys Chem B 126:6820–6834. https://doi.org/10.1021/acs.jpcb.2c04188
8. Jumper J, Evans R, Pritzel A et al (2021) Highly accurate protein structure prediction with AlphaFold. Nature 596:583–589. https://doi.org/10.1038/s41586-021-03819-2
9. Li Z, Jaroszewski L, Iyer M et al (2020) FATCAT 2.0: towards a better understanding of the structural diversity of proteins. Nucleic Acids Res 48:W60–W64. https://doi.org/10.1093/nar/gkaa443
10. Kilinc M, Jia K, Jernigan RL (2023) Improved global protein homolog detection with major gains in function identification. Proc Natl Acad Sci 120:e2211823120. https://doi.org/10.1073/pnas.2211823120
11. Rives A, Meier J, Sercu T et al (2021) Biological structure and function emerge from scaling unsupervised learning to 250 million protein sequences. Proc Natl Acad Sci 118:e2016239118. https://doi.org/10.1073/pnas.2016239118
12. Altschul S (1990) Basic local alignment search tool. J Mol Biol 215:403–410. https://doi.org/10.1006/jmbi.1990.9999
13. van Kempen M, Kim S, Tumescheit C et al (2023) Fast and accurate protein structure search with Foldseek. Nat Biotechnol 1–4. https://doi.org/10.1038/s41587-023-01773-0
14. Bateman A, Martin M-J, Orchard S et al (2022) UniProt: the universal protein knowledgebase in 2023. Nucleic Acids Res 51:D523–D531. https://doi.org/10.1093/nar/gkac1052
15. Zhang Y (2005) TM-align: a protein structure alignment algorithm based on the TM-score. Nucleic Acids Res 33:2302–2309. https://doi.org/10.1093/nar/gki524
16. Kilinc M, Jia K, Jernigan RL (2023) JSONWP: a static website generator for protein bioinformatics research. Bioinformatics advances 3. https://doi.org/10.1093/bioadv/vbad154
17. Jia K, Jernigan RL (2021) New amino acid substitution matrix brings sequence alignments into agreement with structure matches. Proteins: structure. Function, and Bioinformatics 89:671–682. https://doi.org/10.1002/prot.26050
18. Rose AS, Bradley AR, Valasatava Y et al (2018) NGL viewer: web-based molecular graphics for large complexes. Bioinformatics 34:3755–3758. https://doi.org/10.1093/bioinformatics/bty419
19. Yang J, Suzuki M, McCarty DR (2016) Essential role of conserved DUF177A protein in plastid 23S rRNA accumulation and plant embryogenesis. J Exp Bot 67:5447–5460
20. Slupska MM, Chiang J-H, Luther WM et al (2000) Genes involved in the determination of the rate of inversions at short inverted repeats. Genes Cells 5:425–437
21. Olivella M, Gonzalez A, Pardo L, Deupi X (2013) Relation between sequence and structure in membrane proteins. Bioinformatics 29:1589–1592. https://doi.org/10.1093/bioinformatics/btt249

Chapter 10

DescribePROT Database of Residue-Level Protein Structure and Function Annotations

Bi Zhao, Sushmita Basu, and Lukasz Kurgan

Abstract

DescribePROT is a freely available online database of structural and functional descriptors of proteins at the amino acid level. It provides access to 13 diverse descriptors that include sequence conservation, putative secondary structure, solvent accessibility, intrinsic disorder, and signal peptides, and putative annotations of residues that interact with proteins, peptides and nucleic acids. These data can be used to elucidate protein functions, to support efforts to develop therapeutics, and to develop and evaluate future predictors of protein structure and function. DescribePROT includes 7.8 billion predictions for 1.4 million proteins from 83 complete proteomes of popular model organisms. This information can be downloaded at multiple levels of scope (entire database, specific organisms, and individual proteins) and can be interacted with using a graphical interface that simultaneously displays data on multiple descriptors. We describe the contents of this resource, provide directions on how to use its interface, and offer instructions on how to obtain and interact with the underlying data. Moreover, we briefly discuss plans for a future expansion of this database. DescribePROT is available at http://biomine.cs.vcu.edu/servers/DESCRIBEPROT/.

Key words Database, Protein structure, Protein function, Intrinsic disorder, Secondary structure, Solvent accessibility, Protein, ligand binding, Amino acids, Prediction, Machine learning

1 Introduction

The protein sequence space is expanding rapidly. The primary resource that provides access to the protein sequences, the UniProtKB/TrEMBL database [1], has grown from 29.3 million sequences a decade ago (release 2013_01), to 107.6 million 5 years ago (release 2018_01), and to 245.9 million this year (2023_01 release). Correspondingly, scientists face a substantial challenge to characterize the structures and functions of these proteins. This information can be annotated at three levels of resolution: atomic, amino-acid (AA), and whole-protein. The main databases for the atomic-level data include Protein Data Bank (PDB) [2, 3], which focuses on the experimentally solved structures, and AlphaFold DB [4, 5], which stores structures

Andrzej Kloczkowski et al. (eds.), *Prediction of Protein Secondary Structure*, Methods in Molecular Biology, vol. 2867, https://doi.org/10.1007/978-1-0716-4196-5_10,

predicted with AlphaFold [6, 7]. As of early 2023, PDB covers about 200 thousand structures of proteins while AlphaFold DB includes about 200 million putative structures. On the other end of the spectrum, the protein-level data is available from the manually reviewed Swiss-Prot [8] and the computationally annotated TrEMBL [1], which collectively contain about 250 million proteins. The intermediate residue-level annotations, also called one-dimensional descriptors [9, 10], characterize structural and functional features of AAs that compose protein chains. Some of the commonly used residue-level descriptors of structure include secondary structure, torsion angles, solvent accessibility, half-sphere exposure, intrinsic disorder and flexibility. The residue-level annotations of function encompass protein domains, catalytic residues, linkers, signal peptides, and residues that interact with partner molecules, such as proteins, peptides, RNAs, DNA, nucleotides, lipids, and a variety of small ligands. Although the residue-level annotations can be computed from the experimental and putative atomic-resolution structures and extracted from the Swiss-Prot and TrEMBL files, these sources cover a relatively small subset of proteins (PDB and Swiss-Prot), do not include functional annotations (AlphaFold DB), and are limited to a small subset of AAs (Swiss-Prot and TrEMBL).

The relatively low amounts of the AA-level annotations that can be obtained from the above resources can be supplemented with the assistance of computational methods that accurately predict these structural and functional annotations from protein sequences. Hundreds of these predictors have been developed over the last few decades [11–38]. For instance, there are over 60 predictors of secondary structure [13, 22, 23] and over 100 for the intrinsic disorder prediction [34, 36, 39]. Several large-scale empirical evaluations show that some of these tools provide accurate predictions [40–47]. Consequently, a number of them are heavily utilized by the research community, as evidenced by their citation records. For instance, SignalP 4.0 [48], PSIPRED [49], and IUPred [50], which predict signal peptides, secondary structure, and intrinsic disorder, were cited 9634, 6663, and 2176 times, respectively [source: Google Scholar as of May 2023].

While some of these methods are publicly and freely available as webservers and/or standalone software, use of these resources is rather challenging when making predictions for large collections of proteins and/or multiple structural and functional characteristics. First, appropriate (i.e., available, accurate, and/or sufficiently fast) tools for a desired selection of characteristics must be identified by the user. Second, submitting and assembling their results requires substantial amount of effort since different tools typically use different formats for their inputs (i.e., sequences and parameters of predictors) and outputs (i.e., predicted structure/function). Third, the webservers are typically limited to predicting one or a few proteins at a time, which makes working with large protein sets rather challenging. Fourth, the use of downloadable programs often involves complex installation processes (i.e., many of these tools use multiple third-party applications that must be co-installed), and users must have access to computer hardware and operating systems that are compatible with the selected programs, where different applications may require different setups. These difficulties motivate the development of large platforms that combine predictions of multiple structural and functional aspects of proteins. These platforms include PSIPRED workbench [51, 52], PredictProtein [53, 54], MULTICOM [55, 56], and DEPICTER [57, 58]. However, they are typically limited to predicting individual proteins and focus on a specific subset of the residue-level descriptors. For example, MULTICOM and PSIPRED workbench address the prediction of the structural descriptors, while DEPICTER is constrained to the prediction of intrinsic disorder and its functions.

We describe the recently released DescribePROT (**D**atabas**e** of **s**tru**c**tu**r**e and funct**i**on residue-**b**ased pr**e**dictions of **PROT**eins), a comprehensive database of the residue-level predictions of structural and functional characteristics of proteins [59, 60]. DescribePROT offers fast and convenient access to pre-computed predictive results eliminating issues related to the selection of suitable predictors, input/output formatting, limited scope of predictions, and the installation process. Access to these results also prevents wasteful replication of predictions, where the same method is used to predict the same protein by different users. At the same time, this resource is restricted to the proteins that it covers. Therefore, the end users still must use the above-mentioned methods and platforms to collect predictions for the other proteins (*see* **Note 1**). This chapter describes the DescribePROT's scope and contents and provides informative and practical instructions on how to use its webpage interface and how to collect, read, and interpret the predictive results that it includes. We conclude with a brief discussion of plans for a future expansion of this database.

2 DescribePROT Database

2.1 Scope

The DescribePROT database covers 1.36 million proteins and nearly 600 million AAs from 83 complete proteomes that were picked from the UniProt's reference proteomes [59]. The primary selection criterion was to focus on popular model species, such as human, mouse, zebrafish, fruit fly, yeast, *C. elegans*, *A. thaliana*, *E. coli*, and a selection of prevalent viruses, such as influenza, measles, mumps, herpes, papilloma, corona, ebola, and HIV1. This selection also ensures that these organisms cover the four taxonomic

Table 1
Contents of the DescribePROT database

Taxonomic kingdom	Number of proteomes/ species	Number of protein sequences	Number of residue-level predictions
Eukaryotes	56	1,320,498	7,610,642,897
Bacteria	8	31,453	131,841,112
Archaea	6	13,155	48,420,892
Virus	13	840	2,793,518
Total	**83**	**1,365,946**	**7,793,698,419**

kingdoms: Eukaryota, Bacteria, Archaea, and Viruses; Table 1 provides a more detailed breakdown.

The database provides access to a broad spectrum of 10 residue-level descriptors that include four structural and six functional characteristics (Table 2). The structural characteristics include solvent accessibility, disordered linkers, secondary structure, and intrinsic disorder. The functional aspects covered by DescribePROT are sequence alignment profile that includes sequence conservation, signal peptides, and annotations of putative RNA-, DNA-, protein-, and peptide-binding residues. These data are predicted/derived from the protein sequences using a selection of ten popular and accurate methods that we list in Table 2. This table also details specific types of stored data, which include real-valued propensities and binary scores, where the latter are derived from the propensities using a threshold, for example, a given residue is predicted as disordered in binary when its putative real-valued propensity is above a specific threshold. In total, DescribePROT stores nearly 7.8 billion residue-level predictions/annotations. Moreover, for each protein, this database also includes the taxonomy identifier, protein name, UniProt's accession number, and sequence. The accession numbers facilitate linking the protein data from DescribePROT with the corresponding UniProt entries.

2.2 Accessing Data

Users can access DescribePROT using a user-friendly web interface at http://biomine.cs.vcu.edu/servers/DESCRIBEPROT/ (*see* **Note 2**). The main database page provides four options to obtain the underlying data (Fig. 1):

1. Download the contents of the entire database using the download link (middle of Fig. 1) (*see* **Note 3**).
2. Download datasets of proteins that are combined at the proteome/species level. The http://biomine.cs.vcu.edu/servers/DESCRIBEPROT/download.html page, which is linked at the top of the main server page (Fig. 1), provides a list of files

Table 2
Structural and functional characteristics included in the DescribePROT database

Type of characteristic	Name of characteristic (type of data stored)	Predictor [Refs]
Structural	Putative solvent accessibility (real-valued raw solvent accessibility; real-valued relative solvent accessibility; binary: 1 = buried, 0 = not buried)	ASAquick [61, 62]
	Putative disordered linkers (real-valued propensity; binary: 1 = part of a disordered linker, 0 = not part of a disordered linker)	DFLpred [63]
	Putative secondary structure (3-state assignment: H=Helix E=strand C=Coil, with corresponding 3 real-valued propensities)	PSIPRED [49, 64]
	Putative intrinsic disorder (real-valued propensity; binary: 1 = disordered, 0 = structured)	VSL2B [65, 66]
Functional	DNA-binding residues (real-valued propensity; binary: 1 = binding DNA, 0 = not binding DNA)	DisoRDPbind [67–69] DRNApred [70]
	RNA-binding residues (real-valued propensity; binary: 1 = binding RNA, 0 = not binding RNA)	DisoRDPbind [67–69] DRNApred [70]
	Protein-binding residues (real-valued propensity; binary: 1 = binding protein, 0 = not binding protein)	DisoRDPbind [67–69] SCRIBER [71]
	Peptide-binding residues (real-valued propensity; binary: 1 = binding peptide, 0 = not binding peptide)	MoRFchiBi [72]
	Sequence conservation and sequence alignment profile (real-valued conservation score; PSSM (position specific scoring matrix) scores for substitutions to each of the 20 AA types)	MMseqs2 [73, 74]
	Signal peptides (real-valued propensity; binary: 1 = part of a signal peptide, 0 = not part of a signal peptide)	SignalP [75, 76]

that can be downloaded (*see* **Note 3**). The species-level files are offered in two formats: raw per-residue data/predictions and tabulated per-protein summary data (*see* **Note 4**).

3. Search for data for a specific protein. This can be accomplished in three complementary ways: with the UniProt's accession number, with the UniProt's entry (protein) name, and with the protein sequence. This information should be entered using the white entry boxes; the "example" buttons provide a sample of properly formatted inputs (Fig. 1). The first two search options rely on matching the users' input against the list of names and accession numbers of proteins that are included in DescribePROT. The sequence-based option produces a sorted list of proteins in the decreasing order of their similarity to the input protein chain (Fig. 2) (*see* **Note 5**). This result is generated using sequence alignment with the popular

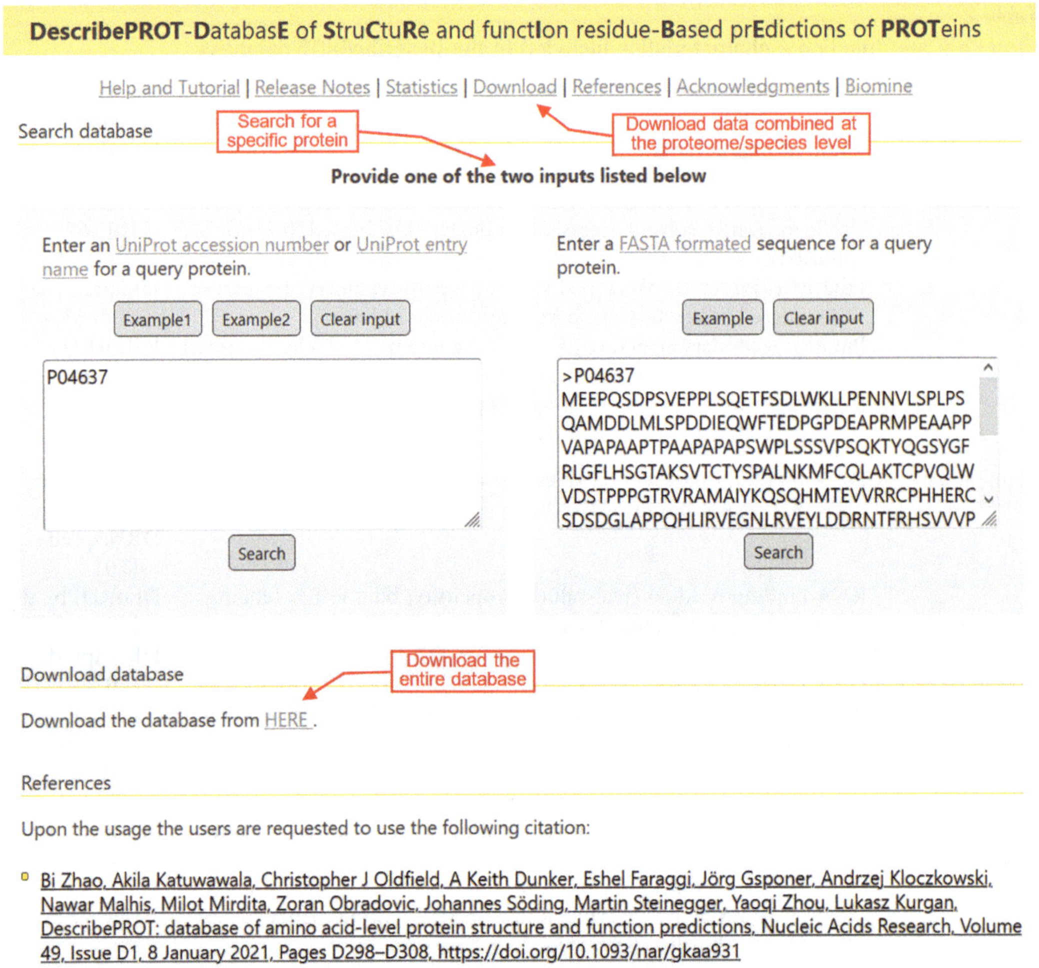

Fig. 1 Home page of the DescribePROT webserver. The web page is set up to collect data for the p53 protein (UniProt ID: P04637)

BLAST program [77, 78]. Users should click on the protein name to access the page with the results/data. Clicking on the UniProt accession number link provides access to the corresponding protein page in UniProt.

4. Directly access data for a specific protein. This can be done based on a direct link that includes the UniProt accession number. For instance, users can obtain results for the p53 protein (UniProt accession number: P04637) using the following direct link: http://biomine.cs.vcu.edu/servers/DESCRIBEPROT/result.php?uniprot=P04637. This option facilitates cross-linking of the data in DescribePROT with the other related databases.

Results for Protein Sequence Query

List of proteins that are similar to the query sequence and ranked by the E-value.

Protein Name	UniProt	E-value	Coverage	Identity	Taxonomy ID
Cellular_tumor_antigen_p53	P04637	0.0	100%	100%	9606
Cellular_tumor_antigen_p53	P56424	0.0	100%	96%	9544
Cellular_tumor_antigen_p53	G5B5D6	0.0	100%	83%	10181
Cellular_tumor_antigen_p53	Q9WUR6	0.0	100%	78%	10141
Cellular_tumor_antigen_p53	Q00366	0.0	100%	78%	10036
Cellular_tumor_antigen_p53	F1PI27	0.0	100%	81%	9615
Cellular_tumor_antigen_p53	P02340	0.0	100%	77%	10090
Cellular_tumor_antigen_p53	P10361	0.0	100%	75%	10116
Cellular_tumor_antigen_p53	P41685	0.0	100%	80%	9685
Cellular_tumor_antigen_p53	M0R497	0.0	66%	98%	10116
Cellular_tumor_antigen_p53	L7N1B2	0.0	98%	74%	59463
Cellular_tumor_antigen_p53_(Fragment)	G1PZ51	1.24e-151	99%	66%	59463
Cellular_tumor_antigen_p53	P10360	1.41e-117	91%	54%	9031
Cellular_tumor_antigen_p53	P79734	2.30e-115	78%	57%	7955

Fig. 2 DescribePROT's page listing results of the sequence similarity search. This page provides partial results for the search for the p53 protein (UniProt ID: P04637)

The last two options redirect the users to a page that provides access to the underlying data and which includes the ability to interact with a graphical visualization of the structural and functional details at the residue level.

2.3 Reading and Interpreting Data for Individual Proteins

Figure 3 shows an example page with the results for a single protein, which in this case is the human p53 (UniProt ID: P04637). The data are available in two complementary ways: as downloadable files and via an interactive graphical interface. The top of the page provides links to the downloadable files that come in two formats (Fig. 3): the comma-separated CSV format (*see* **Note 6**) and the JSON format (*see* **Note 7**). Both types of formats can be relatively easily parsed to obtain the underlying residue-level data. Besides the files, this section of the page includes the protein name and accession number, which double as hyperlinks to the corresponding UniProt record, taxonomy ID, and sequence length.

The middle section of the results page includes color-coded information that is aggregated at the protein level, for example, the overall fraction of the putative disordered residues in the sequence vs. the exact locations of the predicted disordered residues that are shown at the bottom. These results are divided into three distinct sub-parts (Fig. 3):

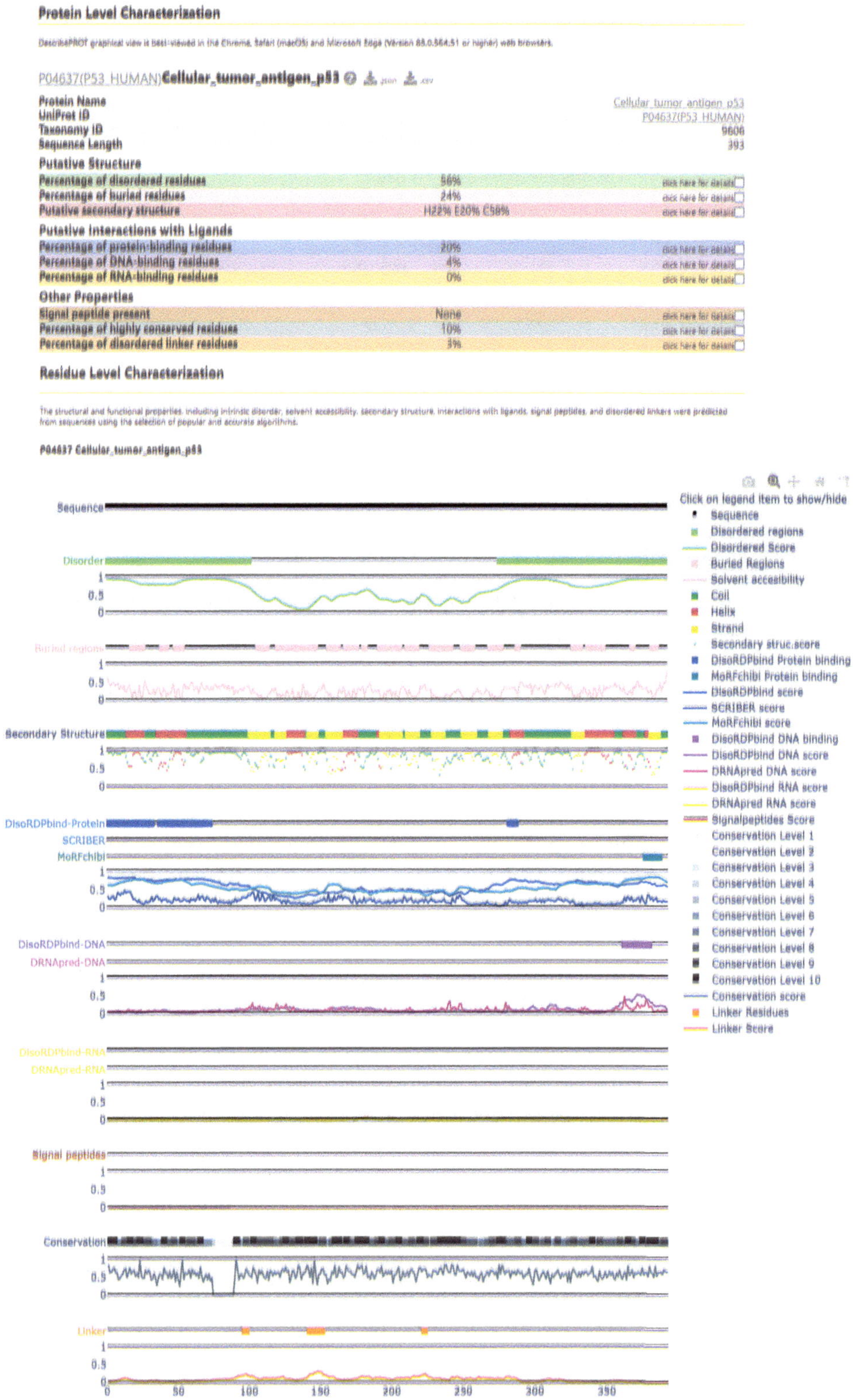

Fig. 3 Results page generated by the DescribePROT webserver. This page provides results for the p53 protein (UniProt ID: P04637)

1. "Putative structure descriptors," which include predictions of intrinsic disorder, solvent accessibility, and secondary structure
2. "Putative interactions with ligands," which cover predictions of peptide and protein-binding residues, DNA-binding residues, and RNA-binding residues
3. "Other properties," which feature sequence conservation, putative disordered linkers, and signal peptides.

The bottom part of the page provides the residue-level data (Fig. 3). These results are shown utilizing an interactive graphical format, which was implemented with the "Feature Viewer" software (DOI: 10.5281/zenodo.345324) (*see* **Note 8**). This graphical report visualizes the real-valued propensities using line plots and the binary scores (labels) using horizontal bars. The actual values of the propensities and the sequence regions covered by the binary predictions are shown on the mouse hover. The graphical plots are interactive. They can be zoomed in and out, paned (moved left and right horizontally), reset to the default view, and annotated with spike lines that provide x and y coordinates for a selected point. This is done by selecting the corresponding operator at the top of the legend on the right side (Fig. 3). Moreover, clicking on an item listed in the legend will hide/unhide it. The color coding used at the bottom panels is consistent with the protein-level data shown in the middle of the page to ease identifying the corresponding data. Furthermore, DescribePROT's website provides tutorial videos that show how to search the database and how to interact with the graphical interface. They can be found using the menu at the top of the main database page (Fig. 1).

3 Future Expansion

The DescribePROT database was initially released in 2020, and it has been continually supported since [59]. The future developments of this resource will focus on three major directions:

1. Enlarging the coverage of the included proteins, with the long-term objective to cover the complete collection of the reference proteomes from the UniProt resource [1].
2. Inclusion of experimental/native annotations of protein structure (secondary structure and solvent accessibility), intrinsic disorder, and binding based on the data available in related databases, such as PDB [79], DisProt [80], and BioLip [81].
3. Inclusion of additional functional descriptors, such as domain boundaries, post-translational modifications, and interactions with small molecule ligands. These will be predicted from protein sequences using a selection of popular, accurate, and high-throughput predictors.

The overarching aim is to further improve the utility of this resource. We note that the original version of DescribePROT was utilized in a wide range of practical applications. Selected recent examples include identification of new biomarkers for Parkinson's disease [82], elucidation of functional roles of intrinsic disorder in the human RNA-binding proteins [83], functional characterization of proteins from the *Anthonomus eugenii* proteome [84], and analysis of conserved motifs that drive attachment of the plant nucleoskeleton [85]. Data from DescribePROT were also used to develop new predictive tools, such as methods that identify putative degrons [86] and gain-of-function and loss-of-function protein variants [87].

4 Notes

1. In case if users would like to collect data for proteins outside of the protein set included in DescribePROT, they will need to utilize the individual tools at the corresponding websites:
   - ASAquick at http://mamiris.com/software.html
   - DFLpred at http://biomine.cs.vcu.edu/servers/DFLpred/
   - DisoRDPbind at http://biomine.cs.vcu.edu/servers/DisoRDPbind/
   - DRNApred at http://biomine.cs.vcu.edu/servers/DRNApred/
   - MMseqs2 at https://github.com/soedinglab/mmseqs2
   - MoRFchibi at https://gsponerlab.msl.ubc.ca/software/morf_chibi/
   - PSIPRED at http://bioinf.cs.ucl.ac.uk/psipred/
   - SCRIBER at http://biomine.cs.vcu.edu/servers/SCRIBER/
   - SignalP at https://services.healthtech.dtu.dk/services/SignalP-5.0/
   - VSL2B at http://www.pondr.com/
2. We recommend using one of the major and popular web browsers, such as Chrome, Firefox, Safari, and Microsoft Edge. This particularly facilitates the ability to work with an interactive interface of the webpage that visualizes results.
3. The aggregate data for the entire database and for the individual proteome/species are in the easy-to-parse JSON format. The size of the complete database file is about 79 Gb.
4. The per-protein tabulated file is in the comma-separated CSV format, where rows are proteins and columns provide protein-level aggregated information. Each file provides data for all

proteins from a given/selected proteome/organism. Each protein/row includes the following information: (1) fraction of putative disordered residues; (2) number of putative long disordered regions (at least 30 consecutive disordered amino acids); (3) fraction of putative helical residues; (4) fraction of putative strand residues; (5) fraction of putative coil residues; (6) fraction of residues that have low sequence conservation; (7) fraction of highly conserved residues; (8) median of residue-level conservation values; (9) presence of putative signal peptides (present = 1; not present = 0); (10) fraction of putative disordered linker residues; (11) fraction of putative buried residues; (12) fraction of putative RNA-binding residues; (13) fraction of putative DNA-binding residues; (14) fraction of putative disordered protein-binding residues; and (15) fraction of putative peptide-binding residues.

5. The protein list is by default sorted by the BLAST-produced E-values. However, users can re-sort this list by the alignment coverage values and the sequence identity values by clicking on the arrow icon at the top (Fig. 2). The list also includes the taxonomy identifier to indicate the taxonomic classification of the listed proteins.
6. The CSV file is organized in a tabular format where rows correspond to individual amino acids and columns to the underlying data, which include:
   - Amino acid type.
   - Binary putative disorder (1 = disordered, 0 – structured).
   - Putative propensity for disorder (0 to 1 range).
   - Putative three-state secondary structure (H=Helix E = strand C=Coil).
   - Putative propensity for each of the three secondary structure states (0 to 100 range).
   - Discretized residue-level conservation score calculated from position specific scoring matrix (PSSM). The range is 0 to 9 where the value corresponds to the corresponding deciles of conservation scores.
   - Conservation score calculated from PSSM.
   - PSSM scores (20 values for amino acids A, C, D, E, F, G, H, I, K, L, M, N, P, Q, R, S, T, V, W, and Y).
   - Binary annotations of signal peptide for the first 70 amino acids (1 = signal peptide, 0 = nonsignal peptide).
   - Propensity for signal peptide for the first 70 amino acids (0 to 1 range).
   - Binary putative disordered flexible linkers (1 = linker, 0 = nonlinker).

- Putative propensity for disordered flexible linkers (0 to 1 range).
- Binary putative solvent accessibility (1 = buried, 0 = nonburied).
- Putative relative solvent accessibility (0 to 1 range).
- Putative raw solvent accessibility.
- Binary putative disordered RNA binding (1 = RNA-binding, 0 = non-RNA-binding).
- Putative propensity for disordered RNA binding (0 to 1 range).
- Binary putative structure-derived RNA binding (1 = RNA-binding, 0 = non-RNA-binding).
- Putative propensity for disordered RNA binding (0 to 1 range).
- Binary putative disordered DNA binding (1 = DNA-binding, 0 = non-DNA-binding).
- Putative propensity for DNA binding (0 to 1 range).
- Binary putative structure-derived DNA binding (1 = DNA-binding, 0 = non-DNA-binding).
- Putative propensity for structure-derived DNA binding (0 to 1 range).
- Binary putative disordered protein binding (1 = protein-binding, 0 = nonprotein-binding).
- Putative propensity for disordered protein binding (0 to 1 range).
- Binary putative structure-derived protein binding (1 = protein binding, 0 = nonprotein-binding).
- Putative propensity for structure-derived protein binding (0 to 1 range).
- Binary putative disordered peptide binding (1 = peptide binding, 0 = nonpeptide binding).
- Putative propensity for disordered peptide binding (0 to 1 range).

7. The JSON file is organized using named fields that provide comma-separable propensity/conservation/solvent accessibility values and binary vectors for the binary annotations. The fields correspond to the list from **Note 6**.
8. The graphical view (bottom of Fig. 3) can be downloaded as an image in the PNG format by clicking the camera icon at the top of the menu on the right-side.

Acknowledgments

This work was supported in part by the Robert J. Mattauch Endowments funds and the National Science Foundation grants (IIS2125218 and DBI2146027) to L.K.

References

1. Uniprot C (2023) UniProt: the universal protein Knowledgebase in 2023. Nucleic Acids Res 51:D523–D531
2. Burley SK, Bhikadiya C, Bi C et al (2023) RCSB Protein Data Bank (RCSB.org): delivery of experimentally-determined PDB structures alongside one million computed structure models of proteins from artificial intelligence/machine learning. Nucleic Acids Res 51:D488–D508
3. Burley SK, Bhikadiya C, Bi C et al (2021) RCSB Protein Data Bank: powerful new tools for exploring 3D structures of biological macromolecules for basic and applied research and education in fundamental biology, biomedicine, biotechnology, bioengineering and energy sciences. Nucleic Acids Res 49:D437–D451
4. David A, Islam S, Tankhilevich E et al (2022) The AlphaFold database of protein structures: a biologist's guide. J Mol Biol 434:167336
5. Varadi M, Anyango S, Deshpande M et al (2022) AlphaFold Protein Structure Database: massively expanding the structural coverage of protein-sequence space with high-accuracy models. Nucleic Acids Res 50:D439–D444
6. Tunyasuvunakool K, Adler J, Wu Z et al (2021) Highly accurate protein structure prediction for the human proteome. Nature 596:590–596
7. Jumper J, Evans R, Pritzel A et al (2021) Highly accurate protein structure prediction with AlphaFold. Nature 596:583–589
8. Boutet E, Lieberherr D, Tognolli M et al (2016) UniProtKB/Swiss-Prot, the manually annotated section of the UniProt Knowledge-Base: how to use the entry view. Methods Mol Biol 1374:23–54
9. Rost B (2003) Prediction in 1D: secondary structure, membrane helices, and accessibility. Methods Biochem Anal 44:559–587
10. Kurgan L, Disfani FM (2011) Structural protein descriptors in 1-dimension and their sequence-based predictions. Curr Protein Pept Sci 12:470–489
11. Si J, Cui J, Cheng J et al (2015) Computational prediction of RNA-binding proteins and binding sites. Int J Mol Sci 16:26303–26317
12. Si J, Zhao R, Wu R (2015) An overview of the prediction of protein DNA-binding sites. Int J Mol Sci 16:5194–5215
13. Oldfield CJ, Chen K, Kurgan L (2019) Computational prediction of secondary and supersecondary structures from protein sequences. Methods Mol Biol 1958:73–100
14. Roche D, Brackenridge DA, Mcguffin LJ (2015) Proteins and their interacting partners: an introduction to protein-ligand binding site prediction methods. Int J Mol Sci 16:29829–29842
15. Maheshwari S, Brylinski M (2015) Predicting protein interface residues using easily accessible on-line resources. Brief Bioinform 16:1025–1034
16. Meng F, Uversky VN, Kurgan L (2017) Comprehensive review of methods for prediction of intrinsic disorder and its molecular functions. Cell Mol Life Sci 74:3069–3090
17. Jiang Q, Jin X, Lee SJ et al (2017) Protein secondary structure prediction: a survey of the state of the art. J Mol Graph Model 76:379–402
18. Katuwawala A, Peng Z, Yang J et al (2019) Computational prediction of MoRFs, short disorder-to-order transitioning protein binding regions. Comput Struct Biotechnol J 17: 454–462
19. Xie J, Ding W, Chen L et al (2015) Advances in protein contact map prediction based on machine learning. Med Chem 11:265–270
20. Lieutaud P, Ferron F, Uversky AV et al (2016) How disordered is my protein and what is its disorder for? A guide through the "dark side" of the protein universe. Intrinsically Disord Proteins 4:e1259708
21. Meng F, Uversky V, Kurgan L (2017) Computational prediction of intrinsic disorder in proteins. Curr Protoc Protein Sci 88:2 16 11–12 16 14
22. Kashani-Amin E, Tabatabaei-Malazy O, Sakhteman A et al (2018) A systematic review on popularity, application and characteristics of protein secondary structure prediction tools. Curr Drug Discov Technol 16:159–172
23. Meng F, Kurgan L (2016) Computational prediction of protein secondary structure from

sequence. Curr Protoc Protein Sci 86:2 3 1–2 3 10

24. Zhang J, Ma Z, Kurgan L (2019) Comprehensive review and empirical analysis of hallmarks of DNA-, RNA- and protein-binding residues in protein chains. Brief Bioinform 20:1250–1268
25. Zhao B, Kurgan L (2023) Machine learning for intrinsic disorder prediction. In: Machine learning in bioinformatics of protein sequences, pp 205–236
26. Basu S, Kihara D, Kurgan L (2023) Computational prediction of disordered binding regions. Comput Struct Biotechnol J 21: 1487–1497
27. Kurgan L, Li M, Li Y (2021) The methods and tools for intrinsic disorder prediction and their application to systems medicine. In: Wolkenhauer O (ed) Systems medicine. Academic Press, Oxford, pp 159–169
28. Wang K, Hu G, Wu Z et al (2020) Comprehensive survey and comparative assessment of RNA-binding residue predictions with analysis by RNA type. Int J Mol Sci 21:6879
29. Wan H, Li JM, Ding H et al (2020) An overview of computational tools of nucleic acid binding site prediction for site-specific proteins and nucleases. Protein Pept Lett 27:370–384
30. Katuwawala A, Kurgan L (2020) Comparative assessment of intrinsic disorder predictions with a focus on protein and nucleic acid-binding proteins. Biomol Ther 10:1636
31. Chen H, Li F, Wang L et al (2020) Systematic evaluation of machine learning methods for identifying human-pathogen protein-protein interactions. Brief Bioinform. https://doi.org/10.1093/bib/bbaa068
32. Barik A, Kurgan L (2020) A comprehensive overview of sequence-based protein-binding residue predictions for structured and disordered regions. In: Protein interactions, pp 33–58
33. Bao Y, Marini S, Tamura T et al (2019) Toward more accurate prediction of caspase cleavage sites: a comprehensive review of current methods, tools and features. Brief Bioinform 20: 1669–1684
34. Zhao B, Kurgan L (2021) Surveying over 100 predictors of intrinsic disorder in proteins. Expert Rev Proteomics 18:1019–1029
35. Li M, Zhang F, Kurgan L (2022) Machine learning methods for predicting protein-nucleic acids interactions. In: Machine learning in bioinformatics of protein sequences, pp 265–287
36. Kurgan L (2022) Resources for computational prediction of intrinsic disorder in proteins. Methods 204:132–141
37. Zhang J, Ghadermarzi S, Kurgan L (2020) Prediction of protein-binding residues: dichotomy of sequence-based methods developed using structured complexes versus disordered proteins. Bioinformatics 36:4729–4738
38. Katuwawala A, Ghadermarzi S, Kurgan L (2019) Computational prediction of functions of intrinsically disordered regions. Prog Mol Biol Transl Sci 166:341–369
39. Zhao B, Kurgan L (2022) Deep learning in prediction of intrinsic disorder in proteins. Comput Struct Biotechnol J 20:1286–1294
40. Zhang H, Zhang T, Chen K et al (2011) Critical assessment of high-throughput standalone methods for secondary structure prediction. Brief Bioinform 12:672–688
41. Zhang J, Kurgan L (2018) Review and comparative assessment of sequence-based predictors of protein-binding residues. Brief Bioinform 19:821–837
42. Yan J, Friedrich S, Kurgan L (2016) A comprehensive comparative review of sequence-based predictors of DNA- and RNA-binding residues. Brief Bioinform 17:88–105
43. Miao Z, Westhof E (2015) A large-scale assessment of nucleic acids binding site prediction programs. PLoS Comput Biol 11:e1004639
44. Necci M, Piovesan D, Predictors C et al (2021) Critical assessment of protein intrinsic disorder prediction. Nat Methods 18:472–481
45. Ruiz-Serra V, Pontes C, Milanetti E et al (2021) Assessing the accuracy of contact and distance predictions in CASP14. Proteins 89: 1888–1900
46. Gallo Cassarino T, Bordoli L, Schwede T (2014) Assessment of ligand binding site predictions in CASP10. Proteins 82(Suppl 2): 154–163
47. Katuwawala A, Oldfield CJ, Kurgan L (2020) Accuracy of protein-level disorder predictions. Brief Bioinform 21:1509–1522
48. Petersen TN, Brunak S, Von Heijne G et al (2011) SignalP 4.0: discriminating signal peptides from transmembrane regions. Nat Methods 8:785–786
49. Jones DT (1999) Protein secondary structure prediction based on position-specific scoring matrices. J Mol Biol 292:195–202
50. Dosztányi Z, Csizmok V, Tompa P et al (2005) IUPred: web server for the prediction of intrinsically unstructured regions of proteins based on estimated energy content. Bioinformatics 21:3433–3434

51. Buchan DWA, Jones DT (2019) The PSIPRED protein analysis workbench: 20 years on. Nucleic Acids Res 47:W402–W407
52. Buchan DWA, Minneci F, Nugent TCO et al (2013) Scalable web services for the PSIPRED protein analysis workbench. Nucleic Acids Res 41:W349–W357
53. Yachdav G, Kloppmann E, Kajan L et al (2014) PredictProtein—an open resource for online prediction of protein structural and functional features. Nucleic Acids Res 42:W337–W343
54. Bernhofer M, Dallago C, Karl T et al (2021) PredictProtein – predicting protein structure and function for 29 years. Nucleic Acids Res 49:W535–W540
55. Cheng J, Li J, Wang Z et al (2012) The MULTICOM toolbox for protein structure prediction. BMC Bioinf 13:65
56. Hou J, Wu T, Guo Z et al (2020) The MULTICOM protein structure prediction server empowered by deep learning and contact distance prediction. Methods Mol Biol 2165:13–26
57. Barik A, Katuwawala A, Hanson J et al (2020) DEPICTER: intrinsic disorder and disorder function prediction server. J Mol Biol 432:3379–3387
58. Basu S, Gsponer J, Kurgan L (2023) DEPICTER2: a comprehensive webserver for intrinsic disorder and disorder function prediction. Nucleic Acids Res. https://doi.org/10.1093/nar/gkad330
59. Zhao B, Katuwawala A, Oldfield CJ et al (2021) DescribePROT: database of amino acid-level protein structure and function predictions. Nucleic Acids Res 49:D298–D308
60. Zhao B, Kurgan L (2023) Databases of protein structure and function predictions at the amino acid level. In: Machine learning in bioinformatics of protein sequences, pp 329–353
61. Faraggi E, Zhou Y, Kloczkowski A (2014) Accurate single-sequence prediction of solvent accessible surface area using local and global features. Proteins 82:3170–3176
62. Faraggi E, Kouza M, Zhou Y et al (2017) Fast and accurate accessible surface area prediction without a sequence profile. Methods Mol Biol 1484:127–136
63. Meng F, Kurgan L (2016) DFLpred: high-throughput prediction of disordered flexible linker regions in protein sequences. Bioinformatics 32:i341–i350
64. Mcguffin LJ, Bryson K, Jones DT (2000) The PSIPRED protein structure prediction server. Bioinformatics 16:404–405
65. Obradovic Z, Peng K, Vucetic S et al (2005) Exploiting heterogeneous sequence properties improves prediction of protein disorder. Proteins 61(Suppl 7):176–182
66. Peng K, Radivojac P, Vucetic S et al (2006) Length-dependent prediction of protein intrinsic disorder. BMC Bioinf 7:208
67. Peng Z, Kurgan L (2015) High-throughput prediction of RNA, DNA and protein binding regions mediated by intrinsic disorder. Nucleic Acids Res 43:e121
68. Peng Z, Wang C, Uversky VN et al (2017) Prediction of disordered RNA, DNA, and protein binding regions using DisoRDPbind. Methods Mol Biol 1484:187–203
69. Oldfield CJ, Peng Z, Kurgan L (2020) Disordered RNA-binding region prediction with DisoRDPbind. Methods Mol Biol 2106:225–239
70. Yan J, Kurgan L (2017) DRNApred, fast sequence-based method that accurately predicts and discriminates DNA- and RNA-binding residues. Nucleic Acids Res 45:e84
71. Zhang J, Kurgan L (2019) SCRIBER: accurate and partner type-specific prediction of protein-binding residues from proteins sequences. Bioinformatics 35:i343–i353
72. Malhis N, Jacobson M, Gsponer J (2016) MoRFchibi SYSTEM: software tools for the identification of MoRFs in protein sequences. Nucleic Acids Res 44:W488–W493
73. Mirdita M, Steinegger M, Soding J (2019) MMseqs2 desktop and local web server app for fast, interactive sequence searches. Bioinformatics 35:2856–2858
74. Steinegger M, Soding J (2017) MMseqs2 enables sensitive protein sequence searching for the analysis of massive data sets. Nat Biotechnol 35:1026–1028
75. Almagro Armenteros JJ, Tsirigos KD, Sonderby CK et al (2019) SignalP 5.0 improves signal peptide predictions using deep neural networks. Nat Biotechnol 37:420–423
76. Nielsen H (2017) Predicting secretory proteins with SignalP. Methods Mol Biol 1611:59–73
77. Camacho C, Coulouris G, Avagyan V et al (2009) BLAST+: architecture and applications. BMC Bioinf 10:421
78. Hu G, Kurgan L (2019) Sequence similarity searching. Curr Protoc Protein Sci 95:e71
79. Wwpdb Consortium (2019) Protein Data Bank: the single global archive for 3D macromolecular structure data. Nucleic Acids Res 47:D520–D528

80. Hatos A, Hajdu-Soltesz B, Monzon AM et al (2020) DisProt: intrinsic protein disorder annotation in 2020. Nucleic Acids Res 48: D269–D276
81. Yang J, Roy A, Zhang Y (2013) BioLiP: a semi-manually curated database for biologically relevant ligand-protein interactions. Nucleic Acids Res 41:D1096–D1103
82. Mackmull MT, Nagel L, Sesterhenn F et al (2022) Global, in situ analysis of the structural proteome in individuals with Parkinson's disease to identify a new class of biomarker. Nat Struct Mol Biol 29:978–989
83. Zhao B, Katuwawala A, Oldfield CJ et al (2021) Intrinsic Disorder in Human RNA-Binding Proteins. J Mol Biol 433: 167229
84. Lechuga-Paredes P, Segura-Leon OL, Cibrian-Tovar J et al (2023) Odorant-binding and chemosensory proteins in Anthonomus eugenii (Coleoptera: Curculionidae) and their tissue expression. Int J Mol Sci 24:3406
85. Mermet S, Voisin M, Mordier J et al (2021) Evolutionary conserved protein motifs drive attachment of the plant nucleoskeleton at nuclear pores. bioRxiv:2021.2003.2020.435662
86. Hou C, Li Y, Wang M et al (2022) Systematic prediction of degrons and E3 ubiquitin ligase binding via deep learning. BMC Biol 20:162
87. Stein D, Bayrak ÇS, Wu Y et al (2022) Genome-wide prediction of pathogenic gain- and loss-of-function variants from ensemble learning of a diverse feature set. bioRxiv:2022.2006.2008.495288

Chapter 11

Structural and Functional Studies on HIV Protease: Mechanism of Action, Subtypes, Inhibitors, and Drug Resistance

Sankaran Venkatachalam, Sowmya Ramaswamy Krishnan, Yasien Sayed, and M. Michael Gromiha

Abstract

Human immunodeficiency virus (HIV) targets the host immune system causing acquired immunodeficiency syndrome (AIDS). Although significant advancements have been made on investigating HIV and related infections, eradicating the virus from the host immune system is still challenging. Nevertheless, the combination therapies using drugs targeting different stages in the viral life cycle are used for treatment in which HIV protease plays a vital role. Hence, it is essential to understand the structure and function of HIV protease. This review focuses on these aspects from different perspectives such as catalytic mechanism, subtypes and role of flaps in drug binding. Further, we highlight the factors affecting drug binding, evolution of drug resistance, and inhibitors reported in the literature using 3D QSAR studies.

Key words HIV protease, Structure and function, Catalytic mechanism, QSAR, Inhibitors

1 Introduction

Acquired immunodeficiency syndrome (AIDS) is an infectious disease caused by the human immunodeficiency virus (HIV). It affects the host immune system by targeting the CD4+ T lymphocytes [1, 2]. Although the first case of AIDS was reported in 1981, extensive understanding of the disease including its epidemiology, mode of transmission, and its impact on the host immune system was reported only in the late 1980s [3]. Despite the long history of AIDS, HIV infection continues to be one of the serious threats to the current global public healthcare systems and importantly to the humankind. This is because the HIV outbreak affects various population from distinct origins with different frequency and behavior [4]. According to a survey from the Joint United Nations Program on HIV/AIDS (UNAIDS), close to 40 million people live with

Andrzej Kloczkowski et al. (eds.), *Prediction of Protein Secondary Structure*, Methods in Molecular Biology, vol. 2867, https://doi.org/10.1007/978-1-0716-4196-5_11,

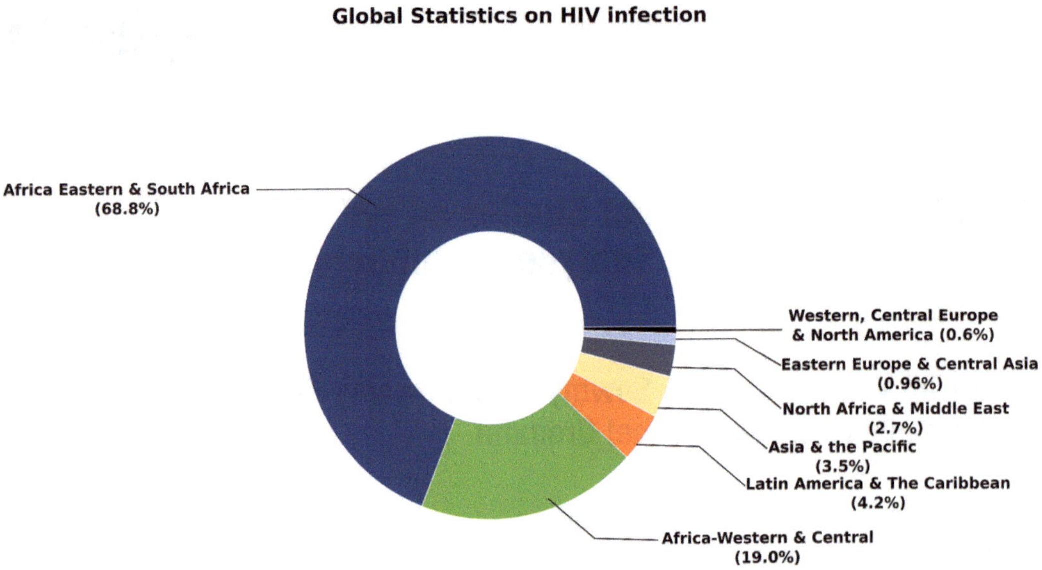

Fig. 1 Infographics on worldwide infection of HIV as presented by UNAIDS

HIV, of which 20.2 million belong to the female population. Further, it is reported that close to 650,000 people died from HIV and related causes in 2021 [5]. Figure 1 shows the infographics on HIV infections across the world. It can be seen that the eastern and southern African regions contribute to ~68.8% of the global HIV infections, making them the most severely affected regions of the world, followed by the western and central African regions accounting for ~19% of the infections.

HIV belongs to the genus *Lentivirus* and family *Retroviridae* and subfamily *Orthoretrovirinae* [6]. Based on the viral antigens and the genetic characteristics, HIV is classified into two types, namely, HIV-1 and HIV-2, where the evolutionary origin of the former is the non-human primate immunodeficiency virus from Central African chimpanzees and the latter is from West African sooty mangabeys. One significant difference observed in these variants is the transmissibility and the likelihood of the disease progression, where it is higher in HIV-1 than in HIV-2. Further, HIV-2 infections are largely confined to the regions of West Africa, whereas the impact of HIV-1 infections is worldwide [7]. Due to this reason, HIV-1 remains more well-studied than HIV-2.

HIV-1 is a retrovirus, i.e., it has the ability to do reverse transcription from DNA to RNA using the Reverse Transcriptase (RT) enzyme. It comprises two identical copies of single-stranded RNA molecules encapsulated within the viral core. The propagation of HIV genome into the host cell occurs by a series of steps, such as the reverse transcription of the viral RNA genome into DNA, followed by RNA degradation and integration of the double-stranded HIV DNA with the human genome. The

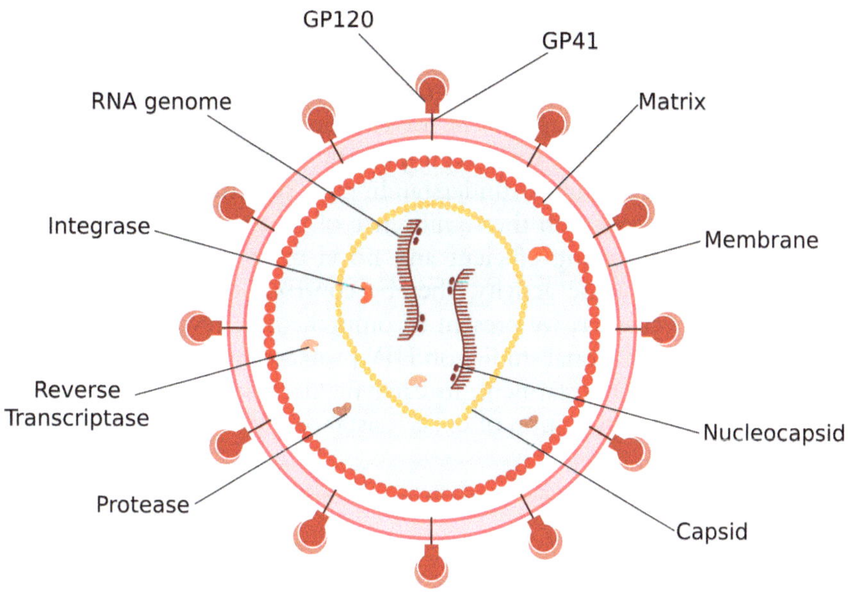

Fig. 2 Structure of HIV

diagrammatic representation of the structure of HIV is shown in Fig. 2. HIV, when matured, is round in shape and measures approximately 100 nm in diameter. The structure of HIV is composed of three major parts, namely, the outer lipid membrane, the viral matrix, and the viral core. The outer lipid membrane acts as an envelope for the viral matrix. The envelope consists of 72 knobs composed of trimers of the envelope (env) proteins. GP120, a surface protein, is anchored to the outer lipid membrane via the transmembrane protein GP41.

The entry of the virus into the host cell is carried out by the binding of the surface envelope proteins to the host CD4 receptor followed by the binding of co-receptor, most commonly CCR5 and CXCR4. Binding to the co-receptor induces conformational changes in the env protein [8, 9]. Upon entering the host cell, the virus replicates to infect the host immune system, causing AIDS. In spite of the advancements in the therapeutics and intense research on HIV and related infections [10], strategies to completely eradicate HIV infection still remain as a challenge. Modern-day anti-retroviral therapies and drugs are designed to primarily target various stages of the viral life cycle including cell fusion, entry into the membrane, and activity of the three important viral enzymes, namely, protease, reverse transcriptase, and integrase [11]. Currently, a strategy called highly active antiretroviral therapy (HAART), a combination therapy, is widely used to treat HIV and related infections. HAART uses a combination of the inhibitors targeting the HIV life cycle [12, 13]. However, the disadvantage is that they are not curative and the patient has to take these therapies throughout their lifetime [14]. Nevertheless, the combination of protease inhibitors with reverse transcriptase inhibitors and integrase inhibitors has been proven to be the most effective treatment for HIV infection. Among these, the protease inhibitors form an important constituent of the HAART therapy [15]. Hence, understanding the structural nuances of HIV protease along with the significance of protease inhibitors is important for designing efficient and novel drug molecules for mitigating the protease activity, thereby controlling the HIV infection. In this chapter, we present a comprehensive review of the structural and functional studies on HIV protease, including its significance in the AIDS treatment, its catalytic mechanism, subtypes, inhibitors, and the evolution of drug resistance to protease inhibitors.

2 HIV Protease

HIV protease (HIV PR) is crucial for producing matured infectious viral particles. It is a 99-residue protein (60% hydrophobic), whose gene is located immediately upstream to the reverse transcriptase (RT) gene in the HIV genome (Fig. 3a) [16]. The protease is translated through ribosomal frameshifting that occurs in the 5′ end of the *gag* and *pol* overlap region in the *pol* open reading frame [17]. HIV protease has a molecular weight and molecular size of 10,792 and 11.5 kDa, respectively [16]. The protease cleaves the gag and gag-pol polyproteins at a minimum of nine distinct sites, enabling the production of mature viral proteins [18, 19]. Due to this reason, HIV protease is recognized as one of the potent targets for treating HIV infections.

2.1 HIV Protease Subgroups

The diversity in the geographic locations and numerous factors give rise to different subgroups in HIV-1. Nevertheless, majority of the infections are caused by the HIV virus belonging to the group M (Major group), accounting for ~90% of the infections worldwide. Further, the HIV-1 group M includes some prominent subtypes, namely, subtypes B, F, C, and D, and other recombinant forms. In general, the sequence level variation between these subtypes ranges from 25 to 30%, and the sequence level variations within the subtypes range from 15% to 20%. The sequence alignment of subtypes C, F, and B is shown in Fig. 4. However, these sequence level variations do not have a great impact on the structure of HIV-1 protease [20–22]. The detailed description of the HIV-1 protease structure and its catalytic mechanism is discussed in Subheadings 2.2 and 2.3.

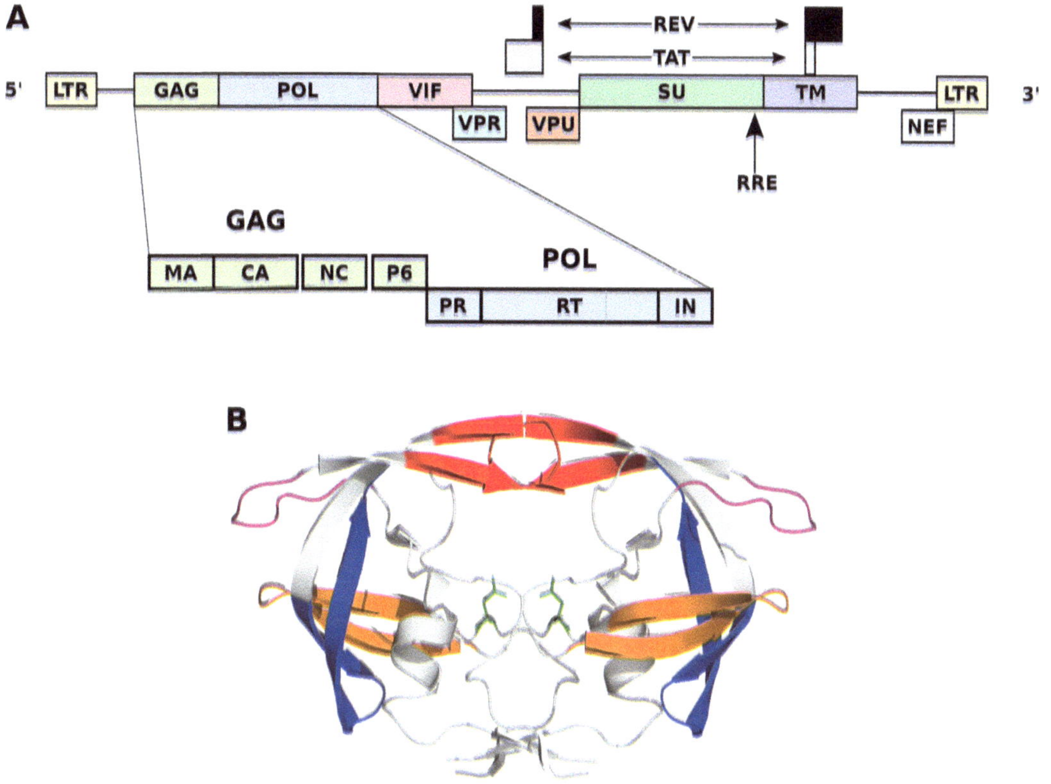

Fig. 3 (**a**) Structural organization of HIV-1 genome. LTR long terminal repeat, gag group-specific antigen, pol polymerase, env envelope, SU surface protein, TM transmembrane protein, *Tat and rev* are gene regulators. Tat transcription protein, rev RNA splicing regulator, nef negative regulating factor, vif viral infectivity protein, vpr virus protein r, vpu virus protein unique. (**b**) 3D structure of HIV protease highlighting flaps (red), hinge (magenta), cantilever (blue), and fulcrum (blue). Asp25/Asp25′ shown in sticks

2.2 Structure of HIV Protease

Various studies in the literature reported the genetic location, structure of HIV-1 PR and its cleavage sites [16, 19, 23, 24]. The first crystal structure of HIV-1 PR was reported in the year 1989 [25–27], and the 3D structure of HIV-1 PR is shown in Fig. 3b. HIV-1 PR is a homodimeric structure with the active site present in the dimer interface. As in other lentivirus proteases, HIV-1 PR belongs to the class of aspartic proteases characterized by the presence of the conserved catalytic triad, Asp-Thr/Ser-Gly. Each monomeric subunit of the dimer contains one copy of the catalytic triad. These active sites help PR to recognize the substrate at the different cleavage sites in the gag and gag-pol polyproteins [16, 25, 28]. Apart from the active site, each subunit of the homodimer includes a number of other functionally important regions. This includes the hinge (residues 35–42), flap (residues 46–54), fulcrum (residues 10–22), and cantilever (residues 62–78). Each of these

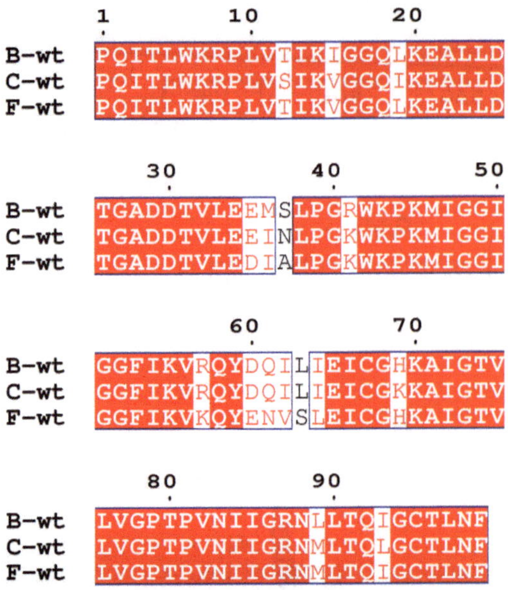

Fig. 4 Sequence alignment of the HIV-1 protease subtypes B, C, and F. Mature wild-type PR often undergoes rapid self-degradation due to the presence of sites that are sensitive to PR autoproteolysis. To enhance the stability, several modifications to the PR sequences have been carried out. One such modification includes Q7K substitution. Sequence containing K at the seventh position is used for the alignment

regions has been validated/identified/established to play a significant role in driving the catalytic mechanism of HIV-1 PR [29].

2.3 Catalytic Mechanism of HIV-1 Protease and Its Specificity

HIV-1 PR, when bound to the substrate, catalyzes the hydrolysis of the peptide bond via nucleophilic attack by the active site water molecule at the carbonyl group of the scissile amide bond [30]. It adopts a closed conformation, where the substrate residues from the gag and gag-pol polyproteins establish interactions with the active site residues. Further, the protease cleaves the substrate at multiple sites enabling the maturation of the viral polyproteins. The first cleavage happens in the C-terminal followed by subsequent cleavages at multiple sites. Upon cleavage, the flaps move away from the active site adopting a free open conformation enabling the product release [31].

HIV-1 PR recognizes its substrate based on their asymmetrical shape rather than the specific amino acid sequence. Hence, despite the differences in the sequences, the protease can cleave its substrate, that fit the active site. Earlier, a study on HIV-1 PR-substrate complexes [32] elucidated the important factors that are responsible for the HIV-1 PR substrate specificity. The study showed that the substrate specificity of the protease depends on four important factors including the conserved hydrogen bonds, substrate-specific hydrogen bonds, shape complementarity, and interaction with water molecules. Since the substrate peptides bind primarily to the beta strand conformation of the protease, the hydrogen bonds are formed between protein and the substrate backbones. Since majority of the conserved hydrogen bonds are formed by the backbone residues, the protease is able to recognize and cleave substrates at different cleavage sites, i.e., it is independent of the substrate sequence. There were eight completely conserved hydrogen bonds (with HIV-1 PR residues Asn25/25′, Gly27/27′, Asp 29/29′, Gly48/48′) and six partially conserved hydrogen bonds. Further, hydrogen bonds that are specific/unique to the substrate of interest were also observed. Besides the conserved and substrate-specific hydrogen bonds, the shape complementarity of the substrates, i.e., the extent to which the solvent-accessible surface area (SASA) changes when the protease binds to the substrate, is said to be less specific, yet important for achieving the specificity. Similar to the conserved hydrogen bonds between the protease and the substrate, the conserved hydrogen bonds between the protease backbone and the water in the active site are also involved in substrate recognition. Further, these water molecules are claimed to be important for product release upon substrate cleavage [32].

2.4 Flap Residues: A Gateway for Entry and Exit

Next to the active site of the HIV-1 PR, the movement of flaps is crucial for enzymatic activity [33, 34]. In HIV-1 PR, the flap constitutes the beta-hairpin structure that forms the part of substrate binding [34]. They act as gates to the active site by governing the entry and exit of the ligand. Upon binding the substrate, the flap region closes owing to the catalytic activity and opens again to release the product [26, 33]. Thus, the opening and closing of flaps are critical for the HIV-1 PR mechanism. A study [35] showed that the transition between the closed and the open conformations is governed by two crucial intermediate states, namely, the curled and the semi-open states. During the transition from closed to curled conformation, the hydrophobic interaction between the two flap regions is reduced, which triggers the opening of flaps. Further, the transition from curled to semi-open conformation involves upward motion of flaps. Meanwhile, the curled flap returns back to the uncurled conformation. Transition to the fully open state from the semi-open conformation requires the breaking of hydrophobic interactions between flaps, making the structure fully open, which is thermodynamically unfavorable and kinetically inaccessible. The transition between states observed in the crystal structure of HIV-1 PR type B is shown in Fig. 5.

The motion and stability of the flap regions are said to be highly dependent on the conformational flexibility of the hinge regions [36]. Further, the movement of the flap is said to be the result of concerted movements of the hinge, fulcrum, and cantilever regions. For example, the curling of the cantilever tip residues

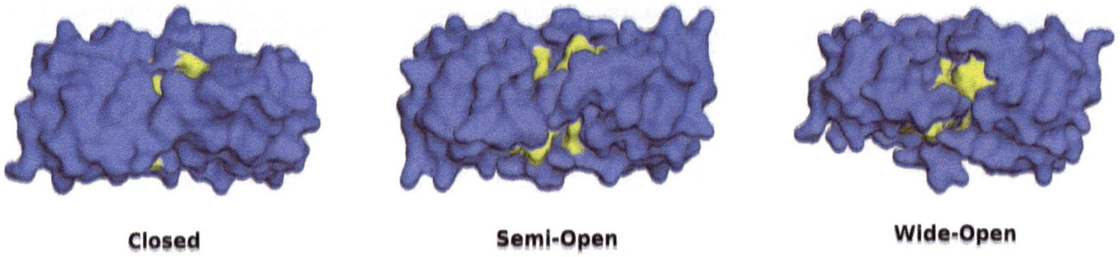

Fig. 5 Crystal structure of three different states of the HIV-1 subtype B protease, namely, closed (PDB ID: 1G6L), semi-open (PDB ID: 3HVP), and wide-open (PDB ID: 2PCO) states (left to right). The figure illustrates the stages of flap opening in the HIV protease and revealing access to the active site. The HIV PR is shown in surface representation, and the active site residues are colored yellow. Water molecules are not shown for clarity. (Figure was created using *Inkscape and PyMol*)

(Ile66/66′, Cys67/67′, Gly68/68′) is inversely correlated with the distance between active site residues and flap tips. The flaps adopt a semi-open conformation when the cantilever tip C_α angle is ≤ 100 (curled in conformation) and vice versa [29]. Similarly, the hinge of the protease significantly contributes to its dynamics [37, 38], and the activity of the protease depends on its conformational dynamics, which in turn depends on its functionally important sites (mentioned above). The perturbation to the HIV-1 PR conformational dynamics has a significant impact on its activity. Hence, restricting the conformational dynamics of HIV-1 PR will lead to loss of function, thereby mitigating the viral replication. However, the perturbations to such restrictions can arise through mutations, and they often lead to the emergence of drug resistance. In the next section, we will discuss about the evolution of resistance in the HIV-1 PR.

2.5 Drug Resistance in HIV Protease

The emergence of drug resistance is a serious problem in HIV treatment [39]. Due to the high frequency of recombination and high mutation rate of the viral genome, HIV can quickly develop resistance to protease inhibitors (PIs) [40, 41]. As a consequence, resistance to almost all the PIs has been reported at a global scale [42]. Currently, ten Food and Drug Administration (FDA)-approved HIV protease inhibitors are available. Figures 6 and 7 show the chemical structure of the ten FDA-approved drugs in 2D and 3D space with hydrogen bond interactions at the active site, respectively. One of the important reasons for the emergence of resistant HIV strains is the evolution of viral quasispecies [15]. Viral quasispecies is defined as the emergence of a massive number of genetically distinct viral variants due to factors such as absence of effective treatment in the infected patients, lack of proofreading by the viral reverse transcriptase, and continuous levels of high replication and recombination [43]. The genetic flexibility of the viral quasispecies together with the selection pressure give rise to the evolution of drug-resistant HIV variants [15]. Interestingly, it was

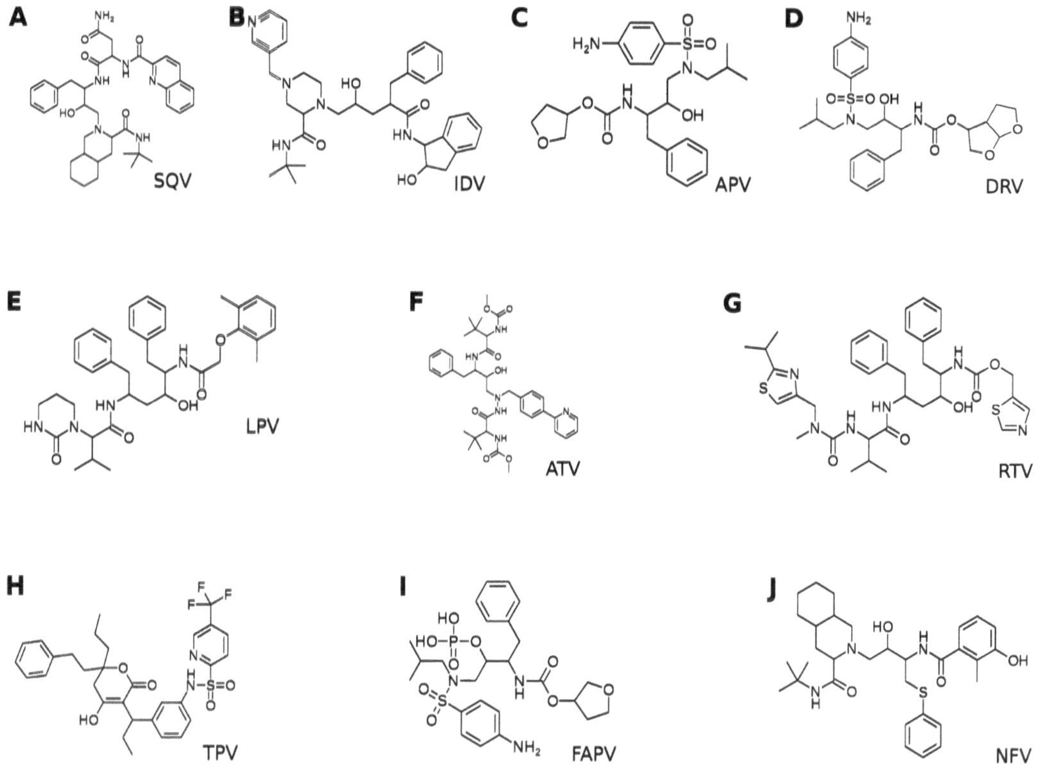

Fig. 6 2D structure of ten FDA-approved protease inhibitors. (**a**) Saquinavir (SQV), (**b**) Indinavir (IDV), (**c**) Amprenavir (APV), (**d**) Darunavir (DRV), (**e**) Lopinavir (LPV), (**f**) Atazanavir (ATZ), (**g**) Ritonavir (RTV), (**h**) Tipranavir (TPV), (**i**) Fosamprenavir (FAPV), (**j**) Nelfinavir (NFV). (Figure was created using *Inkscape and JChemPaint*)

observed that the average number of HIV variants that are involved in the HIV infection is relatively smaller than the population of variants that produce non-infectious viral particles [44, 45].

Focusing on the HIV-1 PR, it confers resistance in a step-wise manner, primed by the amino acid substitutions in the binding site. These mutations, termed as the "primary" or "major" mutations [46], result in overall enlargement of the protease active site. This expansion leads to a decrease in susceptibility to the protease inhibitors. In some cases, the primary mutations can also lead to a decrease in specificity for the natural substrate, leading to decrease in viral replication. The second stage of resistance in the HIV-1 PR is the secondary mutations or the minor mutations. They are observed away from the protease active site [47–49]. For example, the mutation M36I at the hinge region disrupts the formation of a salt bridge between the residues Glu35 and Arg57 in the wild-type HIV-1 PR, making the hinge more flexible. The increase in the flexibility of the hinge increases the flexibility of the flaps gating the binding site, thereby weakening the interaction of the drug with the protease [50]. Apart from the amino acid substitutions,

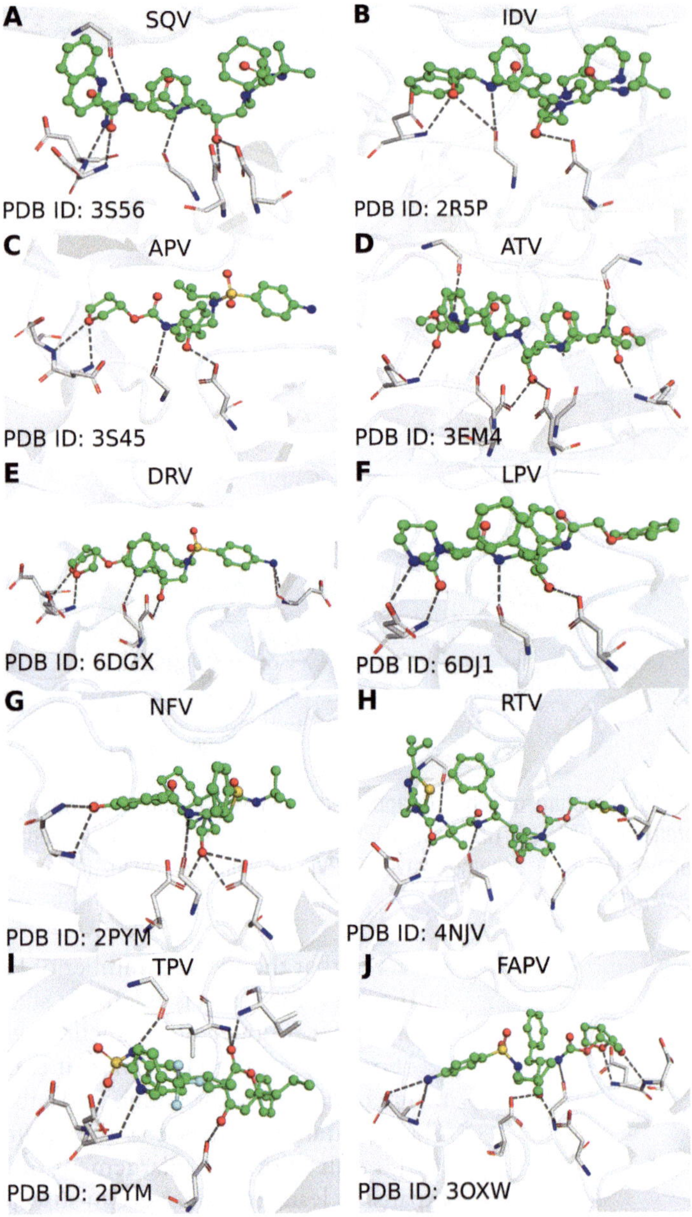

Fig. 7 Representative 3D images showing the hydrogen bond interactions of ten FDA-approved drugs with the active site residues. 2D structure of ten FDA-approved protease inhibitors. (**a**) Saquinavir (SQV), (**b**) Indinavir (IDV), (**c**) Amprenavir (APV), (**d**) Atazanavir (ATZ), (**e**) Darunavir (DRV), (**f**) Lopinavir (LPV), (**g**) Nelfinavir (NFV), (**h**) Ritonavir (RTV), (**i**) Tipranavir (TPV), (**j**) Fosamprenavir (FAPV). (Figure was created using *Inkscape and PyMol*)

Table 1
List of FDA-approved drugs for HIV protease and their corresponding primary mutations

FDA-approved drugs	Major or primary mutation
Atazanavir	I84V, N88S
Darunavir	I47V, I50V, I54M/L, L76V, I84V
Fosamprenavir	I50V, I84V
Indinavir	M46I/L, V82A/F/T, I84V
Lopinavir	V32I, I47V/A, L76V, V82A/F/T/S
Nelfinavir	D30N, L90M
Ritonavir	V82A/F/T/S, I84V
Saquinavir	G48V, L90M
Tipranavir	I47V, Q58E, T74P V82L/T, N83D, 84 V

Data are adapted from [67]

insertion of residues was also observed [51, 52]. Residues positioned between 32 and 42 are reported to be more prone to insertions. In our previous study [52], we reported that the insertion of the residues in the hinge region of the HIV-1 PR subtype C leads to the expansion of the binding site and decreases drug–protease interactions. The list of major mutations conferring resistance to the drug molecules is shown in Table 1. Despite the substitutions in the protease, amino acid substitutions were also observed in the substrate of the viral protease, i.e., in the gag protein. They are thought to adapt the virus through alteration of the active site of the mutant protease, resulting in drug resistance [53]. Numerous databases and computational tools have been developed to facilitate the interpretation of drug-resistant mutations [42]. Nevertheless, selection and design of novel inhibitors are adequate for the treatment of AIDS and related infections. In this context, structure-based drug design has proven to be an effective method for designing novel and adequate anti-retroviral inhibitors [54, 55].

2.6 Application of 3D-QSAR to Study HIV Protease–Inhibitor Interactions

Three-dimensional quantitative structure–activity relationship (3D-QSAR) studies have significantly shaped the drug design process [56]. It uses interdisciplinary approaches such as supervised machine learning techniques and multidimensional data analytics to establish a relationship between target molecular properties with computationally derived molecular chemical structures by permitting the prediction of molecular features in space [56–58]. Aside from designing completely novel inhibitors, 3D-QSAR has been used to improve the efficiency of the existing compounds aiding the

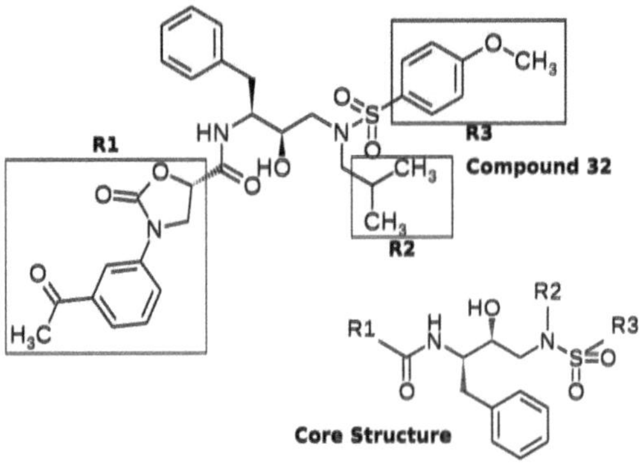

Fig. 8 2D structure of the Compound 32 and its core scaffold as given in the study [59]. (Figure was created using *Inkscape and JChemPaint*)

design of novel derivatives [59]. In recent years, computational studies such as docking, molecular dynamics (MD), and 3D-QSAR have been combined to design new derivatives of HIV-PIs [54, 60].

Ul-Haq et al. [61] used a dataset of 102 darunavir (1 of the 10 FDA-approved second-generation protease inhibitors) derivatives to predict novel HIV-PIs using 3D-QSAR models based on the predicted IC_{50} (pIC_{50}) values. Further, the binding affinities of the modelled structures computed using docking studies revealed that compound 32 (Fig. 8) was most active against the protease binding site. The methoxybenzene group present at the R3 position and the core scaffold of the compound established hydrophobic interactions with Pro81 and Ile50, respectively. Further, the carbonyl oxygen near the R1 position formed a strong hydrogen bond interaction with the Asp29′ and Asp30′ backbone residues. Similarly, other studies based on several HIV-PI derivatives, namely, 3(S)-amino-2-(S)-hydroxyl-4-phenylbutanic acid [62], fullerene [63], chromone [64], cyclic urea [65], and hydroxyethylamines [66], are also reported. One such recent 3D-QSAR study was conducted by Tong et al. [65], where they have used a set of 34 cyclic urea derivatives to investigate the relationship between the inhibitor–protease binding. CoMSIa-based 3D-QSAR method has been adopted to determine the significance of five different properties, namely, electrostatic, steric, hydrophobic, and hydrogen bond donor and acceptor groups of cyclic urea derivatives on HIV inhibitory activity. In addition, the 3D-QSAR results suggested that the steric interaction on R/R0 groups and the electronegativity on R/P2 groups may enhance the inhibitory activity of the enzyme. Further, hydrogen bonding between the hydroxyl group of the ligand and the carboxylate group of Asp25 in the protease was found to be important for its activity.

3 Conclusions

In this review, we focus on the structure, mechanism of action, importance of flaps in drug binding, and evolution of drug resistance in the HIV protease. Specifically, extensive details are provided on (i) the importance of HIV protease as drug target, (ii) different subgroups and structure of HIV protease, (iii) catalytic mechanism and the influence of the flap regions in drug binding, (iv) emergence of drug resistance in HIV protease, and (v) 3D-QSAR studies that are aimed at developing novel and effective drugs for mitigating HIV infections. Taken together, this information will be helpful in gaining deep insights when designing therapeutic strategies.

Acknowledgments

We thank the Department of Biotechnology and Indian Institute of Technology Madras for computational facilities. The work is partially supported by the Department of Science and Technology, Government of India, under BRICS project (DST/ICD/BRICS/Call-3/HIV protease/2019) to MMG, South African National Research Foundation (NRF), via the BRICS multilateral joint science and technology research collaboration (grant number 120452) to YS and Conselho Nacional de Desenvolvimento Científico e Tecnológico (CNPq), Brazil (grant number 402817/2019-2) to VA.

References

1. Barré-Sinoussi F, Chermann JC, Rey F et al (1983) Isolation of a T-lymphotropic retrovirus from a patient at risk for acquired immune deficiency syndrome (AIDS). Science 220(4599):868–871
2. Gallo RC, Sarin PS, Gelmann EP, Robert-Guroff M et al (1983) Isolation of human T-cell leukemia virus in acquired immune deficiency syndrome (AIDS). Science 220(4599): 865–867
3. Sharp PM, Hahn BH (2011) Origins of HIV and the AIDS pandemic. Cold Spring Harb Perspect Med 1(1):a006841
4. Haacker M (2004) The demographic impact of HIV/AIDS. In: The macroeconomics of HIV/AIDSUSA, International Monetary Fund. Available via DIALOG. https://www.elibrary.imf.org/display/book/9781589063600/ch01.xml. Accessed 17 May 2024
5. Joint United Nations Programme on HIV/Aids (UNAIDS) (2022) Global HIV & AIDS statistics-Factssheet. https://www.unaids.org/en/resources/fact sheet#:~:text=39%20million%20%5B33.1%20million%E2%80%9345.7,AIDS%2Drelated%20illnesses%20in%202022. Accessed 15 Sept 2024
6. IARC Working Group on the Evaluation of Carcinogenic Risks to Humans (1996) Human immunodeficiency viruses and human T-cell lymphotropic viruses. IARC monographs on the evaluation of carcinogenic risks to humans, France
7. Nyamweya S, Hegedus A, Jaye A, Rowland-Jones S et al (2013) Comparing HIV-1 and HIV-2 infection: lessons for viral immunopathogenesis. Rev Med Virol 23(4):221–240
8. Choe H, Farzan M, Sun Y, Sullivan N, Rollins B et al (1996) The β-chemokine receptors

CCR3 and CCR5 facilitate infection by primary HIV-1 isolates. Cell 85(7):1135–1148

9. Alkhatib G, Ahuja SS, Light D, Mummidi S, Berger EA, Ahuja SK (1997) CC chemokine receptor 5-mediated signaling and HIV-1 co-receptor activity share common structural determinants: critical residues in the third extracellular loop support HIV-1 *fusion*. J Biol Chem 272(32):19771–19776
10. Maina EK, Adan AA, Mureithi H, Muriuki J et al (2021) A review of current strategies towards the elimination of latent HIV-1 and subsequent HIV-1 cure. Curr HIV Res 19(1): 14–26
11. Tozser J (2003) Stages of HIV replication and targets for therapeutic intervention. Curr Top Med Chem 3(13):1447–1457
12. Lu DY, Wu HY, Yarla NS, Xu B et al (2018) HAART in HIV/AIDS treatments: future trends. Infect Disord Drug Targets (Formerly Current Drug Targets-Infectious Disorders) 18(1):15–22
13. Eggleton JS, Nagalli S (2023) Highly active antiretroviral therapy (HAART). In: StatPearls. StatPearls Publishing
14. Deeks SG, Archin N, Cannon P, Collins S, Jones RB et al (2021) Research priorities for an HIV cure: international AIDS society global scientific strategy 2021. Nat Med 27(12): 2085–2098
15. Wensing AM, van Maarseveen NM, Nijhuis M (2010) Fifteen years of HIV protease inhibitors: raising the barrier to resistance. Antivir Res 85(1):59–74
16. Copeland TD, Oroszlan S (1988) Genetic locus, primary structure, and chemical synthesis of human immunodeficiency virus protease. Gene Anal Tech 5(6):109–115
17. Jacks T, Power MD, Masiarz FR, Luciw PA, Barr PJ, Varmus HE (1988) Characterization of ribosomal frameshifting in HIV-1 gag-pol expression. Nature 331(6153):280–283
18. Weber IT, Wang YF, Harrison RW (2021) HIV protease: historical perspective and current research. Viruses 13(5):839
19. Henderson LE, Sowder RC, Copeland TD, Oroszlan S, Benveniste RE (1990) Gag precursors of HIV and SIV are cleaved into six proteins found in the mature virions. J Med Primatol 19(3–4):411–419
20. Simon V, Ho DD, Karim QA (2006) HIV/AIDS epidemiology, pathogenesis, prevention, and treatment. Lancet 368(9534): 489–504
21. Lessells RJ, Katzenstein DK, De Oliveira T (2012) Are subtype differences important in HIV drug resistance? Curr Opin Virol 2(5): 636–643
22. Sanches M, Krauchenco S, Martins NH, Gustchina A, Wlodawer A, Polikarpov I (2007) Structural characterization of B and non-B subtypes of HIV-protease: insights into the natural susceptibility to drug resistance development. J Mol Biol 369(4):1029–1040
23. Wondrak EM, Louis JM, Mora PT, Oroszlan S (1991) Purification of HIV-1 wild-type protease and characterization of proteolytically inactive HIV-1 protease mutants by pepstatin A affinity chromatography. FEBS Lett 280(2): 347–350
24. Louis JM, Wondrak EM, Copeland TD, Smith CD, Mora PT, Oroszlan S (1989) Chemical synthesis and expression of the HIV-1 protease gene in *Ecoli*. Biochem Biophys Res Commun 159(1):87–94
25. Lapatto R, Blundell T, Hemmings A, Overington J, Wilderspin A, Wood S et al (1989) X-ray analysis of HIV-1 proteinase at 2.7 Å resolution confirms structural homology among retroviral enzymes. Nature 342(6247): 299–302
26. Wlodawer A, Miller M, Jaskólski M, Sathyanarayana BK, Baldwin E et al (1989) Conserved folding in retroviral proteases: crystal structure of synthetic HIV-1 protease. Science 245(4918):616–621
27. Navia MA, Fitzgerald PM, McKeever BM, Leu CT, Heimbach JC, Herber WK et al (1989) Three-dimensional structure of aspartyl protease from human immunodeficiency virus HIV-1. Nature 337(6208):615–620
28. Simoes I, Faro C (2004) Structure and function of plant aspartic proteinases. Eur J Biochem 271(11):2067–2075
29. Sherry D, Worth R, Ismail ZS, Sayed Y (2021) Cantilever-centric mechanism of cooperative non-active site mutations in HIV protease: implications for flap dynamics. J Mol Graph Model 106:107931
30. James MN, Hsu IN, Delbaere LT (1977) Mechanism of acid protease catalysis based on the crystal structure of penicillopepsin. Nature 267(5614):808–813
31. Deshmukh L, Tugarinov V, Louis JM, Clore GM (2017) Binding kinetics and substrate selectivity in HIV-1 protease–Gag interactions probed at atomic resolution by chemical exchange NMR. Proc Natl Acad Sci 114(46): E9855–E9862
32. Prabu-Jeyabalan M, Nalivaika E, Schiffer CA (2002) Substrate shape determines specificity of recognition for HIV-1 protease: analysis of

crystal structures of six substrate complexes. Structure 10(3):369–381

33. Scott WR, Schiffer CA (2000) Curling of flap tips in HIV-1 protease as a mechanism for substrate entry and tolerance of drug resistance. Structure 8(12):1259–1265
34. Freedberg DI, Ishima R, Jacob J, Wang YX, Kustanovich I, Louis JM, Torchia DA (2002) Rapid structural fluctuations of the free HIV protease flaps in solution: relationship to crystal structures and comparison with predictions of dynamics calculations. Protein Sci 11(2): 221–232
35. Yu Y, Wang J, Chen Z, Wang G, Shao Q, Shi J, Zhu W (2017) Structural insights into HIV-1 protease flap opening processes and key intermediates. RSC Adv 7(71):45121–45128
36. Gustchina A, Weber IT (1990) Comparison of inhibitor binding in HIV-1 protease and in non-viral aspartic proteases: the role of the flap. FEBS Lett 269(1):269–272
37. Badaya A, Sasidhar YU (2020) Inhibition of the activity of HIV-1 protease through antibody binding and mutations probed by molecular dynamics simulations. Sci Rep 10(1):5501
38. Sankaran SV, Krishnan SR, Sayed Y, Gromiha MM (2024) Mechanism of drug resistance in HIV-1 protease subtype C in the presence of Atazanavir. Curr Res Struct Biol 7:100132
39. Pennings PS (2013) HIV drug resistance: problems and perspectives. Infect Dis Rep 5(Suppl 1):e5
40. Svicher V, Ceccherini-Silberstein F, Erba F, Santoro M, Gori C, Bellocchi MC et al (2005) Novel human immunodeficiency virus type 1 protease mutations potentially involved in resistance to protease inhibitors. Antimicrob Agents Chemother 49(5):2015–2025
41. Hoffman NG, Schiffer CA, Swanstrom R (2003) Covariation of amino acid positions in HIV-1 protease. Virology 314(2):536–548
42. Sankaran SV, Phogat A, Gromiha MM (2024) Computational resources for understanding disease-causing mutations in proteins: applications to HIV. In: Gromiha MM (ed) Protein mutations: consequence on structure, function and diseases. World Scientific, Singapore, pp 210–228
43. Domingo E, Escarmís C, Sevilla N, Moya A, Elena SF, Quer J, Holland JJ (1996) Basic concepts in RNA virus evolution. FASEB J 10(8):859–864
44. Brown AJL (1997) Analysis of HIV-1 env gene sequences reveals evidence for a low effective number in the viral population. Proc Natl Acad Sci 94(5):1862–1865
45. Nijhuis M, Boucher CA, Schipper P, Leitner T, Schuurman R, Albert J (1998) Stochastic processes strongly influence HIV-1 evolution during suboptimal protease-inhibitor therapy. Proc Natl Acad Sci 95(24):14441–14446
46. Johnson VA, Brun-Vézinet F, Clotet B, Gunthard HF, Kuritzkes DR, Pillay D, Richman DD (2008) Update of the drug resistance mutations in HIV-1: spring 2008. Top HIV Med 16(1):62 68
47. Mammano F, Trouplin V, Zennou V, Clavel F (2000) Retracing the evolutionary pathways of human immunodeficiency virus type 1 resistance to protease inhibitors: virus fitness in the absence and in the presence of drug. J Virol 74(18):8524–8531
48. Mammano F, Petit C, Clavel F (1998) Resistance-associated loss of viral fitness in human immunodeficiency virus type 1: phenotypic analysis of protease and gag coevolution in protease inhibitor-treated patients. J Virol 72(9):7632–7637
49. Nijhuis M, Schuurman R, De Jong D, Erickson J, Gustchina E, Albert J, Boucher CA (1999) Increased fitness of drug resistant HIV-1 protease as a result of acquisition of compensatory mutations during suboptimal therapy. AIDS 13(17):2349–2359
50. Ahmed SM, Kruger HG, Govender T, Maguire GE, Sayed Y, Ibrahim MA, Soliman ME (2013) Comparison of the molecular dynamics and calculated binding free energies for nine FDA-approved HIV-1 PR drugs against subtype B and C-SA HIV PR. Chem Biol Drug Des 81(2):208–218
51. Maseko SB, Padayachee E, Govender T, Sayed Y, Kruger G, Maguire GE, Lin J (2017) I36T↑ T mutation in South African subtype C (C-SA) HIV-1 protease significantly alters protease-drug interactions. Biol Chem 398(10):1109–1117
52. Venkatachalam S, Murlidharan N, Krishnan SR, Ramakrishnan C, Setshedi M, Pandian R, Gromiha MM (2023) Understanding drug resistance of wild-type and L38HL insertion mutant of HIV-1 C protease to Saquinavir. Genes 14(2):533
53. Doyon L, Croteau G, Thibeault D, Poulin F, Pilote L, Lamarre D (1996) Second locus involved in human immunodeficiency virus type 1 resistance to protease inhibitors. J Virol 70(6):3763–3769
54. da Cunha EFF, Sippl W, de Castro RT, Antunes OAC, de Alencastro RB, Albuquerque MG (2009) 3D-QSAR CoMFA/CoMSIA models based on theoretical active conformers of HOE/BAY-793 analogs derived from HIV-1

protease inhibitor complexes. Eur J Med Chem 44(11):4344–4352

55. Sippl W (2002) Development of biologically active compounds by combining 3D QSAR and structure-based design methods. J Comput Aided Mol Des 16:825–830
56. Verma J, Khedkar VM, Coutinho EC (2010) 3D-QSAR in drug design-a review. Curr Top Med Chem 10(1):95–115
57. Akamatsu M (2002) Current state and perspectives of 3D-QSAR. Curr Top Med Chem 2(12):1381–1394
58. Olotu FA, Agoni C, Soremekun O, Soliman ME (2020) The recent application of 3D-QSAR and docking studies to novel HIV-protease inhibitor drug discovery. Expert Opin Drug Discov 15(9):1095–1109
59. Ojha Lokendra K, Rachana S, Rani BM (2013) Modern drug design with advancement in QSAR: a review. Int J Res Biosci 2:1–12
60. Ferreira GL, Leitao A, Montanari AC, Andricopulo DA (2011) Comparative molecular field analysis of a series of inhibitors of HIV-1 protease. Med Chem 7(2):71–79
61. Ul-Haq Z, Usmani S, Shamshad H, Mahmood U, Halim SA (2013) A combined 3D-QSAR and docking studies for the In-silico prediction of HIV-protease inhibitors. Chem Cent J 7:1–12
62. Huang X, Xu L, Luo X, Fan K, Ji R, Pei G, Jiang H (2002) Elucidating the inhibiting mode of AHPBA derivatives against HIV-1 protease and building predictive 3D-QSAR models. J Med Chem 45(2):333–343
63. Durdagi S, Mavromoustakos T, Papadopoulos MG (2008) 3D QSAR CoMFA/CoMSIA, molecular docking and molecular dynamics studies of fullerene-based HIV-1 PR inhibitors. Bioorg Med Chem Lett 18(23):6283–6289
64. Nunthanavanit P, Anthony NG, Johnston BF, Mackay SP, Ungwitayatorn J (2008) 3D-QSAR studies on chromone derivatives as HIV-1 protease inhibitors: application of molecular field analysis. Archiv der Pharmazie Int J Pharm Med Chem 341(6):357–364
65. Tong J, Wu Y, Bai M, Zhan P (2017) 3D-QSAR and molecular docking studies on HIV protease inhibitors. J Mol Struct 1129: 17–22
66. Amin SA, Adhikari N, Bhargava S, Jha T, Gayen S (2018) Structural exploration of hydroxyethylamines as HIV-1 protease inhibitors: new features identified. SAR QSAR Environ Res 29(5):385–408
67. Baxter JD, Chasanov WM, Adams JL (2016) An update on HIV-1 protease inhibitor resistance. J AIDS Clin Res 7:581

Chapter 12

Accurate and Fast Prediction of Intrinsic Disorder Using flDPnn

Kui Wang, Gang Hu, Zhonghua Wu, and Lukasz Kurgan

Abstract

Intrinsically disordered proteins (IDPs) that include one or more intrinsically disordered regions (IDRs) are abundant across all domains of life and viruses and play numerous functional roles in various cellular processes. Due to a relatively low throughput and high cost of experimental techniques for identifying IDRs, there is a growing need for fast and accurate computational algorithms that accurately predict IDRs/IDPs from protein sequences. We describe one of the leading disorder predictors, flDPnn. Results from a recent community-organized Critical Assessment of Intrinsic Disorder (CAID) experiment show that flDPnn provides fast and state-of-the-art predictions of disorder, which are supplemented with the predictions of several major disorder functions. This chapter provides a practical guide to flDPnn, which includes a brief explanation of its predictive model, descriptions of its web server and standalone versions, and a case study that showcases how to read and understand flDPnn's predictions.

Key words Intrinsic disorder, Intrinsically disordered protein, Prediction, flDPnn, Deep learning, Deep neural network, Machine learning, Web server

1 Introduction

Recent advancements in structural bioinformatics and deep learning, exemplified by the release of AlphaFold2 [1] and RoseTTAfold [2], made it possible to accurately predict protein structures and protein–protein interactions from sequences. However, many proteins include intrinsically disordered regions (IDRs), which are sequence regions that lack a well-defined structure under physiological conditions [3–7]. While AlphaFold2 is capable of predicting IDRs, recent research demonstrates that these regions can be more accurately predicted with dedicated sequence-based predictors of IDRs [8–10]. The intrinsically disordered proteins (IDPs) that include one or more IDRs contribute to the formation of the "dark" proteomes [11, 12], which are far less studied and understood when compared to the structured proteins. IDPs are prevalent across all domains of life, and they are particularly common in

Andrzej Kloczkowski et al. (eds.), *Prediction of Protein Secondary Structure*, Methods in Molecular Biology, vol. 2867, https://doi.org/10.1007/978-1-0716-4196-5_12,

the eukaryotic organisms [13–15]. Bioinformatics studies suggest that over one-third of eukaryotic proteins contain at least one long IDR (i.e., regions with 30 or more consecutive residues) [13, 14, 16], with this rate reaching over 40% in the human proteome [4, 11, 17]. IDPs contribute to many cellular functions, with a few examples including molecular recognition, molecular assembly, cell cycle regulation, transcription, translation, and signaling [18–29]. Recently, IDPs were found in a number of viral proteomes [30–36] and were shown to be involved in liquid–liquid phase separation, driving stability and composition of membraneless organelles [37–40]. However, only a small subset of IDPs was characterized so far, with the leading database of IDRs, DisProt, containing data for only about 2700 proteins as of June 2023 [41]. The scarcity of these data motivates the development of accurate algorithms for predicting IDRs/IDPs from the protein sequences.

Motivated by the low disorder annotation coverage and the fact that the disordered nature of IDRs is encoded in the underlying sequence [5, 42, 43], over 100 sequence-based predictors of IDRs/IDPs were developed [44–55]. Several community-organized assessments, including CASP (Critical Assessment of protein Structure Prediction) [56, 57] and CAID (Critical Assessment of Intrinsic Disorder prediction) [58], comparatively assessed the predictive quality of these tools. The most recent CAID assessment evaluated 43 predictors and identified several methods that at this point provide the most accurate results [58]. The authors said that "across the different performance measures, the methods SPOT-Disorder2 [59], flDPnn [60], RawMSA [61] and AUCpreD [62] are consistently found among the top five. While the ordering changes for different measures and reference sets, and the differences among them are not statistically significant, these methods can be seen broadly as performing consistently well." Among these four tools, flDPnn is the fastest (i.e., flDPnn takes about 15 s to predict a protein compared to 100 or more seconds for the other three predictors), and it additionally predicts functions of the putative IDRs including protein-binding, DNA-binding, RNA-binding, and linkers [58, 60]. We also note that these four accurate tools rely exclusively on deep neural network models, and a recent study has found that the deep neural network-based predictors outperform methods that utilize other types of predictive models [63]. This chapter provides a practical introduction to the flDPnn tool inspired by its high levels of accuracy, low runtime, and unique capability to predict disorder functions. We briefly discuss flDPnn's deep neural network-based model and introduce and explain various ways in which flDPnn can be accessed and used, including web server and standalone versions. Moreover, we provide a case study to explain how to read and understand predictions generated by flDPnn.

2 Methods

FlDPnn is a powerful predictor of the intrinsic disorder and several disorder functions for the putative disordered residues [60, 64]. It relies on a prediction process that consists of three main steps that are executed in a sequential fashion (Fig. 1):

1. *Generate sequence profile.* The first step converts the input protein sequence into a profile that covers a variety of sequence-derived information that is relevant to the disorder and disorder function predictions. flDPnn uses a carefully curated collection of fast and accurate third-party tools to generate this profile including PSI-BLAST [65, 66], IUPred [67, 68], PSIPRED [69, 70], DisoRDPbind [71–73], DFLpred [74], and fMoRFpred [20, 75]. PSI-BLAST generates the position-specific scoring matrix (PSSM) that captures evolutionary information, including the PSSM-derived sequence conservation that is quantified at the residue level (*see* **Note 1**). PSIPRED predicts the secondary structure, and IUPred provides residue-level intrinsic disorder prediction; the latter predictions are refined and improved by the flDPnn's

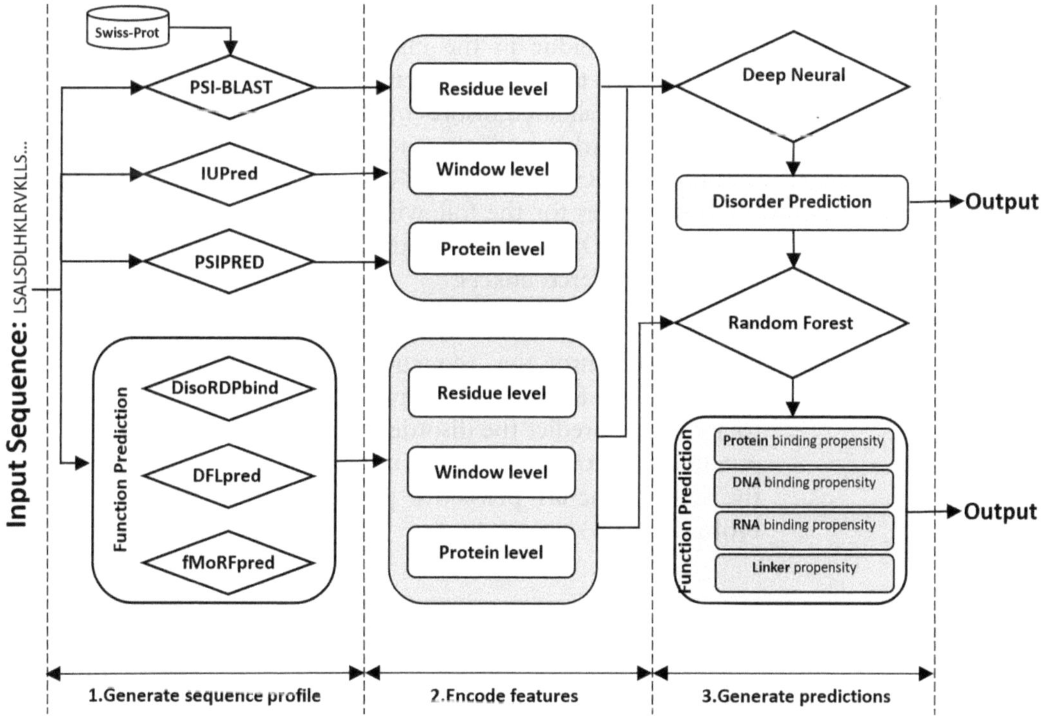

Fig. 1 The flDPnn's architecture. The input to the Random Forest (RF) models includes the putative disorder propensity generated by the deep neural network model and the encoding features. There are four RF models, each dedicated to predicting a specific disorder function

model. The above characteristics are commonly used to make disorder predictions. The profile also incorporates several new characteristics that capture functional aspects of IDRs. They include protein-binding propensities predicted from the sequence by DisoRDPbind and fMoRFpred, DNA-binding and RNA-binding propensities predicted by DisoRDPbind, and putative disordered linker propensities produced by DFLpred (*see* **Note 2**).

2. *Feature encoding.* In the second step, flDPnn uses the sequence profile to encode features that are suitable as inputs for the subsequently used machine learning models. These features are encoded at three different levels: residue, window, and protein level. The residue and window level features describe the local sequence environment of each residue, which is a common practice in the prediction of disorder that relies on machine learning models. One of the innovations introduced in flDPnn is the inclusion of the protein-level features, which aggregate information from the entire sequence. These features include sequence length, distance to each sequence terminus, and sequence-average of the profile values (*see* **Note 3**).
3. *Generate disorder prediction.* In the last step, the encoded features are fed into a feedforward, fully connected deep neural network (DNN) that predicts the putative disorder propensity of each residue in the input protein sequence. This DNN model was trained using a large dataset of proteins with experimentally validated disorder [60]. Furthermore, flDPnn inputs these disorder predictions together with the encoded features into four Random Forest (RF) models that are used to predict propensities for the following four types of functions for the putative IDRs: protein binding, DNA binding, RNA binding, and disordered linkers.

As a few highlights, flDPnn integrates additional functional information into the sequence profile, generates innovative sequence-level features, and applies modern DNN and RF models to accurately predict the disorder propensities and their associated putative functions. The inclusion of these innovations can explain the state-of-the-art predictive performance of flDPnn [60] (*see* **Notes 2** and **3**).

3 Availability of the flDPnn Predictor

The flDPnn predictor can be used in three complementary ways: as a web server, standalone source code, and Docker-based installation. This broad range of options caters to different user needs and usage scenarios. The web server version is the most suitable for users who occasionally need to predict intrinsic disorder and its functions. This version can be used to predict up to 20 proteins at a time where the entire prediction process is done on the server side and without the need to install any software. The results can be accessed through your web browser. We recommend the standalone code version to users who need to run flDPnn for a large collection of proteins and to use it frequently. This version must be installed and run on the user's computer and requires availability and know-how of the Linux or UNIX-like operating system. This option allows users to run flDPnn locally without an Internet connection and to embed this tool into other bioinformatics pipelines. The Docker version also has to be installed and executed locally on the user's computer. This version is arguably easier to install but requires familiarity with the Docker platform (*see* **Note 4**). We note that flDPnn is also available via the DEPICTER2 web server which provides access to multiple predictors of disorder and disorder functions [64]. The following subsections provide in-depth information on how to use the web server and standalone versions of flDPnn.

3.1 Web Server

The flDPnn's web server can be freely accessed at http://biomine.cs.vcu.edu/servers/flDPnn/. Users can do that using any of the popular web browsers (Chrome, Firefox, Edge, IE, etc.) on a computer connected to the Internet. Prior to running flDPnn, users should prepare their query protein sequences in the FASTA format (*see* **Note 5**). The use of flDPnn via the web server involves three simple steps (Fig. 2):

1. Input the FASTA-formatted sequences. Copy and paste the query sequences into the text field (red box 1 in Fig. 2). The number of the query protein sequences is limited to a maximum of 20, and each sequence should be between 21 and 5000 amino acids long (*see* **Note 6**). Users can also click on the "Example" button to see correctly formatted example query sequences.
2. Input your email address (red box 2 in Fig. 2). Users have the option to provide their email address to receive a notification email once the prediction job is completed (*see* **Note 7**).
3. Submit the job and access the prediction results. Click the "Run flDPnn" button (red box 3 in Fig. 2) to start the prediction. The browser will then navigate to a new web page that facilitates monitoring of the prediction process (Fig. 3). If any issues arise, users can email the provided email address and include their job identifier, which is located in the last line in Fig. 3 (*flDPnn:20230530103043*). The processing page refreshes every 10 s until the prediction job is completed (Fig. 4).

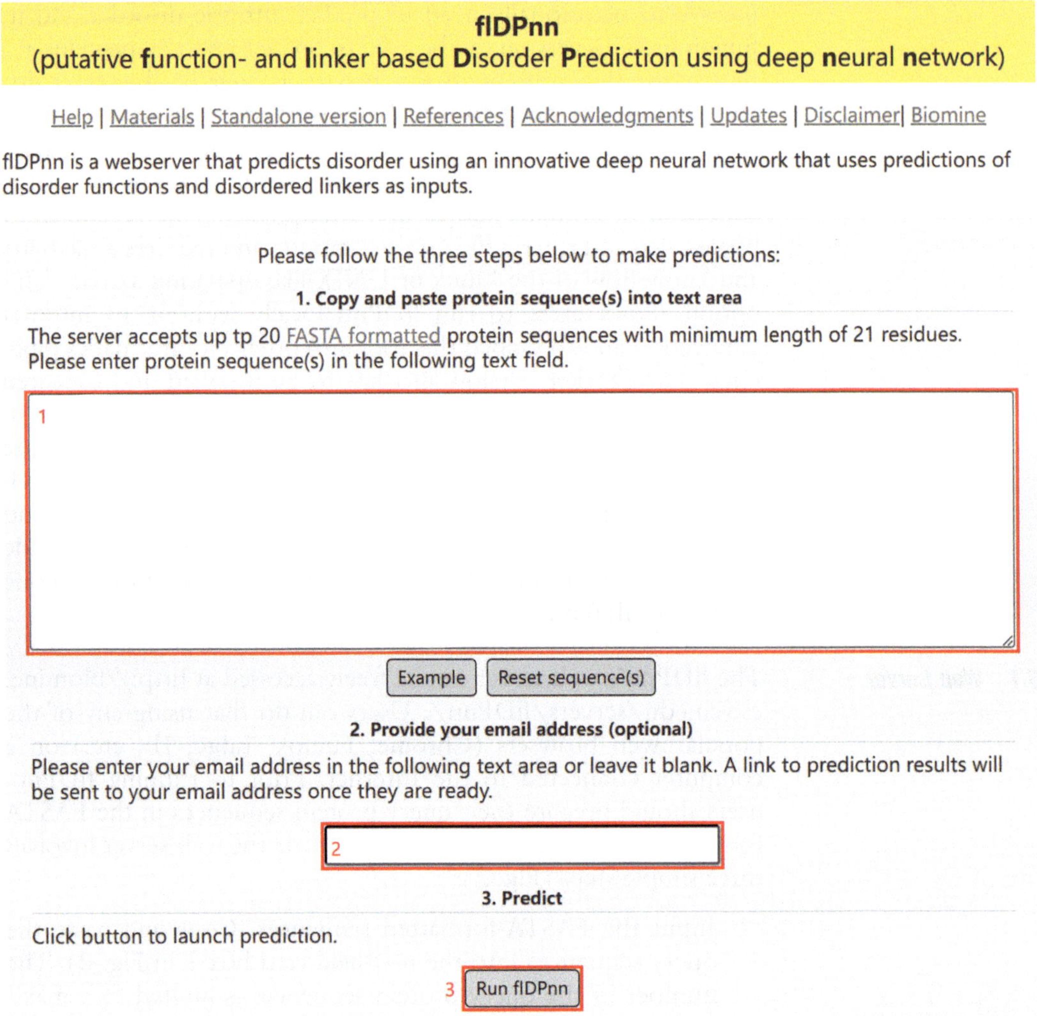

Fig. 2 The main page of the flDPnn's web server

Users who provide the email address will receive a notification email once the job is completed. They can access the results via the links provided in the processing page and in the notification email. These links lead to the CSV-formatted file with the predictions and a web page that provides graphical visualization of the results (Fig. 5).

The flDPnn predictions include disorder propensity scores and the corresponding binary predictions for each of the amino acids in the input sequence. flDPnn outputs a total of five propensity scores, each ranging between 0 and 1. Larger values of the scores suggest higher likelihood that a given residue is intrinsically disordered/binds proteins/binds RNA/binds DNA or is a disordered linker. The propensities are used to derive binary predictions (1 for presence vs. 0 for absence of disorder or disorder function), such

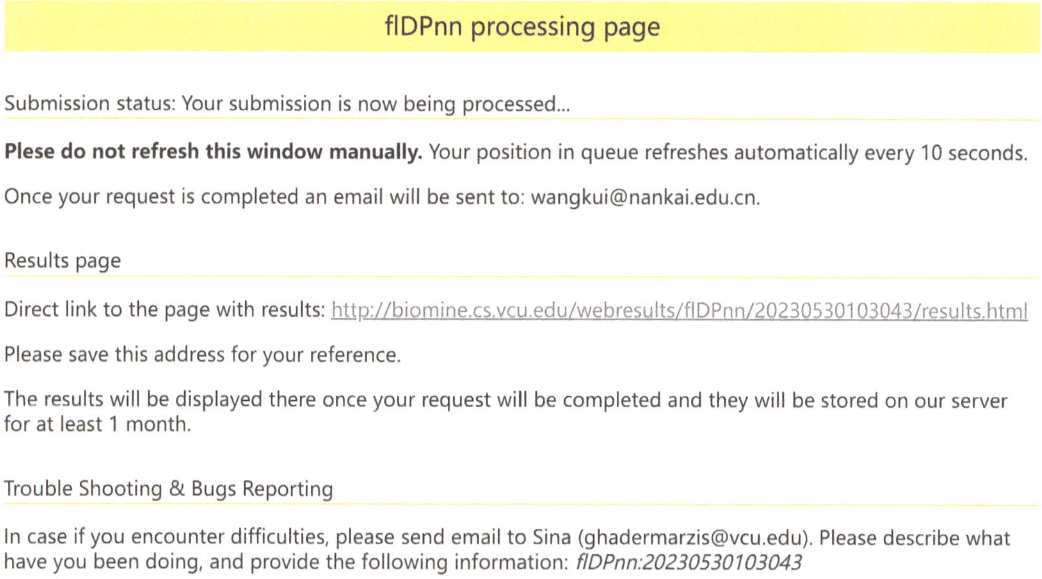

Fig. 3 The flDPnn's processing page. This page refreshes every 10 s

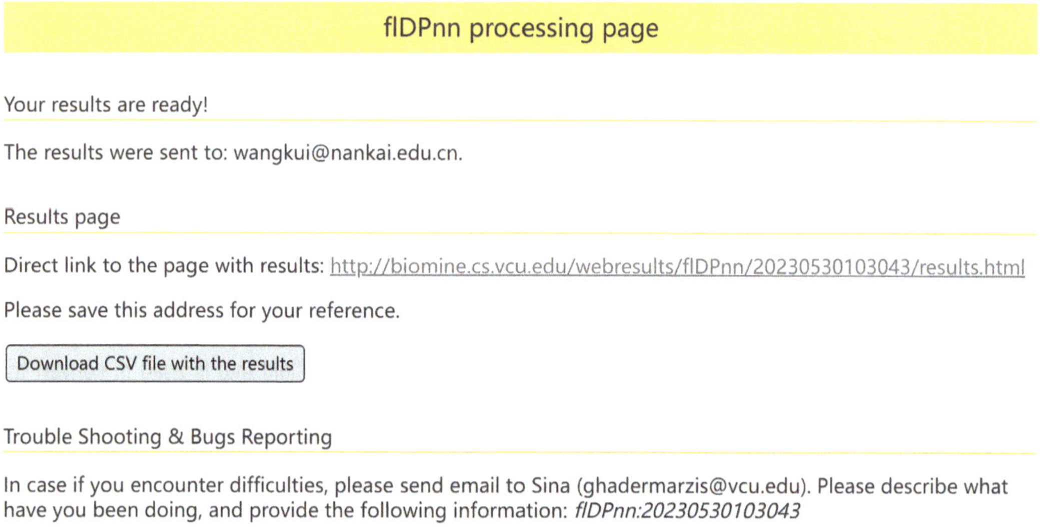

Fig. 4 The flDPnn's processing page that shows when the prediction job is completed. This page provides two options for accessing the results. The "Download CSV file with the results" button offers a link to download the CSV-formatted file containing the predictions. The "Direct Link" directs the user to the graphical visualization of the results in the web browser window

that residues with propensities > a given cut-off value are set to the binary value of 1, while the remaining residues are set to the value of 0. We show an example of this conversion between propensities and binary values for the disorder prediction in Fig. 5.

The comma-separated values (CSV)-formatted results are included in the "results.csv" text file. This file includes the query

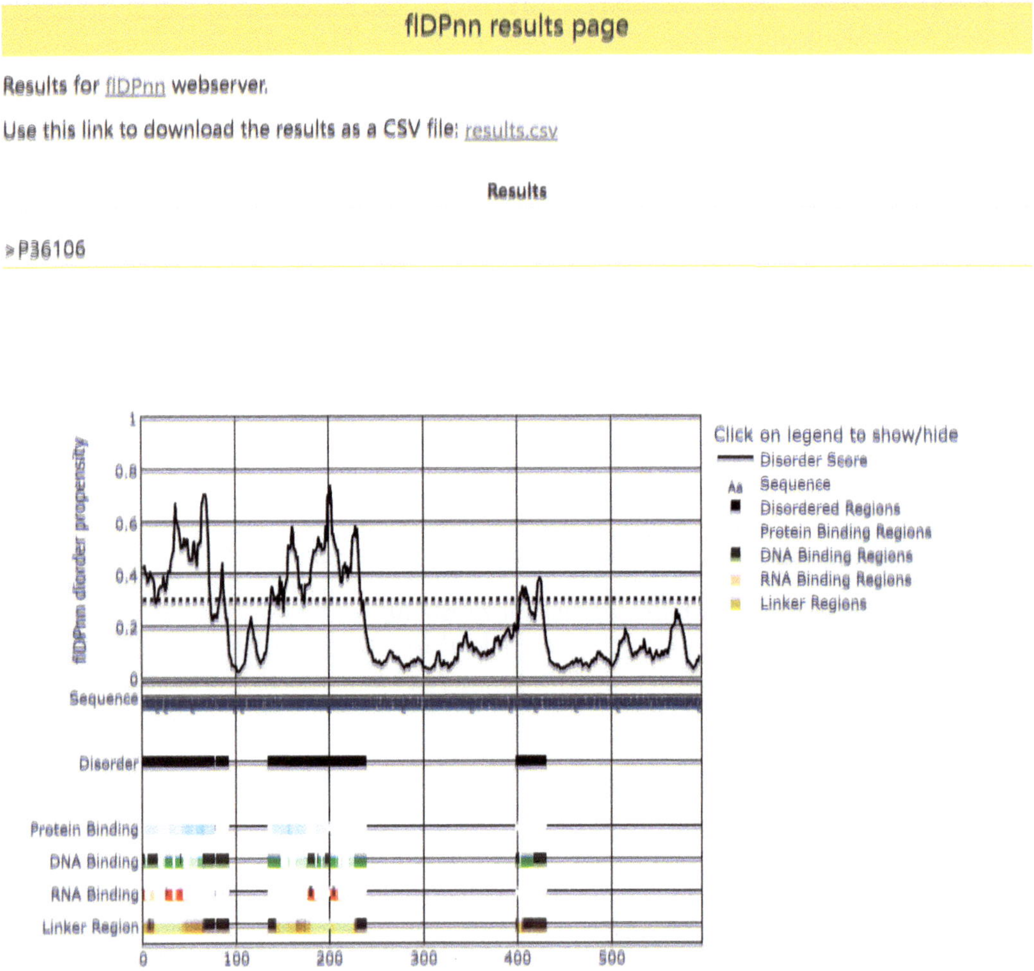

Fig. 5 The web page that visualizes predictions generated by flDPnn. The black line plot at the top represents the disorder propensity scores. The black horizontal dotted line is the threshold that is used to convert the propensities into binary predictions that denote putative IDRs which are shown using black horizontal bars. The color-coded horizontal bars at the bottom show propensities for specific disorder functions

sequence(s) and its/their respective predicted propensity scores and binary predictions. This file can be easily parsed to obtain the predictions for further/subsequent processing (*see* **Note 8**). The graphical visualization of the results is available directly in the browser window (Fig. 5). It shows the disorder propensities using a black line plot, the cut-off using a horizontal dotted line, and the binary disorder predictions using a black horizontal bar in the middle of the plot. The propensities for the four disorder functions are shown at the bottom using color-coded horizontal bars, with blue for protein binding, green for DNA binding, red for RNA binding, and brown for disordered linkers. The color intensity reflects the propensity/likelihood of the corresponding residue for the given type of disorder function, where darker shade corresponds to a higher value. Furthermore, when users hover their mouse pointer over a specific bar/predicted IDRs, additional details, such as position of the selected residue in the sequence and its corresponding propensity score, are displayed. The graphical panel also offers several interactive features. For instance, users can click on specific elements in the figure legend to show or hide these details. They can also zoom in on a specific sequence segment by selecting it when holding the left mouse button, pan the view along the x-axis/sequence, and reset the view to the default zoom level that spans the whole sequence.

3.2 Standalone Versions

Users can also run flDPnn on their own computer hardware. There are two options to do that: the source code from https://gitlab.com/sina.ghadermarzi/fldpnn and the Docker version from https://gitlab.com/sina.ghadermarzi/fldpnn_docker. Users simply need to follow step-by-step instructions provided in these GitLab repositories to install and setup the flDPnn predictor.

While this version is more difficult to use since it requires a lengthy installation process and availability of a suitable computer system, it offers several advantages when compared to the web server. Firstly, the standalone version does not limit the number of input sequences, enabling users to run flDPnn for large datasets of proteins. Secondly, predictions can be made locally without requiring an Internet connection. Thirdly, users can integrate the flDPnn software into their bioinformatics pipelines. However, we note that flDPnn software works only on the Linux operating system and relies on several third-party applications (*see* **Note 9**). To run flDPnn on a local computer, users need the Linux OS (Ubuntu x64 20.04.2), tcsh shell (6.21.00-1), Java Runtime Environment (openjdk 1.0.8), and Python 3 (3.8.5). The following Python packages are also required: plotly (4.14.3), scikit-learn (0.23.1), keras (2.4.3), tensorflow (2.4.1), and pandas (1.2.2). More recent versions of the listed software should be also compatible. Once a user downloads and unpacks the source code and the above requirements are met, (s)he can run flDPnn by following these three steps:

- Go to the directory where flDPnn is located.
- Using the command line run *python3 run_fldpnn.py input_fasta_file*. The input file can include multiple sequences in the FASTA format (*see* **Note 5**). This file might be located in a different folder from flDPnn. For example, command *python3 run_fldpnn.py example/query.fa* should be used if the *query.fa* file is located in the *example* folder.
- Read the results. The standalone version of flDPnn provides the same outputs as the web version. They include the CSV-formatted plain text file named "results.csv" and an

HTML file named "results.html" that includes the graphical view of each query protein (*see* **Note 10**). These two files are generated in the same folder as the query file. We note that running another query in the same folder overwrites the previously generated output files.

4 Case Study

We use a case study to showcase the flDPnn's predictions, explain their format, and provide hints on how to interpret them. The case study concerns the 215 amino acids long multifunctional redox-sensitive HMGB1 protein (UniProt ID: P63159) [76]. In the nucleus, HMGB1 interacts with DNA contributing to transcription, replication, recombination, DNA repair, and chromatin remodeling [76]. In the cytoplasm, this protein acts as a sensor and a chaperone for immunogenic nucleic acids, participating in the activation of TLR9-mediated immune responses and autophagy [77]. Structurally, HMGB1 consists of two HMG domains including domain A (6P to 78Y) which is connected by a basic linker to the domain B (95P to 163R). Furthermore, domain B is connected to an acidic C-terminus tail through another basic linker. This protein has three IDRs (DisProt ID: DP02138): the first basic linker (73R to 99P), the second basic linker (165K to 184K), and the C-terminal tail that is involved in protein binding (184K to 215E) [78]. The latter IDR obstructs the DNA-binding interface by binding to the HMG domains, consequently down-regulating interactions with DNA [78].

We use the HMGB1 sequence and the flDPnn's web server to make the disorder and disorder function predictions. The entire prediction process takes less than 1 min. Figure 6 is the graphical version of the flDPnn's results. The horizontal black bars in the middle represent the binary predictions of disorder, which produce four putative IDRs: 1–13, 32–37, 72–94, and 161–215. These regions were generated from the putative propensities (black line plot at the top of the figure), such that amino acids with values > the cut-off (dotted horizontal line) are assumed to be disordered. The latter two putative IDRs align well with the native IDRs including both disordered linkers (73–99 and 165–184) and the protein-binding IDR (184–215). The other two putative IDRs have lower underlying putative disorder propensities, which suggest that these predictions are less likely to be accurate. More broadly, predictions for residues for which flDPnn generates propensities that are closer to the threshold should be considered as less reliable when compared to the predictions where the propensities are either very high (accurately predicted disorder) or very low (accurately predicted order). Next, we analyze the flDPnn's

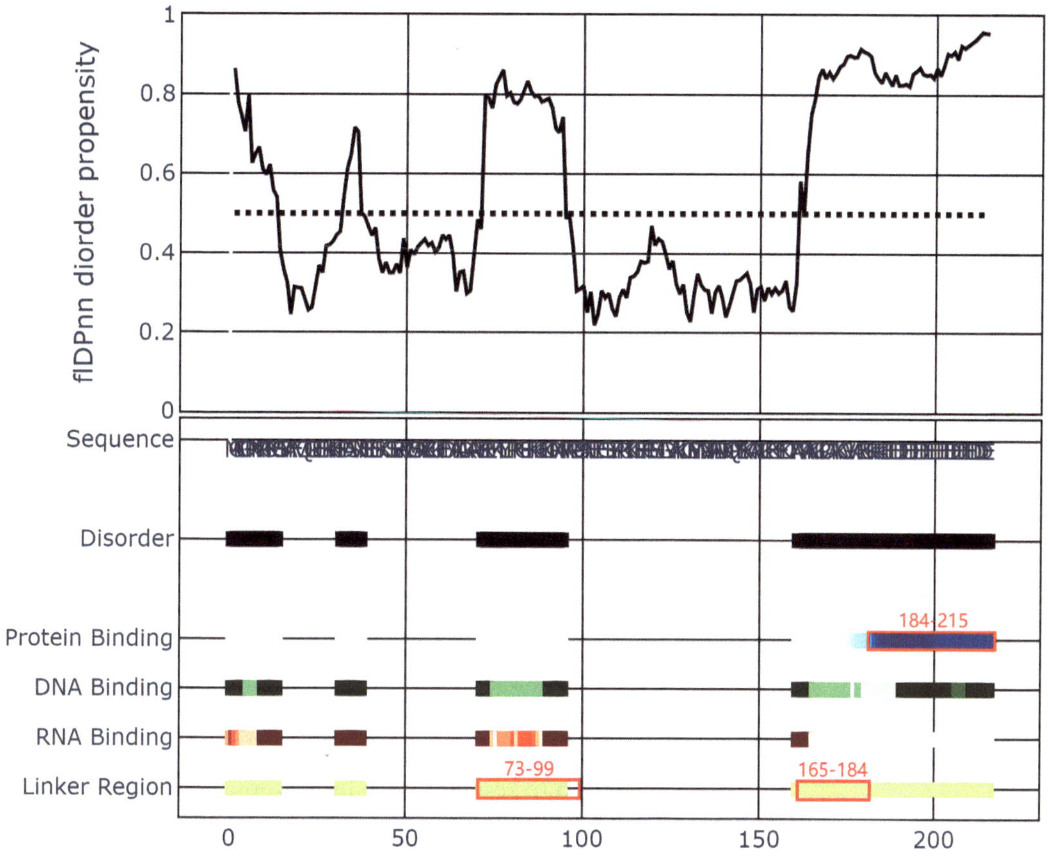

Fig. 6 The flDPnn's prediction for the HMGB1 protein (UniProt ID: P63159; DisProt ID: DP02138). The black line plot at the top gives the disorder propensity scores. The black horizontal dotted line is the threshold that is used to produce binary predictions, which are shown using black horizontal bars. The color-coded horizontal bars at the bottom show propensities for specific disorder functions

predictions of the disorder functions, which are shown as color-coded horizontal bars at the bottom of Fig. 6. The blue bar gives the protein binding propensities for the putative disordered residues. The darker intensity of the color identifies amino acids that are more likely to interact with proteins, suggesting that the part of the putative IDR at the C-terminus binds proteins. This prediction is in good agreement with the native annotations (red box in Fig. 6). Moreover, the horizontal bar at the bottom that visualizes putative propensities for disordered linkers reveals that the two native linkers (red boxes in Fig. 6) are partially identified by flDPnn, although this tool overestimates the linker propensities for the residues near the C-terminus.

5 Notes

1. flDPnn relies on the Position-Specific Scoring Matrix (PSSM) generated by the PSI-BLAST program using a relatively small SwissProt database to produce the sequence profile. The use of this small database facilitates very fast generation of the PSSM but also means that this matrix cannot be generated for some proteins, i.e., those that lack a sufficient number of similar sequences in the SwissProt database. If that is the case, then flDPnn uses a default PSSM as a substitute. The quality of the flDPnn's predictions of disorder might be lower when flDPnn resorts to using the default PSSM. This is denoted by the [WARNING-DEFAULT PSSM] annotation that accompanies the predictions, alerting the end users that the prediction may be less reliable.
2. The addition of the putative functional characteristics of disorder in the flDPnn's sequence profile was empirically shown to substantially improve its predictive performance. More specifically, this resulted in the increase in the F1 measure from 0.60 to 0.63 and in the Matthew's Correlation Coefficient (MCC) metric from 0.43 to 0.49 [60].
3. The inclusion of the protein-level features was empirically demonstrated to result in large improvements in the predictive quality for flDPnn. Empirical results in [60] show that this innovation raised F1 from 0.58 to 0.63 and MCC from 0.41 to 0.49.
4. The Docker software is available at https://www.docker.com/. This software provides a containerized environment that encapsulates all the necessary dependencies and allows for easy deployment and execution of flDPnn. A free version of Docker should be sufficient to install the flDPnn predictor.
5. A FASTA-formatted file is a text file that contains biological sequence data. In the context of flDPnn, the user should input the query protein sequence encoded using single-letter amino acid codes in the input FASTA-formatted file. The header line in the FASTA file should start with the ">" symbol. For more detailed information about the FASTA format, you can refer to https://en.wikipedia.org/wiki/FASTA_format.
6. When providing the query sequences in the FASTA format, we recommend keeping the header line concise/short. Using the UniProt accession number and/or a protein name is a suitable option for the header line, which eases organizing and identifying the results for individual proteins. The upper limit of 20 protein sequences per submission is motivated by the fact that the underlying biomine server provides access to multiple popular bioinformatics tools, which include DescribePROT

[79, 80], PDID [81, 82], DEPICTER [64, 83], qNABpredict [84], CONNECTOR [85], disCoP [86, 87], DRNApred [88], fDETECT [89, 90], MFDp [91, 92], SCRIBER [93], and XRRPred [94]. This limit on the input size ensures that the access to this shared resource is distributed more evenly across users and tools.

7. Once the job is submitted on the website, the computation process will start running in the background, even if the user closes his/her browser or navigates away from the page. Users who do not provide their email address should not close the browser window since otherwise they will not be able to access the results, unless they save the direct link to the results that is available on the processing web page (Fig. 3). Once the prediction job is finished, users will receive a notification email containing the direct links to access the results. These links can be used to access the results page or download the result files, allowing users to conveniently retrieve and view the predictions.
8. The CSV-formatted file with the results generated by flDPnn contains a complete collection of predictions for the query sequences. It includes the propensity scores and binary predictions for both disorder and disorder functions. The format of the CSV file, which is described in the header of the file, is as follows:
   - Line 1: >protein ID. [WARNING-DEFAULT PSSM] note is added in front of the protein identifier if PSSM cannot be calculated for a given input sequence (*see* **Note 1**).
   - Line 2: sequence.
   - Line 3: binary disorder prediction (1 = putative disordered residue; 0 = putative ordered residue).
   - Line 4: putative disorder propensity (higher value denotes higher likelihood that a given residue is disordered).
   - Line 5: binary protein-binding prediction (1 = disordered protein-binding residue; 0 disordered residue that does not bind proteins; X ordered residue).
   - Line 6: protein-binding propensity (higher value denotes higher likelihood that a given residue is disordered and binds proteins; X ordered residue).
   - Line 7: binary DNA-binding prediction (1 = disordered DNA-binding residue; 0 disordered residue that does not bind DNA; X ordered residue).
   - Line 8: DNA-binding propensity (higher value denotes higher likelihood that a given residue is disordered and binds DNA; X ordered residue).

- Line 9: binary RNA-binding prediction (1 = disordered RNA-binding residue; 0 disordered residue that does not bind RNA; X ordered residue).
- Line 10: RNA-binding propensity (higher value denotes higher likelihood that a given residue is disordered and binds RNA; X ordered residue).
- Line 11: binary linker prediction (1 = disordered linker residue; 0 disordered residue that is not a linker; X ordered residue).
- Line 12: linker propensity (higher value denotes higher likelihood that a given residue is the disordered linker; X ordered residue).

9. flDPnn aims to ease the installation process by including all the necessary third-party tools and datasets with its source code. This includes a full version of PSI-BLAST, IUPred (v1), PSIPRED, DisoRDPbind, DFLpred, and fMoRFpred. Additionally, flDPnn also includes the SwissProt dataset that is used by PSI-BLAST. By bundling these tools and datasets, users can install flDPnn without the need to separately install and configure each component.
10. If a user does not need the prediction of disorder functions and only wants to obtain the disorder prediction, (s)he should run the "fldpnn-nofunc.py" Python script using the following command: *python3 fldpnn-nofunc.py input_fasta_file*.

Acknowledgments

This work was supported in part by the National Natural Science Foundation of China (92370128 and 12326611 to GH and KW), the National Science Foundation (DBI2146027 and IIS2125218 to LK), and the Robert J. Mattauch Endowment funds to LK.

References

1. Jumper J, Evans R, Pritzel A et al (2021) Highly accurate protein structure prediction with AlphaFold. Nature 596:583–589
2. Baek M, Dimaio F, Anishchenko I et al (2021) Accurate prediction of protein structures and interactions using a three-track neural network. Science 373:871–876
3. Oldfield CJ, Uversky VN, Dunker AK et al (2019) Introduction to intrinsically disordered proteins and regions. In: Salvi N (ed) Intrinsically disordered proteins. Academic, pp 1–34
4. Van Der Lee R, Buljan M, Lang B et al (2014) Classification of intrinsically disordered regions and proteins. Chem Rev 114:6589–6631
5. Dunker AK, Babu MM, Barbar E et al (2013) What's in a name? Why these proteins are intrinsically disordered. Intrinsically Disord Proteins 1:e24157
6. Habchi J, Tompa P, Longhi S et al (2014) Introducing protein intrinsic disorder. Chem Rev 114:6561–6588
7. Lieutaud P, Ferron F, Uversky AV et al (2016) How disordered is my protein and what is its disorder for? A guide through the "dark side"

of the protein universe. Intrinsically Disord Proteins 4:e1259708
8. Wilson CJ, Choy WY, Karttunen M (2022) AlphaFold2: a role for disordered protein/region prediction? Int J Mol Sci 23:4591
9. Piovesan D, Monzon AM, Tosatto SCE (2022) Intrinsic protein disorder and conditional folding in AlphaFoldDB. Protein Sci 31:e4466
10. Zhao B, Ghadermarzi S, Kurgan L (2023) Comparative evaluation of AlphaFold2 and disorder predictors for prediction of intrinsic disorder, disorder content and fully disordered proteins. Comput Struct Biotechnol J 21: 3248–3258
11. Hu G, Wang K, Song J et al (2018) Taxonomic landscape of the dark proteomes: whole-proteome scale interplay between structural darkness, intrinsic disorder, and crystallization propensity. Proteomics 18:e1800243
12. Kulkarni P, Uversky VN (2018) Intrinsically disordered proteins: the dark horse of the dark proteome. Proteomics 18:e1800061
13. Peng Z, Yan J, Fan X et al (2015) Exceptionally abundant exceptions: comprehensive characterization of intrinsic disorder in all domains of life. Cell Mol Life Sci 72:137–151
14. Xue B, Dunker AK, Uversky VN (2012) Orderly order in protein intrinsic disorder distribution: disorder in 3500 proteomes from viruses and the three domains of life. J Biomol Struct Dyn 30:137–149
15. Ward JJ, Sodhi JS, Mcguffin LJ et al (2004) Prediction and functional analysis of native disorder in proteins from the three kingdoms of life. J Mol Biol 337:635–645
16. Peng Z, Mizianty MJ, Kurgan L (2014) Genome-scale prediction of proteins with long intrinsically disordered regions. Proteins 82:145–158
17. Tunyasuvunakool K, Adler J, Wu Z et al (2021) Highly accurate protein structure prediction for the human proteome. Nature 596:590
18. Zhao B, Katuwawala A, Uversky VN et al (2020) IDPology of the living cell: intrinsic disorder in the subcellular compartments of the human cell. Cell Mol Life Sci 78:2371
19. Piersimoni L, Abd El Malek M, Bhatia T et al (2022) Lighting up Nobel Prize-winning studies with protein intrinsic disorder. Cell Mol Life Sci 79:449
20. Yan J, Dunker AK, Uversky VN et al (2016) Molecular recognition features (MoRFs) in three domains of life. Mol BioSyst 12:697–710
21. Peng Z, Mizianty MJ, Xue B et al (2012) More than just tails: intrinsic disorder in histone proteins. Mol BioSyst 8:1886–1901
22. Toth-Petroczy A, Oldfield CJ, Simon I et al (2008) Malleable machines in transcription regulation: the mediator complex. PLoS Comput Biol 4:e1000243
23. Uversky VN, Oldfield CJ, Dunker AK (2005) Showing your ID: intrinsic disorder as an ID for recognition, regulation and cell signaling. J Mol Recognit 18:343–384
24. Shammas SL (2017) Mechanistic roles of protein disorder within transcription. Curr Opin Struct Biol 42:155–161
25. Staby L, O'shea C, Willemoes M et al (2017) Eukaryotic transcription factors: paradigms of protein intrinsic disorder. Biochem J 474: 2509–2532
26. Peng Z, Oldfield CJ, Xue B et al (2014) A creature with a hundred waggly tails: intrinsically disordered proteins in the ribosome. Cell Mol Life Sci 71:1477–1504
27. Wang C, Uversky VN, Kurgan L (2016) Disordered nucleiome: abundance of intrinsic disorder in the DNA- and RNA-binding proteins in 1121 species from Eukaryota, Bacteria and Archaea. Proteomics 16:1486–1498
28. Zhao B, Katuwawala A, Oldfield CJ et al (2021) Intrinsic Disorder in Human RNA-Binding Proteins. J Mol Biol 433: 167229
29. Hu G, Wu Z, Uversky VN et al (2017) Functional analysis of human hub proteins and their interactors involved in the intrinsic disorder-enriched interactions. Int J Mol Sci 18:2761
30. Xue B, Blocquel D, Habchi J et al (2014) Structural disorder in viral proteins. Chem Rev 114:6880–6911
31. Fan X, Xue B, Dolan PT et al (2014) The intrinsic disorder status of the human hepatitis C virus proteome. Mol BioSyst 10:1345–1363
32. Xue B, Mizianty MJ, Kurgan L et al (2012) Protein intrinsic disorder as a flexible armor and a weapon of HIV-1. Cell Mol Life Sci 69: 1211–1259
33. Meng F, Badierah RA, Almehdar HA et al (2015) Unstructural biology of the Dengue virus proteins. FEBS J 282:3368–3394
34. Kumar N, Kaushik R, Tennakoon C et al (2021) Comprehensive intrinsic disorder analysis of 6108 viral proteomes: from the extent of intrinsic disorder penetrance to functional annotation of disordered viral proteins. J Proteome Res 20:2704–2713
35. Cubuk J, Alston JJ, Incicco JJ et al (2021) The SARS-CoV-2 nucleocapsid protein is dynamic, disordered, and phase separates with RNA. Nat Commun 12:1936
36. Giri R, Bhardwaj T, Shegane M et al (2020) Understanding COVID-19 via comparative

analysis of dark proteomes of SARS-CoV-2, human SARS and bat SARS-like coronaviruses. Cell Mol Life Sci. https://doi.org/10.1007/s00018-020-03603-x

37. Wang J, Shi CR, Xu Q et al (2021) SARS-CoV-2 nucleocapsid protein undergoes liquid-liquid phase separation into stress granules through its N-terminal intrinsically disordered region. Cell Discov 7:5
38. Darling AL, Liu Y, Oldfield CJ et al (2018) Intrinsically disordered proteome of human membrane-less organelles. Proteomics 18: e1700193
39. Ibrahim AY, Khaodeuanepheng NP, Amarasekara DL et al (2023) Intrinsically disordered regions that drive phase separation form a robustly distinct protein class. J Biol Chem 299:102801
40. Uversky VN (2017) Protein intrinsic disorder-based liquid-liquid phase transitions in biological systems: complex coacervates and membrane-less organelles. Adv Colloid Interf Sci 239:97–114
41. Quaglia F, Meszaros B, Salladini E et al (2022) DisProt in 2022: improved quality and accessibility of protein intrinsic disorder annotation. Nucleic Acids Res 50:D480–D487
42. Zhao B, Kurgan L (2022) Compositional bias of intrinsically disordered proteins and regions and their predictions. Biomol Ther 12:888
43. Campen A, Williams RM, Brown CJ et al (2008) TOP-IDP-scale: a new amino acid scale measuring propensity for intrinsic disorder. Protein Pept Lett 15:956–963
44. Zhao B, Kurgan L (2021) Surveying over 100 predictors of intrinsic disorder in proteins. Expert Rev Proteomics 18:1019–1029
45. Basu S, Kihara D, Kurgan L (2023) Computational prediction of disordered binding regions. Comput Struct Biotechnol J 21: 1487–1497
46. Kurgan L (2022) Resources for computational prediction of intrinsic disorder in proteins. Methods 204:132–141
47. Katuwawala A, Kurgan L (2020) Comparative assessment of intrinsic disorder predictions with a focus on protein and nucleic acid-binding proteins. Biomol Ther 10:1636
48. Liu Y, Wang X, Liu B (2019) A comprehensive review and comparison of existing computational methods for intrinsically disordered protein and region prediction. Brief Bioinform 20: 330–346
49. Katuwawala A, Peng Z, Yang J et al (2019) Computational prediction of MoRFs, short disorder-to-order transitioning protein binding regions. Comput Struct Biotechnol J 17:454–462
50. Meng F, Uversky VN, Kurgan L (2017) Comprehensive review of methods for prediction of intrinsic disorder and its molecular functions. Cell Mol Life Sci 74:3069–3090
51. Punta M, Simon I, Dosztanyi Z (2015) Prediction and analysis of intrinsically disordered proteins. Methods Mol Biol 1261:35–59
52. Deng X, Eickholt J, Cheng J (2012) A comprehensive overview of computational protein disorder prediction methods. Mol BioSyst 8:114–121
53. He B, Wang K, Liu Y et al (2009) Predicting intrinsic disorder in proteins: an overview. Cell Res 19:929–949
54. Zhao B, Kurgan L (2023) Machine learning for intrinsic disorder prediction. In: Machine learning in bioinformatics of protein sequences, pp 205–236
55. Meng F, Uversky V, Kurgan L (2017) Computational prediction of intrinsic disorder in proteins. Curr Protoc Protein Sci 88:2 16 11–12 16 14
56. Monastyrskyy B, Kryshtafovych A, Moult J et al (2014) Assessment of protein disorder region predictions in CASP10. Proteins 82(Suppl 2): 127–137
57. Monastyrskyy B, Fidelis K, Moult J et al (2011) Evaluation of disorder predictions in CASP9. Proteins 79(Suppl 10):107–118
58. Necci M, Piovesan D, Predictors C et al (2021) Critical assessment of protein intrinsic disorder prediction. Nat Methods 18:472–481
59. Hanson J, Paliwal KK, Litfin T et al (2020) SPOT-Disorder2: improved protein intrinsic disorder prediction by ensembled deep learning. Genomics Proteomics Bioinformatics. https://doi.org/10.1016/j.gpb.2019.01.004
60. Hu G, Katuwawala A, Wang K et al (2021) flDPnn: accurate intrinsic disorder prediction with putative propensities of disorder functions. Nat Commun 12:4438
61. Mirabello C, Wallner B (2019) rawMSA: end-to-end deep learning using raw multiple sequence alignments. PLoS One 14:e0220182
62. Wang S, Ma JZ, Xu JB (2016) AUCpreD: proteome-level protein disorder prediction by AUC-maximized deep convolutional neural fields. Bioinformatics 32:672–679
63. Zhao B, Kurgan L (2022) Deep learning in prediction of intrinsic disorder in proteins. Comput Struct Biotechnol J 20:1286–1294
64. Basu S, Gsponer J, Kurgan L (2023) DEPICTER2: a comprehensive webserver for intrinsic

disorder and disorder function prediction. Nucleic Acids Res 51:W141

65. Hu G, Kurgan L (2019) Sequence similarity searching. Curr Protoc Protein Sci 95:e71
66. Altschul SF, Madden TL, Schaffer AA et al (1997) Gapped BLAST and PSI-BLAST: a new generation of protein database search programs. Nucleic Acids Res 25:3389–3402
67. Dosztanyi Z (2018) Prediction of protein disorder based on IUPred. Protein Sci 27:331–340
68. Dosztanyi Z, Csizmok V, Tompa P et al (2005) IUPred: web server for the prediction of intrinsically unstructured regions of proteins based on estimated energy content. Bioinformatics 21:3433–3434
69. Buchan DWA, Jones DT (2019) The PSIPRED protein analysis workbench: 20 years on. Nucleic Acids Res 47:W402–W407
70. Buchan DWA, Minneci F, Nugent TCO et al (2013) Scalable web services for the PSIPRED protein analysis workbench. Nucleic Acids Res 41:W349–W357
71. Oldfield CJ, Peng Z, Kurgan L (2020) Disordered RNA-binding region prediction with DisoRDPbind. Methods Mol Biol 2106:225–239
72. Peng Z, Wang C, Uversky VN et al (2017) Prediction of disordered RNA, DNA, and protein binding regions using DisoRDPbind. Methods Mol Biol 1484:187–203
73. Peng Z, Kurgan L (2015) High-throughput prediction of RNA, DNA and protein binding regions mediated by intrinsic disorder. Nucleic Acids Res 43:e121
74. Meng F, Kurgan L (2016) DFLpred: high-throughput prediction of disordered flexible linker regions in protein sequences. Bioinformatics 32:i341–i350
75. Disfani FM, Hsu WL, Mizianty MJ et al (2012) MoRFpred, a computational tool for sequence-based prediction and characterization of short disorder-to-order transitioning binding regions in proteins. Bioinformatics 28:i75–i83
76. Stros M (2010) HMGB proteins: interactions with DNA and chromatin. Biochim Biophys Acta 1799:101–113
77. Li GQ, Liang XY, Lotze MT (2013) HMGB1: the central cytokine for all lymphoid cells. Front Immunol 4:68
78. Stott K, Watson M, Howe FS et al (2010) Tail-mediated collapse of HMGB1 is dynamic and occurs via differential binding of the acidic tail to the A and B domains. J Mol Biol 403:706–722
79. Zhao B, Kurgan L (2023) Databases of protein structure and function predictions at the amino acid level. In: Machine learning in bioinformatics of protein sequences, pp 329–353
80. Zhao B, Katuwawala A, Oldfield CJ et al (2021) DescribePROT: database of amino acid-level protein structure and function predictions. Nucleic Acids Res 49:D298–D308
81. Wang C, Brylinski M, Kurgan L (2019) PDID: database of experimental and putative drug targets in human proteome. In: Roy K (ed) In silico drug design. Academic, pp 827–847
82. Wang C, Hu G, Wang K et al (2015) PDID: database of molecular-level putative protein-drug interactions in the structural human proteome. Bioinformatics. https://doi.org/10.1093/bioinformatics/btv597
83. Barik A, Katuwawala A, Hanson J et al (2020) DEPICTER: intrinsic disorder and disorder function prediction server. J Mol Biol 432: 3379–3387
84. Wu Z, Basu S, Wu X et al (2023) qNABpredict: quick, accurate, and taxonomy-aware sequence-based prediction of content of nucleic acid binding amino acids. Protein Sci 32:e4544
85. Wang C, Kurgan L (2019) Review and comparative assessment of similarity-based methods for prediction of drug-protein interactions in the druggable human proteome. Brief Bioinform 20:2066–2087
86. Oldfield CJ, Fan X, Wang C et al (2020) Computational prediction of intrinsic disorder in protein sequences with the disCoP meta-predictor. Methods Mol Biol 2141:21–35
87. Fan X, Kurgan L (2014) Accurate prediction of disorder in protein chains with a comprehensive and empirically designed consensus. J Biomol Struct Dyn 32:448–464
88. Yan J, Kurgan L (2017) DRNApred, fast sequence-based method that accurately predicts and discriminates DNA- and RNA-binding residues. Nucleic Acids Res 45: e84
89. Meng F, Wang C, Kurgan L (2018) fDETECT webserver: fast predictor of propensity for protein production, purification, and crystallization. BMC Bioinf 18:580
90. Mizianty MJ, Fan X, Yan J et al (2014) Covering complete proteomes with X-ray structures: a current snapshot. Acta Crystallogr D Biol Crystallogr 70:2781–2793
91. Mizianty MJ, Peng ZL, Kurgan L (2013) MFDp2: accurate predictor of disorder in proteins by fusion of disorder probabilities, content and profiles. Intrinsically Disord Proteins 1:e24428

92. Mizianty MJ, Stach W, Chen K et al (2010) Improved sequence-based prediction of disordered regions with multilayer fusion of multiple information sources. Bioinformatics 26: i489–i496
93. Zhang J, Kurgan L (2019) SCRIBER: accurate and partner type-specific prediction of protein-binding residues from proteins sequences. Bioinformatics 35:i343–i353
94. Ghadermarzi S, Krawczyk B, Song J et al (2021) XRRpred: accurate predictor of crystal structure quality from protein sequence. Bioinformatics 37:4366

Chapter 13

Prediction of Disordered Linkers Using APOD

Zhenling Peng, Haiyan Wu, Yuxian Luo, and Lukasz Kurgan

Abstract

Intrinsically disordered linkers (DLs) connect protein domains and structural elements within domains and facilitate allosteric regulation. Computational studies suggest that thousands of proteins have DLs. Since there are only about 250 proteins with manually curated DL annotations (DisProt database ver. 9.3), computational approaches that make accurate predictions of DLs from the protein sequences are essential for reducing this annotation gap. To this end, we recently released the Accurate Predictor Of DLs (APOD) method. Empirical tests show that APOD achieves Area Under the ROC Curve (AUC) of 0.82 and Matthews Correlation Coefficient (MCC) of 0.42 on a low-similarity test dataset. We implement APOD as a freely available and convenient web server at https://yanglab.qd.sdu.edu.cn/APOD/. This web server takes a protein sequence as the input and outputs an easy-to-parse prediction result, with the entire prediction process done on the server side. We also provide a standalone version of APOD for users who want to process large datasets of sequences. This version must be installed and run locally on the end user's computer. In this chapter, we overview APOD, explain how to locate and use the web server and the standalone implementation, and discuss how to read and interpret APOD's outputs. We also demonstrate utility of APOD based on a case study protein.

Key words Intrinsic disorder, Intrinsically disordered protein, Disordered linker, Web server, Support vector machine, Machine learning

1 Introduction

Intrinsically disordered proteins (IDPs) include one or more intrinsically disordered regions (IDRs), which are defined as sequence segments that lack stable three-dimensional structure under physiological conditions [1–4]. Studies have shown that the IDPs are abundant in nature, particularly in eukaryotes, and carry out a variety of cellular functions that include regulation, signaling, translation, and transcription [5–17]. They contribute to the formation of dark proteomes [18–20], are found across several subcellular compartments [21], and were recently identified as key drivers of phase separation [22, 23].

Andrzej Kloczkowski et al. (eds.), *Prediction of Protein Secondary Structure*, Methods in Molecular Biology, vol. 2867, https://doi.org/10.1007/978-1-0716-4196-5_13,

DisProt database is the primary source of the experimentally and functionally annotated IDRs [24, 25]. The most commonly annotated disorder function of IDRs in the current version 9.3 of this database is entropic chain [25], where 81% of the entropic chains are disordered linker (DL) regions. DLs typically serve as linkers between protein domains and between structural elements/ constituents that make up domains [26]. They were found to be involved in peptide aggregation [27], connecting disordered protein binding regions [28], facilitating movement of structured domains between catalytic sites [29], and in allosteric regulation [30]. DLs are different from linker regions in multiple ways [31], such as the degree of their conformational flexibility (disordered vs. ordered), the region length (DLs tend to be longer), and localization (inter- and intra-domain vs. inter-domain) [32]. A computational analysis suggests about 7% of human proteins are likely to have at least one long putative DL (20 or more consecutive residues in length) and 10% of human proteins have at least 30% of their residues predicted as DLs [32]. While these computational studies suggest that thousands of proteins have DLs, only a relative handful of these annotations is available, i.e., approximately 250 proteins with the curated DL annotations are available in the DisProt database version 9.3. Consequently, computational methods that make accurate predictions of DLs from protein sequences are needed to reduce this annotation gap.

Recent studies demonstrate that machine learning-based approaches can be used to develop accurate predictors of intrinsic disorder and various disorder functions [32–42]. To this end, we recently released APOD, an accurate machine learning-based predictor of DLs from protein sequences [42]. The article that introduces APOD shows that this tool offers state-of-the-art predictive performance. Using a low-similarity test dataset, APOD secures Area Under the ROC Curve (AUC) of 0.82 and Matthews Correlation Coefficient (MCC) of 0.42. APOD is freely available as a convenient web server at https://yanglab.qd.sdu.edu.cn/APOD/ that automates the entire prediction process on the server side. We also provide a standalone software package for APOD at https://yanglab.qd.sdu.edu.cn/download/APOD_20230321.tar.gz, which should be used to make predictions on a larger scale. This package must be installed on the user's computer and can make predictions for multiple protein sequences at the time. This chapter overviews architecture of the predictive model implemented in APOD, details how to access and run the user-friendly web server version of APOD and how to install the standalone version, discusses how to read and understand the outputs generated by APOD, and illustrates these predictions with a case study.

2 Methods

2.1 Architecture of the APOD Predictor

APOD takes a protein sequence as the input and processes it in three steps to produce the DL prediction (Fig. 1). In the first step, we apply three third-party programs including PSI-BLAST [43], DISOPRED3 [44], and SPARKS-X [45], to generate an information-rich *sequence profile* that characterizes relevant (to the DL prediction) putative properties of the input chain. In the second step, we convert this sequence profile into a vector of numeric features that can be used as an input for the predictive model. More precisely, we compute the amino acid composition (AAC) directly from the input sequence, calculate residue-level sequence conservation (CONS) values from the outputs of PSI-BLAST generated using the non-redundant (NR) sequence database [43], produce putative secondary structure (SS) content and relative solvent accessibility (RSA) from the SPARKS-X's predictions [45], and derive sliding window-based representation of the intrinsic disorder (Swdis) and protein-level value of the intrinsic disorder content (DisCon) from the results of DISOPRED3 [44]. We note that these features cover protein-level features (that quantify the disorder content) and sequence window-level features; Subheading 2.3 in the original publication details these features [42]. In the third step, we input these features into the Support Vector Machine (SVM) model that predicts DLs. We designed this SVM model

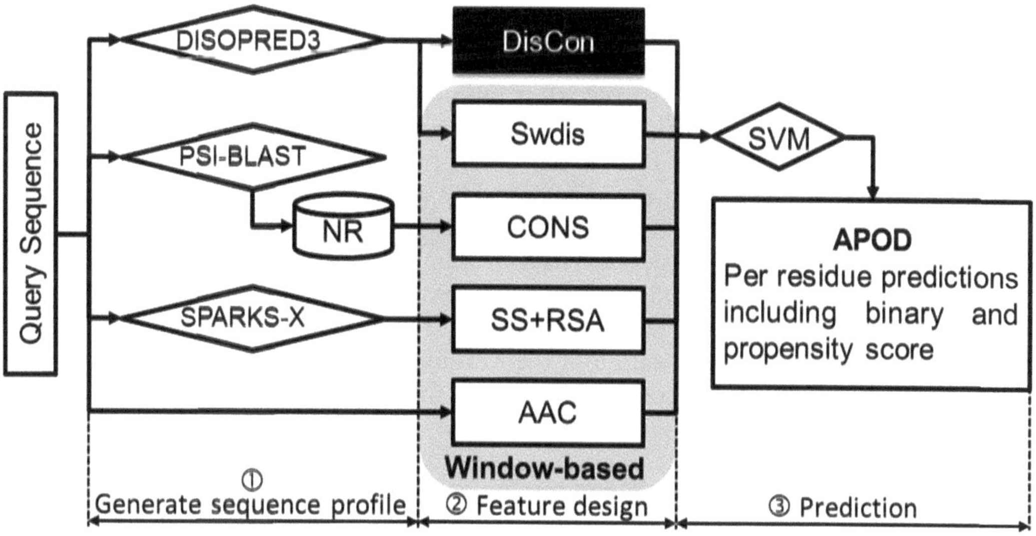

Fig. 1 Predictive architecture of APOD. Gray and black highlights identify the protein-level and the sequence window-based features, respectively. Abbreviations: amino acid composition (AAC), sequence conservation (CONS), secondary structure (SS), relative solvent accessibility (RSA), sliding window-based representation of the intrinsic disorder (Swdis), protein-level value of the intrinsic disorder content (DisCon), and Support Vector Machine (SVM)

using the machine learning approach, where internal parameters of the model are estimated from a training dataset to minimize differences between the model's outputs and the known ground truth annotations of DLs. The predictions consist of real-value propensity values for every amino acid in the input sequence. These values quantify likelihood that a given residue is in a DL region, and they are used to derive binary scores that categorize whether a given residue is located in DL or not (*see* **Note 1**).

2.2 Predictive Quality and Runtime of the APOD Predictor

The predictive performance of APOD was empirically evaluated using a test dataset with 82 proteins that includes 2223 DL residues. This test dataset was generated from the version 8.0 of the DisProt database [24] and is available at https://yanglab.qd.sdu.edu.cn/APOD/. Importantly, the test proteins are highly dissimilar, with <25% sequence similarity, to the data that was used to train/derive the SVM model; additional details are in the original publication [42].

A side-by-side comparison of APOD to its predecessor DFLpred [32], which is the only other sequence-based predictor of DLs, reveals that APOD provides substantially higher levels of predictive performance with MCC (that focuses on the binary predictions) equal 0.42 vs 0.15 and AUC (that evaluates the propensity values) equal 0.82 vs 0.64 [42]. We also quantify and report the runtime of APOD. This analysis was done on a modern server with Intel Xeon Platinum 8260 CPUs at 2.40GHz. We measured the runtime for each test protein. Next, we sorted these proteins by their sequence length and divided them into 10 equally sized subsets of protein with the increasing length values. This allows us to properly average the runtime values (for each subset) and study relation between the runtime and protein size. Figure 2 summarizes the average per-protein runtime values measured in seconds. This analysis reveals that the predictions generated by APOD take between about 60 and 500 s (between 1 and 9 min) for protein sequences with lengths that range between approximately 60 and 1000 residues. The dotted line in Fig. 2 denotes a linear fit into the empirically derived measurements of runtime in the function of the protein sequence length n. This function provides a good quality approximation of the runtime values, revealing that the execution time increases linearly with the protein size (*see* **Note 2**).

3 APOD Web Server and Standalone Implementation

APOD can be accessed by the end users in two complementary ways: a web server and standalone code. The web server is more suitable for small jobs (one or a few proteins) that are submitted by less computer-savvy users, given that it does not require installation or running any software. The standalone

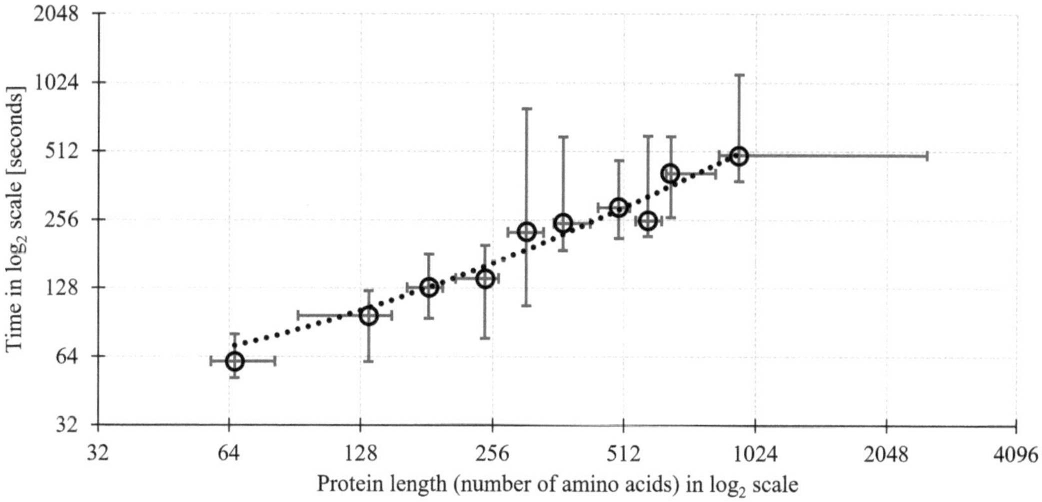

Fig. 2 The runtime of APOD measured on proteins from the test dataset. Each data point (black circle marker) is the runtime measured for a disjoint subset of 10% of test proteins sorted by their sequence length. The *y*-axis reports median runtime for each protein subset in seconds using the base 2 logarithmic scale. The *x*-axis shows the corresponding median protein length using the base 2 logarithmic scale. The error bars along both axes denote the 5th and 95th percentiles of the values in a given protein set. The dotted line shows the linear fit into the experimental values

version targets more computer-educated users since it requires installation of multiple pieces of software and allows for larger size jobs.

3.1 Web Server

We provide a convenient and free to use web server for APOD, which is located at https://yanglab.qd.sdu.edu.cn/APOD/. This web server performs all computations on the server side and requires the FASTA-formatted protein sequences as the only required input (*see* **Note 3**). The web server accepts a single protein sequence at the time (*see* **Note 4**). We advise users who want to predict many sequences to install and use APOD on their own computer; we discuss the standalone package in Subheading 3.2. The results are delivered directly in the web browser window and are also sent to the user's email address, if provided (*see* **Note 5**).

Three easy steps, which are highlighted in red font in Fig. 3, are needed to run the APOD web server:

1. Copy and paste the FASTA-formatted protein sequence into the input box, or upload a file with the sequence using the "upload the sequence file" option.
2. Provide the email address where the notification of the completed results will be delivered. This step is optional.
3. Click the "submit" button. The "reset" button clears the input data.

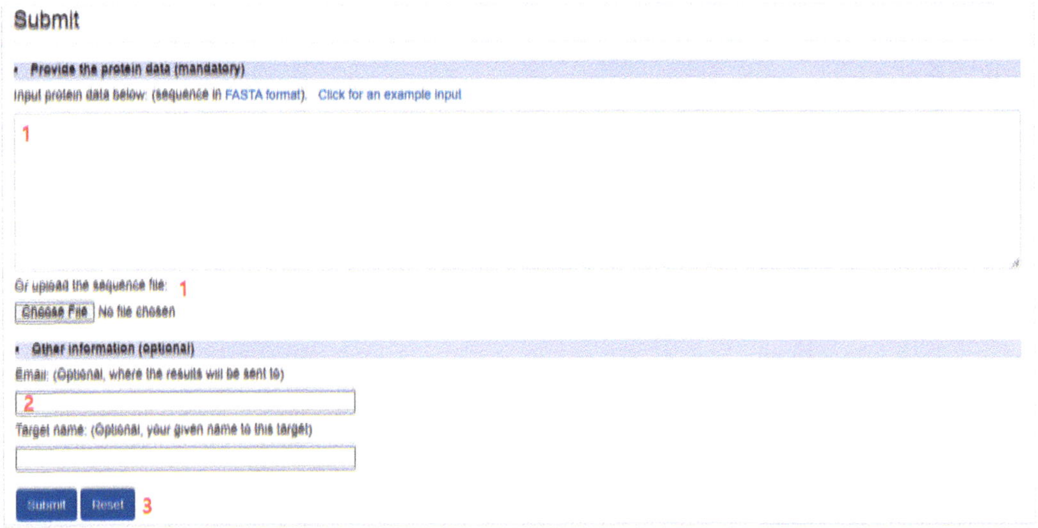

Fig. 3 The input web page of the APOD web server. The numbers in red font identify the three main steps that are needed to set up the prediction

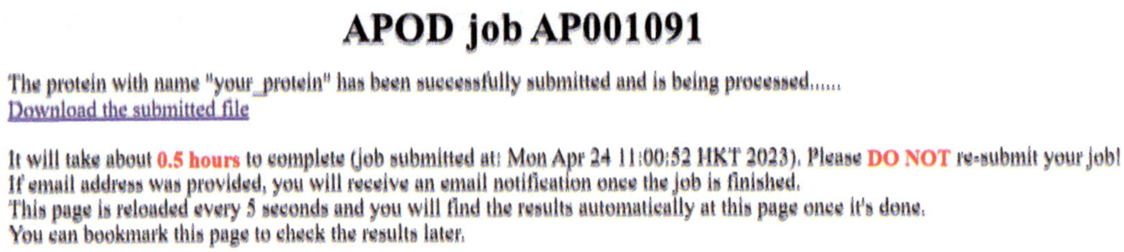

Fig. 4 The job processing web page of the APOD web server

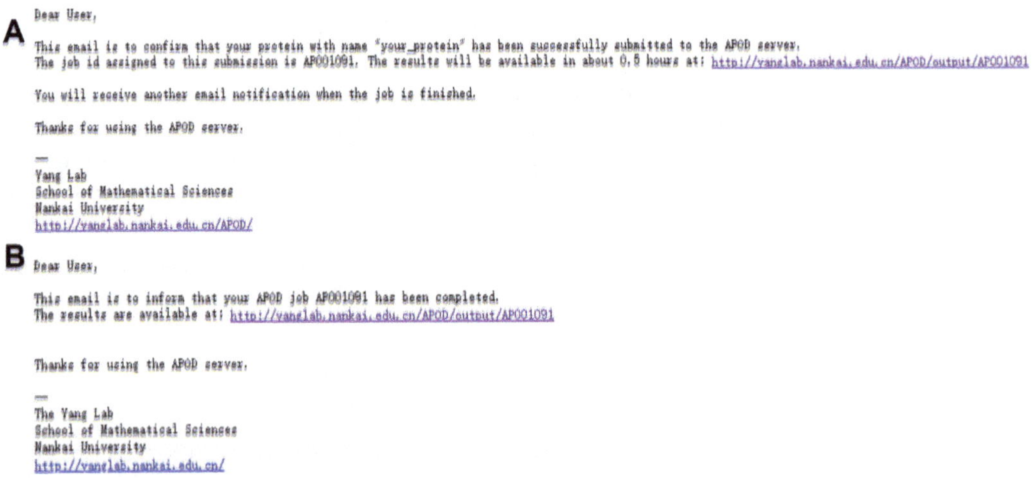

Fig. 5 The notification emails sent by the APOD web server. Users should receive two emails. The first email (Panel A) is sent after the job submission. The second email (Panel B) is sent after the job is completed, notifying the users that the predictions are ready to be downloaded

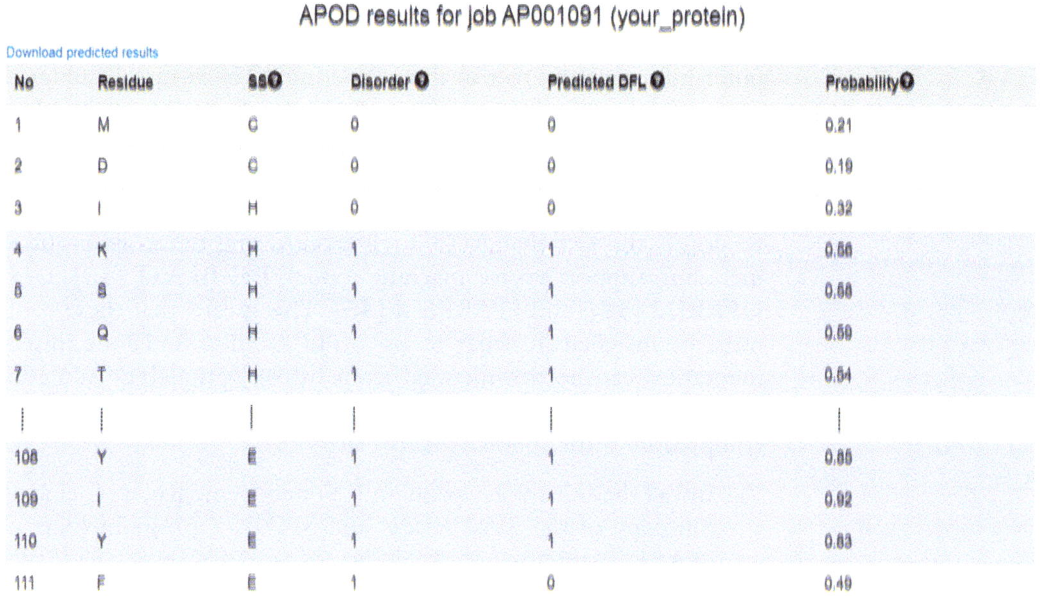

APOD results for job AP001091 (your_protein)

Download predicted results

No	Residue	SS	Disorder	Predicted DFL	Probability
1	M	C	0	0	0.21
2	D	C	0	0	0.19
3	I	H	0	0	0.32
4	K	H	1	1	0.56
5	S	H	1	1	0.58
6	Q	H	1	1	0.59
7	T	H	1	1	0.54
⋮	⋮	⋮	⋮	⋮	⋮
108	Y	E	1	1	0.85
109	I	E	1	1	0.92
110	Y	E	1	1	0.83
111	F	E	1	0	0.49

Fig. 6 Web page of the APOD web server that visualizes prediction results

After submitting the request, the web browser window is redirected to a new web page that shows the processing status for the prediction (Fig. 4). Each request is tagged with a job ID in the AP* format, where * denotes a six-digit number (*see* **Note 6**). The job processing page refreshes automatically until the request is completed, which is when the browser is redirected to the web page with the prediction results. In fact, users who provide email address receive two messages, the first that acknowledges the submission, which is delivered right after the request is submitted (Fig. 5a), and the second that marks completion of the submitted job (Fig. 5b). The second email typically arrives within 24 h of the submission.

The web page that shows outputs of the APOD server (Fig. 6) is organized in a tabular format where rows correspond to individual residues in the input sequence and columns provide information generated by the predictor. This information includes the amino acid type ("Residue" column), its position in the protein chain ("No" column), the putative secondary structure that is produced by SPARKS-X ("SS" column) [45], the putative intrinsic disorder which is generated by DISOPRED3 [44] ("Disorder" column), and the binary prediction ("Predicted DFL" column) and the real-valued propensity ("Probability" column) that are output by APOD. The putative DL regions are highlighted with the purple background. The "?" icons provide a brief explanation for the information in the corresponding column.

3.2 Standalone Implementation

The standalone package of APOD is aimed at users capable of installing and compiling programs and who have their own computer infrastructure to run the predictions. The main advantage is the ability to run the predictions locally and possibly embedding APOD into other bioinformatics platforms. This option also facilitates the ability to perform large jobs for hundreds or thousands of proteins. A particularly challenging aspect of using the standalone version is the fact that APOD relies on several other applications including support vector machine model, PSI-BLAST [43], DISOPRED3 [44], and SPARKS-X [45]. This means that these tools must be installed in order to use APOD. The standalone implementation can be downloaded from https://yanglab.qd.sdu.edu.cn/download/APOD_20230321.tar.gz. The installation requires completion of the following three steps:

1. Install the third-party programs. For convenience, we include a full copy of the SVM model and SPINE-X program (part of SPARKS-X that we use to derive the putative secondary structure) in the APOD's standalone package. Information about how to obtain copies of the DISOPRED3 and PSI-BLAST programs is available at https://yanglab.qd.sdu.edu.cn/download/. Users should follow instructions described on their download pages and/or the "readMe" files to install these applications.
2. Compile the APOD program by using the included scripts. Users should run the Perl script "makefile.pl" to generate the binary files that implement the APOD predictor. Before running APOD, users should double-check the setting of the absolute path for the third-party programs in the "APOD.pl" and "APOD_multi.pl" files to ensure that APOD can correctly call these programs. In this step, users can speed up APOD by adjusting the value of parameter "-a" to run PSI-BLAST (*see* **Note 7**).
3. Run the APOD program. Users should utilize the "APOD.pl" and "APOD_multi.pl" scripts to make DL prediction for a single protein and multiple sequences, respectively. Since "APOD_multi.pl" relies on "APOD.pl," we advise users to use "APOD.pl" to check if it executes correctly.

Users should refer to the "readMe" file in the APOD's standalone package for further details. The APOD package takes the FASTA-formatted protein sequences (*see* **Note 3**) as the input and outputs predictions for each input sequence using the following format:

- Line 1: >protein name
- Line 2: 1-letter encoded amino acid sequence

- Line 3: binary prediction, where 1 and 0 indicate that a given residue is predicted as the DL residue and the non-DL residue, respectively (*see* **Note 1**)
- Line 4: comma-separated propensity scores that quantify likelihood of a given residue to be located in a DL region

4 Case Study

The ADRM1 protein (UniProt ID: Q16186) is part of a multimeric 26S proteasome complex. This complex participates the ATP-dependent degradation of ubiquitinated proteins, plays a key role in the maintenance of protein homeostasis by removing misfolded or damaged proteins, and is involved in several cellular processes, such as DNA damage repair and apoptosis [46–53]. Serving as a proteasomal ubiquitin receptor in 26S proteasome [50], ADRM1 engages and activates the deubiquitinases UCHL5 and PSMD14 during protein degradation [47–49, 51], where both of them are a component of the 26S proteasome, and UCHL5 is associated with the 19S regulatory subunit. Intrinsic disorder is instrumental to these diverse functions. ADRM1 has three IDRs that are annotated in the DisProt database (DisProt ID: DP00839): 1M to 21K, 132P to 285Q, and 386K to 407D. The middle IDR includes a DL region 132P to 252A (red horizontal line in Fig. 7) and a disordered protein binding region 253S to 285Q. Moreover, UniProt [54] identifies two domains: the Pru domain (18A to 131P) and the DEUBAD domain (277P to 391E (orange horizontal bar in Fig. 6). The DL region connects these two domains.

We use the ADRM1 sequence to make predictions with APOD. We show these predictions in Fig. 7 as the blue horizontal bar (binary predictions) and the black line (putative propensity scores) (*see* **Note 1**). We find that the binary predictions (blue bar) align relatively accurately with the location of the native DL (red bar). The APOD also correctly identifies both IDRs at the sequence termini (green bars) as non-DLs, producing low propensities and binary predictions of 0 (no blue bar). This suggests that APOD is able to discriminate between IDRs that are DLs and other types of disordered regions. Moreover, we note that the putative propensities generated by APOD provide useful context for the binary predictions. The residues that are associated with either high (near 1) or low (near 0) propensities are more likely to correspond to accurate predictions when compared to the residues that have propensities close to the 0.5 threshold value. For instance, we observe a relatively high number of residues with the propensities around 0.5 in the 253S to 285Q IDRs that is partly and incorrectly predicted as DL. This suggests that users should use the binary

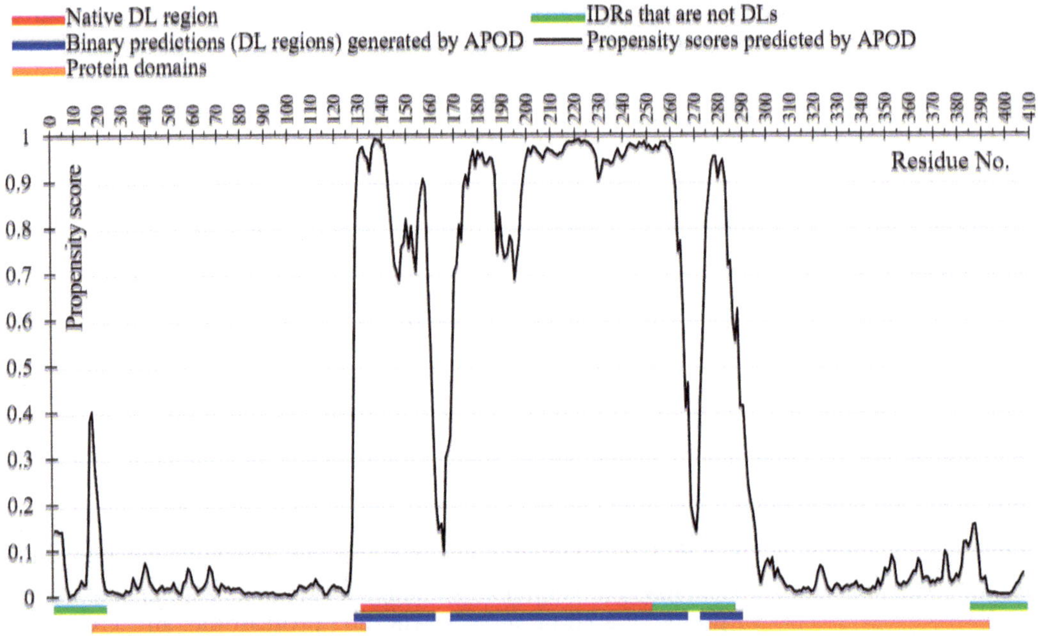

Fig. 7 Native annotations and predictions generated by APOD for the ADRM1 protein (UniProt ID: Q16186; DisProt ID: DP00839). This protein has three native IDRs: 1M to 21K, 132P to 285Q, and 386K to 407D, where the 132P to 252A segment is the DL region and the 253S to 285Q segment is annotated as protein-binding IDR. We show the corresponding native DL region and the other disordered non-DL regions as the red and green horizontal bars at the bottom, respectively. We show the APOD's predictions as black line (propensity scores) and blue horizontal bar (binary scores). We annotate the two known domains for this protein based on UniProt annotations as orange horizontal bar at the bottom

predictions to identify locations of putative DLs and should contextualize this analysis with the predicted propensities that inform about the likelihood that a given prediction is accurate.

5 Notes

1. The binary prediction is generated from the predicted propensity scores using the threshold = 0.5. This means that the residues with the propensity >0.5 are assigned with the binary value 1 (they are part of a DL region) and the remaining residues are assigned with the binary value 0.
2. The runtime t quantified in seconds for a protein with n residues can be estimated using the following formula, $t = 0.4916*n + 39.068$. We establish this formula based on the linear fit into the empirical data from Fig. 2. This formula can be used to estimate the runtime for a protein sequence of any size. For protein with $n = 300$ residues, the estimated runtime = 187 seconds = 3.11 minutes. For proteins with

$n = 1000$ residues, the approximate runtime = 531 seconds = 8.84 minutes.

3. Server accepts the input sequences in the FASTA format, using the single-letter amino acid encoding. This format is described at https://en.wikipedia.org/wiki/FASTA_format.
4. Although APOD only takes about 3 min for an average protein of 300 residues, it may take hours to process this prediction since the underlying web server is used to support many other bioinformatics servers/tools. The jobs/requests for these tools are placed in a single job queue that is processed in the first-in-first-out fashion.
5. Keeping the web browser window open for a long time to collect results could be prohibitive. Therefore, we recommend that the users provide their email addresses where they will be notified when their job/request is completed, with a link to the location of the results. In case if the email is not provided and the browser window is closed/terminated, the users will not be able to obtain the link to the location of the results.
6. Users should store the job ID AP* for future reference, where * is a six-digit number. They can retrieve the prediction results directly from the job-specific page address https://yanglab.qd.sdu.edu.cn/APOD/output/AP*/. The results of the APOD prediction are kept for 2 months under this link. The job ID AP* and the corresponding link are sent via email when the prediction is finished, if the user provides the email address. We encourage users to copy the AP* or the link from the web browser if they did not provide the email address.
7. The standalone package of APOD sets parameter "-a 4" to run PSI-BLAST, i.e., this setting utilizes 4 CPUs to run this program in parallel. If a user has a more powerful server (with more than 4 CPUs), then they should adjust this parameter accordingly. PSI-BLAST is a major bottleneck of the APOD application, and so adjusting this parameter appropriately should have substantial impact on the overall runtime of APOD. Moreover, user can speed up DISOPRED3 in the same way since this application is also using PSI-BLAST.

Acknowledgment

This work was supported in part by the National Natural Science Foundation of China (grants T2222012 and 61873185) to Zhenling Peng.

References

1. Van Der Lee R, Buljan M, Lang B et al (2014) Classification of intrinsically disordered regions and proteins. Chem Rev 114:6589–6631
2. Oldfield CJ, Uversky VN, Dunker AK et al (2019) Introduction to intrinsically disordered proteins and regions. In: Salvi N (ed) Intrinsically disordered proteins. Academic, pp 1–34
3. Dunker AK, Babu MM, Barbar E et al (2013) What's in a name? Why these proteins are intrinsically disordered. Intrinsically Disord Proteins 1:e24157
4. Habchi J, Tompa P, Longhi S et al (2014) Introducing protein intrinsic disorder. Chem Rev 114:6561–6588
5. Xie H, Vucetic S, Iakoucheva LM et al (2007) Functional anthology of intrinsic disorder. 1. Biological processes and functions of proteins with long disordered regions. J Proteome Res 6:1882–1898
6. Peng Z, Yan J, Fan X et al (2015) Exceptionally abundant exceptions: comprehensive characterization of intrinsic disorder in all domains of life. Cell Mol Life Sci 72:137–151
7. Wang C, Uversky VN, Kurgan L (2016) Disordered nucleiome: abundance of intrinsic disorder in the DNA- and RNA-binding proteins in 1121 species from Eukaryota, Bacteria and Archaea. Proteomics 16:1486–1498
8. Lieutaud P, Ferron F, Uversky AV et al (2016) How disordered is my protein and what is its disorder for? A guide through the "dark side" of the protein universe. Intrinsically Disord Proteins 4:e1259708
9. Meng F, Na I, Kurgan L et al (2015) Compartmentalization and functionality of nuclear disorder: intrinsic disorder and protein-protein interactions in intra-nuclear compartments. Int J Mol Sci 17:24
10. Babu MM (2016) The contribution of intrinsically disordered regions to protein function, cellular complexity, and human disease. Biochem Soc Trans 44:1185–1200
11. Kjaergaard M, Kragelund BB (2017) Functions of intrinsic disorder in transmembrane proteins. Cell Mol Life Sci 74:3205–3224
12. Dunker AK, Silman I, Uversky VN et al (2008) Function and structure of inherently disordered proteins. Curr Opin Struct Biol 18: 756–764
13. Chen J, Kriwacki RW (2018) Intrinsically disordered proteins: structure, function and therapeutics. J Mol Biol 430:2275–2277
14. Zhao B, Katuwawala A, Oldfield CJ et al (2021) Intrinsic Disorder in Human RNA-Binding Proteins. J Mol Biol 433: 167229
15. Hu G, Wu Z, Uversky VN et al (2017) Functional analysis of human hub proteins and their interactors involved in the intrinsic disorder-enriched interactions. Int J Mol Sci 18:2761
16. Peng Z, Oldfield CJ, Xue B et al (2014) A creature with a hundred waggly tails: intrinsically disordered proteins in the ribosome. Cell Mol Life Sci 71:1477–1504
17. Peng Z, Mizianty MJ, Xue B et al (2012) More than just tails: intrinsic disorder in histone proteins. Mol BioSyst 8:1886–1901
18. Hu G, Wang K, Song J et al (2018) Taxonomic landscape of the dark proteomes: whole-proteome scale interplay between structural darkness, intrinsic disorder, and crystallization propensity. Proteomics 18:e1800243
19. Kulkarni P, Uversky VN (2018) Intrinsically disordered proteins: the dark horse of the dark proteome. Proteomics 18:e1800061
20. Bhowmick A, Brookes DH, Yost SR et al (2016) Finding our way in the dark proteome. J Am Chem Soc 138:9730–9742
21. Zhao B, Katuwawala A, Uversky VN et al (2020) IDPology of the living cell: intrinsic disorder in the subcellular compartments of the human cell. Cell Mol Life Sci. https://doi.org/10.1007/s00018-020-03654-0
22. Ibrahim AY, Khaodeuanepheng NP, Amarasekara DL et al (2023) Intrinsically disordered regions that drive phase separation form a robustly distinct protein class. J Biol Chem 299:102801
23. Uversky VN (2021) Recent developments in the field of intrinsically disordered proteins: intrinsic disorder-based emergence in cellular biology in light of the physiological and pathological liquid-liquid phase transitions. Annu Rev Biophys 50:135–156
24. Hatos A, Hajdu-Soltesz B, Monzon AM et al (2020) DisProt: intrinsic protein disorder annotation in 2020. Nucleic Acids Res 48: D269–D276
25. Quaglia F, Meszaros B, Salladini E et al (2022) DisProt in 2022: improved quality and accessibility of protein intrinsic disorder annotation. Nucleic Acids Res 50:D480–D487
26. Dunker AK, Brown CJ, Lawson JD et al (2002) Intrinsic disorder and protein function†. Biochemistry 41:6573–6582
27. Shvadchak VV, Subramaniam V (2014) A four-amino acid linker between repeats in the alpha-Synuclein sequence is important for fibril formation. Biochemistry 53:279–281

28. Oldfield CJ, Dunker AK (2014) Intrinsically disordered proteins and intrinsically disordered protein regions. Annu Rev Biochem 83:553–584
29. Anand S, Mohanty D (2012) Inter-domain movements in polyketide synthases: a molecular dynamics study. Mol BioSyst 8:1157–1171
30. Sorensen CS, Kjaergaard M (2019) Effective concentrations enforced by intrinsically disordered linkers are governed by polymer physics. Proc Natl Acad Sci USA 116:23124–23131
31. Chen X, Zaro JL, Shen WC (2013) Fusion protein linkers: property, design and functionality. Adv Drug Deliv Rev 65:1357–1369
32. Meng F, Kurgan L (2016) DFLpred: high-throughput prediction of disordered flexible linker regions in protein sequences. Bioinformatics 32:i341–i350
33. Zhao B, Kurgan L (2023) Machine learning for intrinsic disorder prediction. In: Machine learning in bioinformatics of protein sequences, pp 205–236
34. Zhao B, Kurgan L (2021) Surveying over 100 predictors of intrinsic disorder in proteins. Expert Rev Proteomics 18:1019–1029
35. Basu S, Kihara D, Kurgan L (2023) Computational prediction of disordered binding regions. Comput Struct Biotechnol J 21:1487–1497
36. Kurgan L, Li M, Li Y (2021) The methods and tools for intrinsic disorder prediction and their application to systems medicine. In: Wolkenhauer O (ed) Systems medicine. Academic, Oxford, pp 159–169
37. Katuwawala A, Peng Z, Yang J et al (2019) Computational prediction of MoRFs, short disorder-to-order transitioning protein binding regions. Comput Struct Biotechnol J 17:454–462
38. Kurgan L (2022) Resources for computational prediction of intrinsic disorder in proteins. Methods 204:132–141
39. Liu Y, Wang X, Liu B (2019) A comprehensive review and comparison of existing computational methods for intrinsically disordered protein and region prediction. Brief Bioinform 20:330–346
40. Katuwawala A, Ghadermarzi S, Kurgan L (2019) Computational prediction of functions of intrinsically disordered regions. Prog Mol Biol Transl Sci 166:341–369
41. Morgunov AS, Saar KL, Vendruscolo M et al (2021) New frontiers for machine learning in protein science. J Mol Biol 433:167232
42. Peng Z, Xing Q, Kurgan L (2020) APOD: accurate sequence-based predictor of disordered flexible linkers. Bioinformatics 36:i754–i761
43. Altschul SF, Madden TL, Schaffer AA et al (1997) Gapped BLAST and PSI-BLAST: a new generation of protein database search programs. Nucleic Acids Res 25:3389–3402
44. Meszaros B, Erdos G, Dosztanyi Z (2018) IUPred2A: context-dependent prediction of protein disorder as a function of redox state and protein binding. Nucleic Acids Res 46:W329–W337
45. Yang Y, Faraggi E, Zhao H et al (2011) Improving protein fold recognition and template-based modeling by employing probabilistic-based matching between predicted one-dimensional structural properties of query and corresponding native properties of templates. Bioinformatics 27:2076–2082
46. Jorgensen JP, Lauridsen AM, Kristensen P et al (2006) Adrm1, a putative cell adhesion regulating protein, is a novel proteasome-associated factor. J Mol Biol 360:1043–1052
47. Yao T, Song L, Xu W et al (2006) Proteasome recruitment and activation of the Uch37 deubiquitinating enzyme by Adrm1. Nat Cell Biol 8:994–1002
48. Hamazaki J, Iemura S, Natsume T et al (2006) A novel proteasome interacting protein recruits the deubiquitinating enzyme UCH37 to 26S proteasomes. EMBO J 25:4524–4536
49. Qiu XB, Ouyang SY, Li CJ et al (2006) hRpn13/ADRM1/GP110 is a novel proteasome subunit that binds the deubiquitinating enzyme, UCH37. EMBO J 25:5742–5753
50. Husnjak K, Elsasser S, Zhang N et al (2008) Proteasome subunit Rpn13 is a novel ubiquitin receptor. Nature 453:481–488
51. Jiao L, Ouyang S, Shaw N et al (2014) Mechanism of the Rpn13-induced activation of Uch37. Protein Cell 5:616–630
52. Sahtoe DD, Van Dijk WJ, El Oualid F et al (2015) Mechanism of UCH-L5 activation and inhibition by DEUBAD domains in RPN13 and INO80G. Mol Cell 57:887–900
53. Vander Linden RT, Hemmis CW, Schmitt B et al (2015) Structural basis for the activation and inhibition of the UCH37 deubiquitylase. Mol Cell 57:901–911
54. Uniprot C (2023) UniProt: the universal protein knowledgebase in 2023. Nucleic Acids Res 51:D523–D531

Chapter 14

Computational Prediction of Linear Interacting Peptides

Nawar Malhis and Jörg Gsponer

Abstract

Intrinsically disordered protein regions, IDRs, are observed in many eukaryotic proteins. They play critical roles in essentially all cellular processes because segments of these regions, known as linear interacting peptides (LIPs), are heavily involved in regulatory protein interactions across proteomes. This chapter presents an integrated summary of the results from the last two Critical Assessments of protein Intrinsic Disorder predictions, known as CAID events, on the computational prediction of LIP segments. Because the CAID community questioned the quality of the test dataset used by the first CAID event, we reannotated this dataset using more accurate annotations from the latest DisProt database release. Then, we compared the results of the first CAID with the updated data and the results of the second CAID event. Our comparison highlights the importance of data annotation on the evaluation outcome and provides recommendations for users of LIP predictors.

Key words CAID, Prediction, Intrinsic disorder, MoRF, Linear interacting peptides LIP, Protein protein interactions

1 Introduction

Until the early 1990s, we build our understanding of proteins within the framework of the structure–function paradigm, which states that proteins' tertiary structures determine their function [1]. Today, it is clear that a significant portion of proteins, and protein regions, lack a unique tertiary structure when alone in solution under physiological conditions. Proteins with large intrinsically disordered protein regions (IDRs) constitute about one-third of eukaryotic proteomes. They are heavily involved in different types of cellular processes [2–5]. Fundamental to the contribution of IDRs to cell physiology is often their ability to interact with partner biomolecules such as DNA, RNA, and other proteins [2–5]. Modes of interactions of IDRs with partners are quite diverse, as are the sequence elements that mediate these interactions. ELMs and SLiMs, which refer to Eukaryotic Linear Motifs/Short Linear Motifs, are short conserved sequences mostly

Andrzej Kloczkowski et al. (eds.), *Prediction of Protein Secondary Structure*, Methods in Molecular Biology, vol. 2867, https://doi.org/10.1007/978-1-0716-4196-5_14,

from 3 to 11 amino acids that are able to recognize specific partner domains [6–8]. MoREs and MoRFs, molecular recognition elements/features, are short- to medium-length disordered sequences with up to 70 amino acids that undergo a disorder-to-order transition upon binding to partners [9–11]. Finally, disordered DNA and RNA-binding domains encompass all segments in IDRs that bind to nucleotide sequences. These different types of sequence segments that mediate interactions of IDRs have recently been labeled under the umbrella term of linear interacting peptides (LIPs) [12].

Proteins with IDRs and LIPs are believed to have certain advantages over mainly folded proteins as far as interactions are concerned. The flexibility of IDRs and their LIPs enables them to adopt different secondary and tertiary structures, which can fit the molecular surfaces of diverse biomolecular partners. Individual IDRs have been shown to be able to interact with multiple, structurally quite diverse partners [13, 14]. Interactions with diverse partners are further supported by multiple LIPs in the same protein. Moreover, IDRs are often subject to alternative splicing and post-translational modifications (PTMs), specifically phosphorylation, both cellular processes known to contribute decisively to the regulation and diversification of protein interactions. Recently, it has even been proposed that IDRs are particularly equipped to help in the rewiring of protein interactions in a tissue- and cell-type-specific manner [15]. Again, the conformational flexibility and lack of a unique tertiary structure of IDRs are believed to facilitate alternative splicing and PTMs, because these features of IDRs entail reduced structural constraints. Consequently, proteins with IDRs are over-represented among the hubs of eukaryotic interactomes [1] and major signaling pathways [3].

Given the vital role LIP-mediated interactions play in cellular processes, identifying LIPs from sequence information only has become of significant interest to the computational biology community and led to the development of many computational tools that predict the location of LIPs in IDRs [9, 16–48]. Many of these methods are now broadly used by the scientific community to further our understanding of IDR-mediated interactions, supporting both computational and experimental studies [49–54]. To assess the accuracy of these computational tools, LIP predictors were included in the recent community-driven CAID evaluations. CAID [55], the critical assessment of protein intrinsic disorder predictions, is a community-based effort launched in 2018 to understand and evaluate computational methods that predict IDRs and LIPs. Two CAID events have been completed thus far, CAID1 in 2018 [55] and CAID2 in 2022. In each CAID contest, manually curated sets of protein disorder regions and LIPs are assembled and used as ground truth to evaluate the accuracy of predictors. Predictors are then ranked according to different performance metrics and recommendations for usage provided.

This chapter reviews the CAID results for 15 computational tools that predict protein-binding LIPs and participated in at least one of the CAID events. We discuss metrics used in their evaluation and highlight some challenges facing the prediction of IDR's binding segments.

2 Materials and Methods

2.1 CAID Test Datasets

The two CAID rounds used fair benchmarking data for evaluating and contrasting the participating prediction tools because these benchmarking data were collected after training the tested methods. However, even such community-based assessments have shortcomings, given the complexity of the subject and the difficulties in defining the ground truth for LIPs. For instance, the CAID community highlighted in the comprehensive study published in 2021 that the CAID1 LIP dataset, which was annotated between June and November 2018, is "strongly unbalanced" because most IDRs used in the test set do not include any interacting (LIP) segments, and the LIP segments in the remaining IDRs are often loosely defined [55]. In this chapter, we are using the two CAID test datasets, CAID1 and CAID2. In addition, to quantify the effect of the strong imbalance in the CAID1 data on the results of the first CAID event, we created an updated version of that dataset, uCAID1, based on more accurate annotations.

The first CAID binding dataset, CAID1, encompasses 646 manually curated protein sequences with IDR annotations, including 232 sequences with LIP annotations. Two of these sequences include non-standard amino acids and failed to be processed by some predictors. The remaining 644 sequences have 335,742 residues, including 21,027 residues in 254 LIP segments.

The updated CAID1 binding dataset, uCAID1, utilizes annotations of the CAID1 sequences that we updated using the latest DisProt release (2022_12). In line with the CAID2 dataset, uCAID1 only includes sequences with at least one LIP. This dataset consists of 240 sequences with 133,051 residues, including 22,236 LIP residues in 269 LIP segments and 152 residues belonging to ambiguous LIP segments, which we excluded from the assessment. Table 1 shows the annotation overlap between the CAID1 and uCAID1 datasets. Overall, the annotation of 4175 residues in CAID1 is changed (Table 1), meaning that 1.2% of non-LIP and 1.8% of LIP residues in CAID1 are reannotated.

Unlike the CAID1 dataset, the second CAID binding dataset, CAID2, includes IDRs that contain LIPs exclusively. Only 62 out of the 78 sequences in CAID2 are scored by all the predictors because AlphaFold-binding requires AlphaFold structures. These 62 sequences involve 39,736 amino acids, including 72 LIP segments with 6715 residues.

Table 1
Annotation overlap between CAID1 and uCAID1. X represents residues that we excluded from the assessment because they belong to ambiguous LIP segments

CAID1	uCAID1	Non-LIP	LIP	X
	Non-LIP	110,288	3648	152
	LIP	375	18,588	0

Sequences included in these intersections are limited to those shared by both sets

2.2 Prediction Tools

We grouped prediction methods that participated in the CAID1 competition as first-generation (G1) tools. We did so because these methods were optimized using smaller and lower-quality training data compared to those developed later and participating only in CAID2. In addition, G1 tools were developed before the widespread use of deep learning methodologies and, thus, are based on other machine learning models such as support vector machines (SVM). The G1 group includes ANCHOR2 [35], fMoRFpred [28], MoRFchibi [30], OPAL [36], DISOPRED3 [25], and DisoRDPbind [29]. The second-generation (G2) tools participated in CAID2 but not in CAID1, including AlphaFold-binding [47], CLIP [48], DeepDISOBind [45], DisoBindPred [56], DRPBind [56], and ENSHROUD [56]. While most of these tools are based on machine learning models, ANCHOR2 and AlphaFold-binding are simple statistical models with only a few optimizable parameters. CAID revealed that the performances of tools that predict DNA or RNA binding of LIPs are much poorer than those that predict protein-binding LIPs. The reason for this low performance might be the fact that protein-binding LIPs highly dominate the training and testing datasets. Thus, we excluded these tools from this review.

2.3 Evaluation Metrics

The standard metrics used to evaluate LIP predictions in the first CAID event [55] can be divided into two categories: metrics that exploit the range of predicted propensity scores and those based on binary predictions (LIP vs. non-LIP) obtained by a single classification threshold. The former metrics include the receiver operating characteristic curve (ROC), the area under this ROC curve (AUC), and the precision/recall curve. The latter metrics include the F1-score and the Matthews correlation coefficient (MCC). All these metrics require the following quantities: the true positive (TP) is the number of LIP residues that are accurately predicted, false positive (FP) is the number of non-LIP residues that are mispredicted as LIP, true negative (TN) is the number of non-LIP residues that are accurately predicted as non-LIP, and

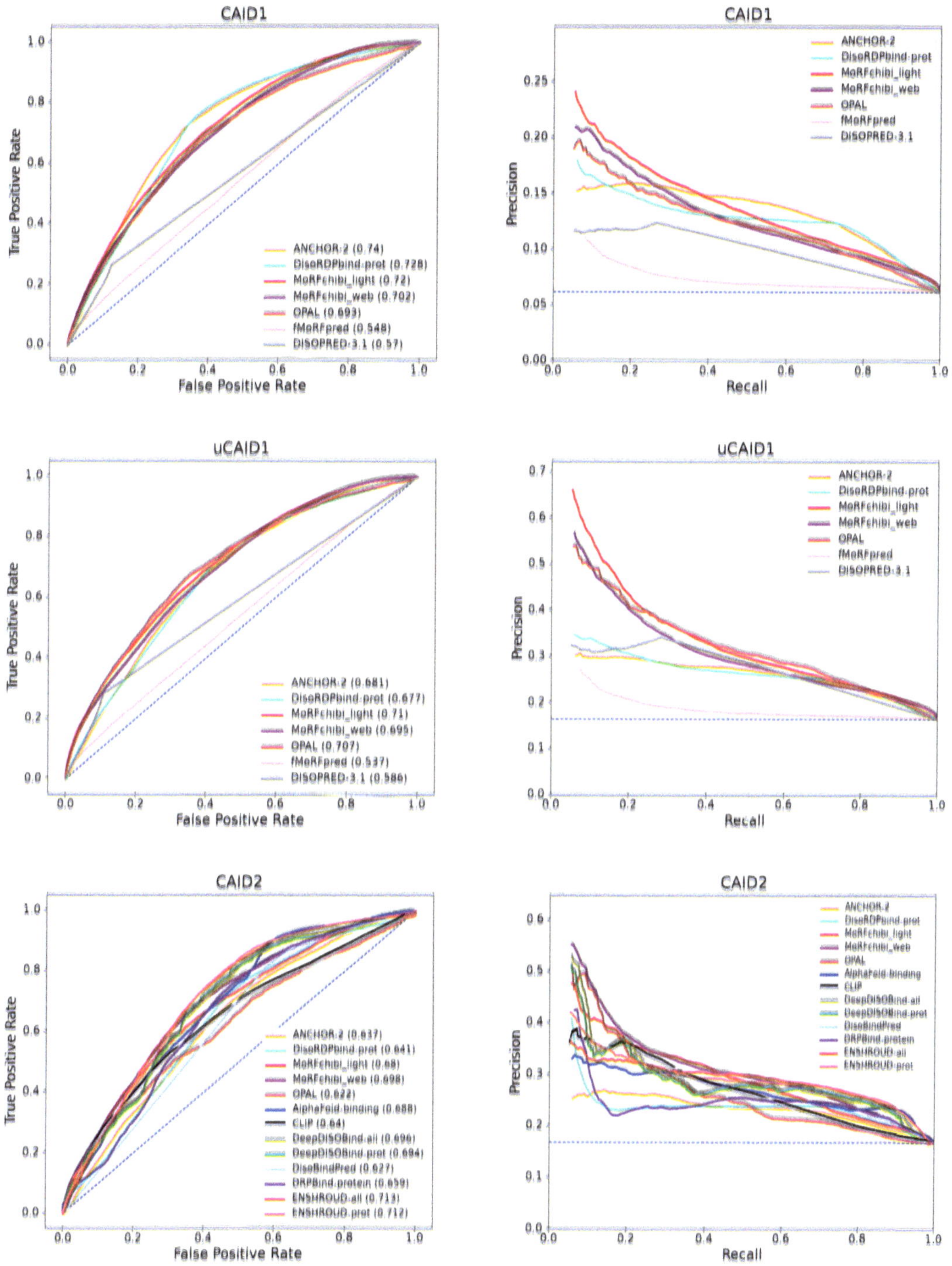

Fig. 1 Results for tested tools using the CAID1 (top), uCAID1 (middle), and CAID2 (lower) datasets. We only included sequences that all predictors process. Left) ROC curves with AUC values in parentheses. Right) Precision–recall curves. Precision scores for recall values less than 0.05 are excluded to reduce noise and improve figure clarity. The dotted line in the ROC curves represents the performance of a naïve/random predictor. The horizontal dotted line in the precision–recall curves represents the precision of a naïve/random predictor, i.e., the percentage of LIP residues in the input data

Table 2
The F1-scores, the MCC values, and the AUC values for the tested tools using the CAID1, the uCAID1, and the CAID2 datasets. We only included sequences that all predictors process

	CAID1 F1	uCAID1 F1	CAID2 F1	CAID1 MCC	uCAID1 MCC	CAID2 MCC	CAID1 AUC	uCAID1 AUC	CAID2 AUC
Sequences	*644*	*238*	*62*	*644*	*238*	*62*	*644*	*238*	*62*
Count LIP segments	*254*	*267*	*72*	*254*	*267*	*72*	*254*	*267*	*72*
Count LIP residues	*21,027*	*21,936*	*6715*	*21,027*	*21,936*	*6715*	*21,027*	*21,936*	*6715*
Count non-LIP residues	*314,715*	*110,059*	*33,021*	*314,715*	*110,059*	*33,021*	*314,715*	*110,059*	*33,021*
ANCHOR-2	0.226	0.357	0.335	0.188	0.194	0.151	0.740	0.681	0.637
DisoRDPbind-prot	0.173	0.244	0.204	0.110	0.133	0.065	0.728	0.677	0.641
MoRFchibi_Light	0.182	0.246	0.201	0.130	0.207	0.153	0.720	0.710	0.680
MoRFchibi_Web	0.185	0.256	0.240	0.128	0.192	0.173	0.702	0.695	0.698
OPAL	0.168	0.372	0.319	0.135	0.221	0.118	0.693	0.707	0.622
fMoRFpred	0.073	0.074		0.054	0.064		0.548	0.537	
DISOPRED-3.1	0.124	0.188		0.062	0.011		0.570	0.586	
AlphaFold-binding			0.371			0.230			0.688
CLIP			0.121			0.102			0.640
DeepDISOBind-all			0.383			0.226			0.696
DeepDISOBind-protein			0.290			0.153			0.694
DisoBindPred			0.349			0.178			0.627
DRPBind-protein			0.323			0.145			0.659
ENSHROUD-all			0.400			0.255			0.713
ENSHROUD-protein			0.399			0.252			0.712

The default cut-off threshold values are used to convert predicted propensity scores to binary values, which are used to compute the F1 scores and the MCC values

false negative (FN) is the number of LIP residues that are wrongly predicted non-LIP.

The ROC curve is used to evaluate the performance of binary classification tools. For every possible discrimination threshold score, the ROC curve charts the relationship between the percentage of the positive class passing that score (true-positive rate, TPR) and the percentage of the negative class passing the score (false-positive rate, FPR). The formal definitions of TPR and FPR are:

$$\mathrm{TPR} = \frac{\mathrm{TP}}{\mathrm{TP}+\mathrm{FN}} \;\&\; \mathrm{FPR} = \frac{\mathrm{FP}}{\mathrm{FP}+\mathrm{TN}}$$

For a random (naive) classifier, at any score, TPR is likely to be equal to FPR; thus, it is represented by the diagonal line in a ROC plot. The better the classifier, the higher the curve is located. Since TPR is a percentage of the positive, and FPR is a percentage of the negative, ROC curves are insensitive to dataset imbalance. ROC curves are often presented as the relationship between sensitivity (equal to TPR) and specificity (equal to 1-FPR). The lower left corner of the ROC curves is the most important, as it represents the prediction performance for residues with high prediction scores. Figure 1 shows the ROC curves for the predictions made by the discussed tools using the CAID1, uCAID1, and CAID2 datasets.

The AUC value is used for evaluating ROC curves. The AUC value is in the range of 0 to 1. For a random (naive) classifier, the AUC is the area under the diagonal line, i.e., AUC = 0.5. The better the classifier is, the higher the AUC value is. Since ROC curves are insensitive to dataset imbalance, so are AUC values. For a randomly selected positive class instance and a randomly selected negative class instance, the AUC is the probability of ranking the positive instance higher than the negative one [57]. Figure 1 and Table 2 show the AUC values for the predictions made by the discussed tools using the CAID1, uCAID1, and CAID2 datasets.

The precision–recall curve charts the relationship between the precision and the recall for every possible threshold score. The precision is the percentage of true positives in the predicted positive, and the recall is the percentage of true positives of the total positive.

$$\mathrm{precision} = \frac{\mathrm{TP}}{\mathrm{TP}+\mathrm{FP}} \;\&\; \mathrm{recall} = \frac{\mathrm{TP}}{\mathrm{TP}+\mathrm{FN}}$$

In contrast to ROC curves, precision–recall curves are sensitive to dataset imbalance. The left side of the precision–recall curves, at low recall values, is most important as it demonstrates the prediction performance for residues with high prediction scores and corresponds to the lower left corner of the ROC curves. Figure 1 shows the precision–recall curves for the predictions made by the discussed tools using the CAID1, uCAID1, and CAID2 datasets.

The F1 score is a single value representing both the precision and the recall at some classification threshold cut-off.

$$\mathrm{F1} = \frac{2 * \mathrm{precision} * \mathrm{recall}}{\mathrm{precision} + \mathrm{recall}}$$

The threshold cut-off value that maximizes the F1 score using some test datasets, F1-max, is often used as the optimal threshold cut-off to convert real propensity scores to binary values. Table 2 shows the F1 scores based on the default threshold values for all tools using the two CAID and the updated uCAID1 datasets.

The Matthews correlation coefficient (MCC) measures the correlation between the binary prediction and the ground truth values. Unlike the F1 score, MCC does not depend on the class that is positive.

$$\mathrm{MCC} = \frac{\mathrm{TP} * \mathrm{TN} + \mathrm{FP} * \mathrm{FN}}{\sqrt{(\mathrm{TP}+\mathrm{FP})(\mathrm{TP}+\mathrm{FN})(\mathrm{TN}+\mathrm{FP})(\mathrm{TN}+\mathrm{FN})}}$$

MCC ranges from −1 to +1, where 1 implies perfect predictions, 0 random predictions, and −1 perfect reverse predictions. Table 2 shows the MCC scores based on the default threshold values for all tools using the two CAID and the updated uCAID1 datasets.

Regarding the use of metrics that employ a single classification threshold, it needs to be stressed that it can be challenging to interpret results or compare methods using this approach. Developers select default thresholds to generate binary predictions that maximize the performance of these predictors on their datasets. In other words, thresholds are selected to optimize performance on sets with a specific content of LIP residues that are limited in size and characteristics. Consequently, the number of predicted LIP residues can vary substantially between predictors; some methods can significantly under-predict, while others over-predict LIPs using these default thresholds. For example, the percentages of residues predicted by ANCHOR-2, DisoRDPbind-prot, and MoRFchibi_Light to be LIP residues in the CAID1 dataset using their default threshold are 24.8%, 8.5%, and 5.8%, while the actual LIP residue content is 6.3%. For CAID2, these numbers are 40.4%, 13.2%, and 5.6%, and the actual rate is about 17%. Thus, results based on binary outputs cannot be reliably compared for two reasons. First, default thresholds for different tools generate different output sizes. Second, optimal threshold values will likely differ between different datasets. To address this second point, the analysis of the first CAID event contrasted predictors based on new threshold values for all tools computed to maximize their performance on the CAID1 dataset. Given these challenges, we will rely in this chapter on the AUC values when contrasting the performance of LIP prediction tools.

3 Results and Discussion

The CAID community analysis concluded that identifying LIPs by computational tools is a challenging, unsolved task. This conclusion was based on the finding that many competing predictors were only marginally better than some binary prediction baselines suggested by the CAID assessors [55]. They also considered the old DisProt LIP annotation used in CAID1 to be imprecise and, thus, contain loosely defined LIPs. That is, the initial mapping of LIPs often originates from experiments in which large segments of a protein IDR are detected, and the resulting deletion construct then assayed for interactions with partners using different biophysical and biochemical means. As a result, LIPs are not precisely mapped, and LIP segments in CAID1 are long, including some non-binding IDR residues (Table 1). After the first CAID event, the organizers of CAID stated that the annotation of future test datasets would be more detailed, thus, less biased toward long-binding segments [55]. Therefore, to quantify the effect of the old annotation strategy on the first CAID evaluation, we compared the results of the five predictors that participated in both CAID events, namely, ANCHOR2, DisoRDPBind, MoRFchibi_Light, MoRFchibi_Web, and OPAL, on all three datasets, CAID1, uCAID1, and CAID2. These five predictors show that the CAID2 dataset is more consistent with the uCAID1 dataset than the CAID1 (Fig. 1 and Table 2).

The AUC ranking of ANCHOR2 and DisoRDPBind is considerably higher when using the CAID1 dataset than uCAID1 and CAID2 (Fig. 1, Table 2). The better performance of ANCHOR2 on longer LIP segments contaminated by non-binding IDR residues might be a result of the simplicity of its model and the fact that its final output is generated using scores averaged over long segments (62 and 80 amino acids long). In this implementation, the effects of small shifts on the amino acid distribution of such large window sizes are minimal. Consistent with the finding that ANCHOR2 performs better on longer protein-binding LIPs, we recently recommended using this predictor for that purpose [58]. In contrast to ANCHOR2, DisoRDPbind utilizes smaller sliding windows (33 amino acids). Thus, in theory, it should be able to identify shorter segments. However, the DisoRDPBind training dataset was collected from DisProt (v5.6 2010). Therefore, its LIP annotation mimics that of the CAID1 test dataset. Apart from ANCHOR2 and DisoRDPBind, the remaining G1 predictors showed little change in their AUC performance between CAID1 and uCAID1 datasets, implying that the 4175 inaccurately annotated residues in CAID1 are scored by these tools neither high nor low. We can also observe a low OPAL prediction quality of the CAID2 dataset compared to uCAID1, which we can only attribute to noise as both datasets are relatively small, which is especially true for CAID2. We can see some inconsistencies between the AUC results and that of F1 and MCC (Table 2), which is expected as the latter represents only a single point in the ROC and the precision-real curves.

The second CAID event shows that G2 tools generally provided higher prediction quality than G1 tools—most prominantly DeepDISOBind (all and protein) and ENSHROUD (all and protein). DeepDISOBind [45] is a multi-task hybrid convolution neural network with multi-kernel sizes and a feed-forward neural network that utilizes 33 input features. ENSHROUD [56] integrates deep learning and other machine learning models such as XGBoost, LightGBM, and LightGBMXT to improve the generalization and robustness. ENSHROUD makes its predictions based on various protein features such as accessible surface area, disordered binding region, secondary structure, and PSSM.

It is important to note that because of the limited number of LIP residues compared to the total residues, estimated to be around 1% in Eukaryota [28], the percentage of the true positive in the predicted LIP residues is usually low, i.e., low precision. However, in many cases, LIP predictors are used to identify the positions of LIP residues in sequences that are assumed to contain LIPs, i.e., the percentage of LIP residues in such arrangements is high, which lessens the effect of this class imbalance. From the precision–recall curves in Fig. 1, we can estimate that using a single high-quality predictor for an input sequence with 18% LIP residues, the prediction precision for high-scored residues is about 40% or higher.

4 Conclusions

We can conclude that the observed differences in prediction quality are heavily influenced by the quality of the test data. When larger test data became available, we believe that separating nucleotide-binding LIPs from those that bind to protein can help improve the evaluation outcome. In addition, multi-model and deep learning used by some predictors, such as MoRFchibi, OPAL, DeepDISO-Bind, and ENSHROUD, provided higher-quality predictions than single weighted-sum models used by some G1 tools. We agree with the CAID community in their assessment that the performance of future predictors is likely to improve as the quality and quantity of the data available for training improve. Because of the significant differences that currently exist between predictors, we recommend using multiple predictors and assessing the outcome manually whenever possible to overcome the high false-positive rate.

References

1. Haynes C, Oldfield CJ, Ji F et al (2006) Intrinsic disorder is a common feature of hub proteins from four eukaryotic interactomes. PLoS Comput Biol 2(8):e100
2. Oldfield CJ, Uversky VN, Dunker AK et al (2019) Chapter 1: Introduction to intrinsically disordered proteins and regions. In: Salvi N (ed) Intrinsically disordered proteins. Academic, Cambridge, MA, pp 1–34
3. Wright PE, Dyson HJ (2015) Intrinsically disordered proteins in cellular signalling and regulation. Nat Rev Mol Cell Biol 16(1): 18–29. PMID: 25531225
4. Oldfield CJ, Dunker AK (2014) Intrinsically disordered proteins and intrinsically disordered protein regions. Annu Rev Biochem 83:553–584. PMID: 24606139

5. Van der Lee R, Buljan M, Lang B et al (2014) Classification of intrinsically disordered regions and proteins. Chem Rev 114(13): 6589–6631. PMID: 24773235
6. Neduva V, Russell RB (2005) Linear motifs: evolutionary interaction switches. FEBS Lett 579:3342–3345
7. Davey NE, Van Roey K, Weatheritt RJ et al (2012) Attributes of short linear motifs. Mol BioSyst 8:268–281
8. Dinkel H, Van Roey K, Michael S et al (2016) ELM 2016—data update and new functionality of the eukaryotic linear motif resource. Nucleic Acids Res 44:D294–D300
9. Oldfield CJ, Cheng Y, Cortese MS et al (2005) Coupled folding and binding with alpha-helix-forming molecular recognition elements. Biochemistry 44(37):12454–12470. https://doi.org/10.1021/bi050736e
10. Mohan A, Oldfield CJ, Radivojac P et al (2006) Analysis of molecular recognition features (MoRFs). J Mol Biol 362:1043–1059
11. Vacic V, Oldfield CJ, Mohan A et al (2007) Characterization of molecular recognition features, MoRFs, and their binding partners. J Proteome Res 6:2351–2366
12. Piovesan D, Necci M, Escobedo N et al (2021) MobiDB: intrinsically disordered proteins in 2021. Nucleic Acids Res 49(D1):D361–D367. https://doi.org/10.1093/nar/gkaa1058
13. Babu MM, van der Lee R, de Groot NS et al (2011) Intrinsically disordered proteins: regulation and disease. Curr Opin Struct Biol 21(3):432–440. https://doi.org/10.1016/j.sbi.2011.03.011. Epub 2011 Apr 20. PMID: 21514144
14. Cumberworth A, Lamour G, Babu MM et al (2013) Promiscuity as a functional trait: intrinsically disordered regions as central players of interactomes. Biochem J 454(3):361–369. https://doi.org/10.1042/BJ20130545. PMID: 23988124
15. Holguin-Cruz JA, Foster LJ, Gsponer J (2022) Where protein structure and cell diversity meet. Trends Cell Biol 32(12):996–1007. https://doi.org/10.1016/j.tcb.2022.04.004. Epub 2022 May 7. PMID: 35537902
16. Cheng Y, Oldfield CJ, Meng J et al (2007) Mining alpha-helix-forming molecular recognition features with cross species sequence alignments. Biochemistry 46(47): 13468–13477. https://doi.org/10.1021/bi7012273. Epub 2007 Nov 1. PMID: 17973494; PMCID: PMC2570644
17. Dosztányi Z, Mészáros B, Simon I (2009) ANCHOR: web server for predicting protein binding regions in disordered proteins. Bioinformatics 25(20):2745–2746. https://doi.org/10.1093/bioinformatics/btp518. Epub 2009 Aug 28. PMID: 19717576; PMCID: PMC2759549
18. Mészáros B, Simon I, Dosztányi Z (2009) Prediction of protein binding regions in disordered proteins. PLoS Comput Biol 5(5): e1000376. https://doi.org/10.1371/journal.pcbi.1000376. Epub 2009 May 1. PMID: 19412530; PMCID: PMC2671142
19. Xue B, Dunker AK, Uversky VN (2010) Retro-MoRFs: identifying protein binding sites by normal and reverse alignment and intrinsic disorder prediction. Int J Mol Sci 11(10): 3725–3747. https://doi.org/10.3390/ijms11103725. PMID: 21152297; PMCID: PMC2996789
20. Miri Disfani F, Hsu WL, Mizianty MJ et al (2012) MoRFpred, a computational tool for sequence-based prediction and characterization of short disorder-to-order transitioning binding regions in proteins. Bioinformatics 28(12):i75–i83
21. Mooney C, Pollastri G, Shields DC et al (2012) Prediction of short linear protein binding regions. J Mol Biol 415(1):193–204. https://doi.org/10.1016/j.jmb.2011.10.025. Epub 2011 Oct 21. PMID: 22079048
22. Khan W, Duffy F, Pollastri G et al (2013) Predicting binding within disordered protein regions to structurally characterised peptide-binding domains. PLoS One 8(9):e72838. https://doi.org/10.1371/journal.pone.0072838. PMID: 24019881; PMCID: PMC3760854
23. Fang C, Noguchi T, Tominaga D et al (2013) MFSPSSMpred: identifying short disorder-to-order binding regions in disordered proteins based on contextual local evolutionary conservation. BMC Bioinf 14:300. https://doi.org/10.1186/1471-2105-14-300. PMID: 24093637; PMCID: PMC3853019
24. Malhis N, Gsponer J (2015) Computational identification of MoRFs in protein sequences. Bioinformatics 31(11):1738–1744. PMID: 25637562
25. Jones DT, Cozzetto D (2015) DISOPRED3: precise disordered region predictions with annotated protein-binding activity. Bioinformatics 31(6):857–863. https://doi.org/10.1093/bioinformatics/btu744. Epub 2014 Nov 12. PMID: 25391399; PMCID: PMC4380029
26. Malhis N, Wong TCE, Nassar R et al (2015) Computational identification of MoRFs in protein sequences using hierarchical application of Bayes rule. PLoS One. https://doi.org/10.

1371/journal.pone.0141603. PMID: 26517836

27. Palopoli N, Lythgow KT, Edwards RJ (2015) QSLiMFinder: improved short linear motif prediction using specific query protein data. Bioinformatics 31(14):2284–2293. https://doi.org/10.1093/bioinformatics/btv155. Epub 2015 Mar 19. PMID: 25792551; PMCID: PMC4495300
28. Yan J, Dunker AK, Uversky VN et al (2016) Molecular recognition features (MoRFs) in three domains of life. Mol BioSyst 12(3): 697–710. https://doi.org/10.1039/c5mb00640f. PMID: 26651072
29. Peng Z, Kurgan LA (2015) High-throughput prediction of RNA, DNA and protein binding regions mediated by intrinsic disorder. Nucleic Acids Res 43(18):e121
30. Malhis N, Jacobson M, Gsponer J (2016) MoRFchibi SYSTEM: software tools for the identification of MoRFs in protein sequences. Nucleic Acids Res. https://doi.org/10.1093/nar/gkw409. PMID: 27174932
31. Sharma R, Kumar S, Tsunoda T et al (2016) Predicting MoRFs in protein sequences using HMM profiles. BMC Bioinf 17(Suppl 19):504. https://doi.org/10.1186/s12859-016-1375-0. PMID: 28155710; PMCID: PMC5259822
32. Krystkowiak I, Davey NE (2017) SLiMSearch: a framework for proteome-wide discovery and annotation of functional modules in intrinsically disordered regions. Nucleic Acids Res 45 (W1):W464–W469. https://doi.org/10.1093/nar/gkx238. PMID: 28387819; PMCID: PMC5570202
33. Peng Z, Wang C, Uversky VN et al (2017) Prediction of disordered RNA, DNA, and protein binding regions using DisoRDPbind. Methods Mol Biol 1484:187–203. https://doi.org/10.1007/978-1-4939-6406-2_14. PMID: 27787828
34. Sharma R, Bayarjargal M, Tsunoda T et al (2018) MoRFPred-plus: computational identification of MoRFs in protein sequences using physicochemical properties and HMM profiles. J Theor Biol 21(437):9–16. https://doi.org/10.1016/j.jtbi.2017.10.015. Epub 2017 Oct 16. PMID: 29042212
35. Mészáros B, Erdos G, Dosztányi Z (2018) IUPred2A: context-dependent prediction of protein disorder as a function of redox state and protein binding. Nucleic Acids Res 46 (W1):W329–W337. https://doi.org/10.1093/nar/gky384
36. Sharma R, Raicar G, Tsunoda T et al (2018) OPAL: prediction of MoRF regions in intrinsically disordered protein sequences. Bioinformatics 34(11):1850–1858. https://doi.org/10.1093/bioinformatics/bty032. PMID: 29360926
37. Sharma R, Sharma A, Raicar G et al (2019) OPAL+: length-specific MoRF prediction in intrinsically disordered protein sequences. Proteomics 19(6):e1800058. https://doi.org/10.1002/pmic.201800058. Epub 2018 Nov 2. PMID: 30324701
38. Fang C, Moriwaki Y, Tian A et al (2019) Identifying short disorder-to-order binding regions in disordered proteins with a deep convolutional neural network method. J Bioinforma Comput Biol 17(1):1950004. https://doi.org/10.1142/S0219720019500045. PMID: 30866736
39. He H, Zhao J, Sun G (2019) Computational prediction of MoRFs based on protein sequences and minimax probability machine. BMC Bioinf 20(1):529. https://doi.org/10.1186/s12859-019-3111-z. PMID: 31660849; PMCID: PMC6819637
40. Fang C, Moriwaki Y, Li C et al (2019) MoRFPred_en: sequence-based prediction of MoRFs using an ensemble learning strategy. J Bioinforma Comput Biol 17(6):1940015. https://doi.org/10.1142/S0219720019400158. PMID: 32019410
41. He H, Zhao J, Sun G (2019) Prediction of MoRFs in protein sequences with MLPs based on sequence properties and evolution information. Entropy (Basel) 21(7):635. https://doi.org/10.3390/e21070635. PMID: 33267349; PMCID: PMC7515128
42. Hanson J, Litfin T, Paliwal K et al (2020) Identifying molecular recognition features in intrinsically disordered regions of proteins by transfer learning. Bioinformatics 36(4): 1107–1113. https://doi.org/10.1093/bioinformatics/btz691. PMID: 31504193
43. He H, Zhou Y, Chi Y et al (2021) Prediction of MoRFs based on sequence properties and convolutional neural networks. BioData Min 14(1):39. https://doi.org/10.1186/s13040-021-00275-6. PMID: 34391457; PMCID: PMC8364704
44. Hu G, Katuwawala A, Wang K et al (2021) flDPnn: accurate intrinsic disorder prediction with putative propensities of disorder functions. Nat Commun 12(1):4438. https://doi.org/10.1038/s41467-021-24773-7. PMID: 34290238; PMCID: PMC8295265
45. Zhang F, Zhao B, Shi W et al (2022) DeepDISOBind: accurate prediction of RNA-, DNA- and protein-binding intrinsically disordered residues with deep multi-task learning. Brief

Bioinform 23(1):bbab521. https://doi.org/10.1093/bib/bbab521. PMID: 34905768

46. Katuwawala A, Zhao B, Kurgan L (2021) DisoLipPred: accurate prediction of disordered lipid-binding residues in protein sequences with deep recurrent networks and transfer learning. Bioinformatics 38(1):115–124. https://doi.org/10.1093/bioinformatics/btab640. PMID: 34487138
47. Varadi M, Anyango S, Deshpande M et al (2022) AlphaFold protein structure database: massively expanding the structural coverage of protein-sequence space with high-accuracy models. Nucleic Acids Res 50:D439–D444
48. Peng Z, Li Z, Meng Q et al (2023) CLIP: accurate prediction of disordered linear interacting peptides from protein sequences using co-evolutionary information. Brief Bioinform 24(1):bbac502. https://doi.org/10.1093/bib/bbac502. PMID: 36458437
49. Sun X, Malhis N, Zhao B et al (2020) Computational disorder analysis in ethylene response factors uncovers binding motifs critical to their diverse functions. Int J Mol Sci 21(1):74. https://doi.org/10.3390/ijms21010074
50. Kuechler ER, Budzyńska PM, Bernardini JP et al (2020) Distinct features of stress granule proteins predict localization in Membraneless organelles. J Mol Biol 432(7):2349–2368
51. Bhardwaj T, Saumya KU, Kumar P et al (2020) Japanese encephalitis virus—exploring the dark proteome and disorder–function paradigm. FEBS J 287(17):3751–3776. https://doi.org/10.1111/febs.15427
52. Hassan SS, Attrish D, Ghosh S et al (2021) Notable sequence homology of the ORF10 protein introspects the architecture of SARS-CoV-2. Int J Biol Macromol 181:801–809. https://doi.org/10.1016/j.ijbiomac.2021.03.199
53. Hou C, Li Y, Wang M et al (2022) Systematic prediction of degrons and E3 ubiquitin ligase binding via deep learning. BMC Biol 20:162. https://doi.org/10.1186/s12915-022-01364-6
54. Mignon J, Mottet D, Leyder T et al (2022) Structural characterisation of amyloidogenic intrinsically disordered zinc finger protein isoforms DPF3b and DPF3a. Int J Biol Macromol 218:57–71. https://doi.org/10.1016/j.ijbiomac.2022.07.102
55. Necci M, Piovesan D, Hoque T et al (2021) Critical assessment of protein intrinsic disorder prediction. Nat Methods 18(5):472–481. https://doi.org/10.1038/s41592-021-01117-3
56. Del Conte A, Bouhraoua A, Mehdiabadi M et al (2023) CAID prediction portal: a comprehensive service for predicting intrinsic disorder and binding regions in proteins. Nucleic Acids Res. https://doi.org/10.1093/nar/gkad430
57. Fawcett T (2006) An introduction to ROC analysis. Pattern Recogn Lett 27:861–874
58. Kurgan L, Hu G, Wang K et al (2023) Tutorial: a guide for the selection of fast and accurate computational tools for the prediction of intrinsic disorder in proteins. Nat Protoc. https://doi.org/10.1038/s41596-023-00876-x

Chapter 15

Accurate Prediction of Protein-Binding Residues in Protein Sequences Using SCRIBER

Jian Zhang, Feng Zhou, Xingchen Liang, and Lukasz Kurgan

Abstract

Deciphering molecular-level mechanisms that govern protein–protein interactions (PPIs) relies in part on the accurate prediction of protein-binding partners and protein-binding residues. These predictions can be used to support a wide spectrum of applications that include development of PPI networks and protein docking programs, drug design studies, and investigations of molecular details that underlie certain diseases. Computational methods that predict protein-binding residues offer convenient, inexpensive, and relatively accurate data that can aid these efforts. We introduce and describe a user-friendly webserver for the SCRIBER method that conveniently provides state-of-the-art predictions of protein-binding residues and that minimizes cross-predictions, i.e., incorrect prediction of residues that bind other/non-protein ligands as protein binding. SCRIBER relies on a two-layer architecture that is specifically designed to reduce the cross-predictions. We motivate and explain this predictive architecture. We describe how to use the webserver, interact with its web interface, and collect, read, and understand results generated by SCRIBER. The SCRIBER webserver is available at http://biomine.cs.vcu.edu/servers/SCRIBER/.

Key words Protein-protein interactions, Protein-binding residues, Prediction, Cross-prediction, Web server, SCRIBER, Machine learning, Logistic regression

1 Introduction

Proteins interact with a variety of molecules including proteins, DNAs, RNAs, lipids, and a number of small ligands [1–8]. Protein–protein interactions (PPIs) are instrumental for a wide spectrum of cellular functions, with a few examples of regulation of cell cycle, signal transduction, and metabolism [9–11]. Besides facilitating efforts to decipher cellular processes, knowledge of PPIs has numerous practical applications in the construction of PPI networks, development of protein docking programs, improving our understanding of molecular mechanisms that underpin certain diseases, and in rational drug design [12–20]. Data on PPIs can be found in several databases including molecular-level details in the Protein Data Bank (PDB) [21], amino acid-level annotations in

Andrzej Kloczkowski et al. (eds.), *Prediction of Protein Secondary Structure*, Methods in Molecular Biology, vol. 2867,
https://doi.org/10.1007/978-1-0716-4196-5_15,

DisProt [22] and BioLiP [23], and protein-level information in multiple resources including STRING [24], mentha [25], and BioGRID [26]. Although billions of PPIs are annotated at the protein level [24], the additional details of the interacting residues in the protein sequences are available for a relatively small fraction of these interactions. This motivates the development of computational methods that support efforts to annotate PPIs.

The computational predictors include protein-level and residue-level methods [27–31]. The former tools predict pairs of protein sequences that interact with each other, while the latter identify putative protein-binding residues in a given protein sequence. Some of the recently published protein-level methods include CNN-FSRF [32], NVDT [33], ProteinPrompt [34], MM-StackEns [35], and MARPPI [36]. A representative collection of modern residue-level predictors includes SCRIBER [37], DeepPPISP [38], PROBselect [39], ProNA2020 [40], HybridPBRpred [41], DELPHI [42], ProB-site [43], PITHIA [44], DeepDISOBind [45], and DeepPRObind [46]. These tools can be used to accurately predict PPIs with high throughput and at a low cost. In particular, the residue-level tools provide additional details for the large number of known protein-level annotations of PPIs [24]. However, several studies identified a substantial cross-prediction problem that many of these methods suffer. This means that amino acids that interact with other types of partners (e.g., nucleic acid or lipids) are cross-predicted as interacting with proteins, in essence resulting in a partner-agnostic prediction of binding residues [27, 39, 41, 47–49]. This drawback likely stems from the fact that these methods are typically designed using training datasets that focus solely on the protein-binding proteins, underrepresenting or entirely missing proteins that interact with the other partner types.

A recently released residue-level predictor, SCRIBER (**S**ele**C**tive p**R**ote**I**n-**B**inding r**E**sidue p**R**edictor), addresses this drawback by accurately predicting protein-binding residues while also minimizing the cross-predictions [37]. The latter is due to the use of a high-quality training dataset that covers DNA-, RNA- and small ligand-binding proteins and an innovative predictive architecture. This architecture includes models that predict binding with DNA, RNA, and small ligands and uses their predictions to reduce cross-predictions generated by the main model that predicts protein-binding residues. SCRIBER's predictive performance was assessed and compared to several other predictors on a relatively large test dataset in the original article [37] (*see* **Note 1**). The results reveal that SCRIBER provides relatively accurate predictions, with the area under the ROC curve (AUC) of 0.715, Matthews correlation coefficient (MCC) = 0.23, and area under the cross-prediction curve (AUCPC) = 0.28 [37]. To compare, the next best method that was evaluated in this comparison, SSWRF [50], secures

AUC = 0.687, MCC = 0.18, and AUCPC = 0.42 [37]. We note that larger values of AUCPC indicate higher levels of cross-predictions (worse quality) and 0.50 corresponds to the partner-type agnostic prediction (i.e., protein-binding residues cannot be differentiated from DNA-/RNA−/small ligand-binding residues).

This chapter describes the SCRIBER's architecture, details its prediction process, and provides informative and practical instructions that explain how to use its webserver. Moreover, we briefly discuss how to read and interpret the predictive results generated by SCRIBER.

2 Methods

SCRIBER relies on an innovative two-layer design that we summarize in Fig. 1. The first layer generates a profile that covers a comprehensive collection of characteristics that are relevant to the prediction of binding residues. This profile includes several major elements that are predicted/derived from the input protein sequence: (1) relative amino acid propensities (RAAPs), which are computed using an approach from [51] and which quantify propensity of amino acids for binding to various partner types; (2) relative solvent accessibility (RSA) predicted with an accurate and fast

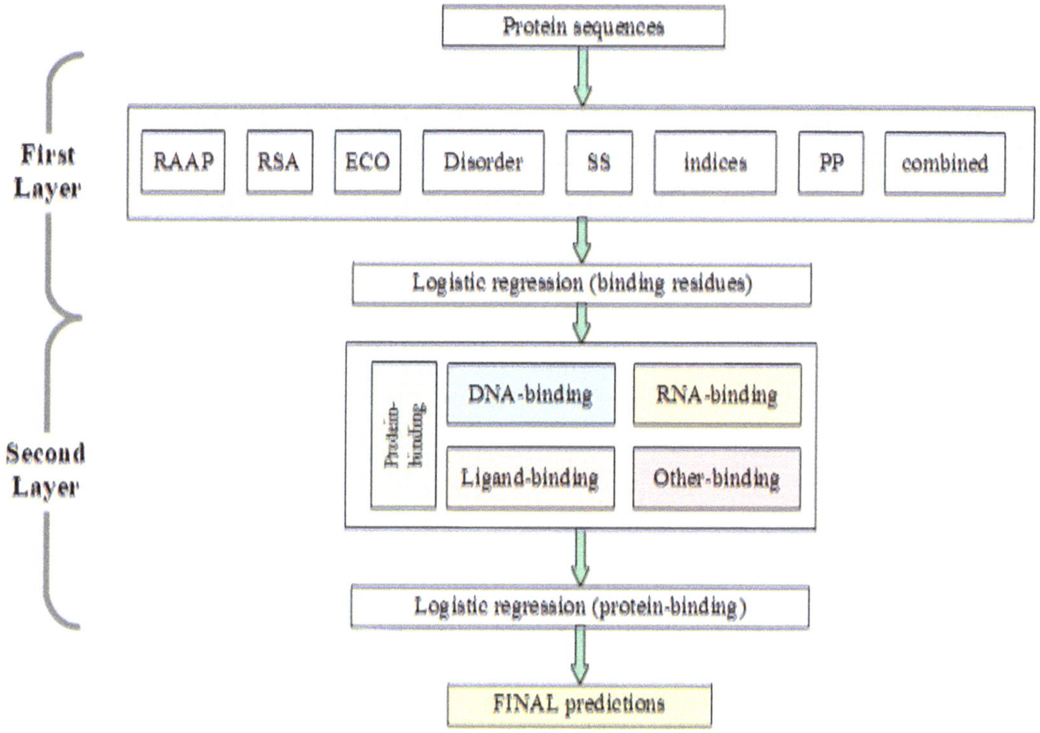

Fig. 1 Flowchart of the prediction process utilized by the SCRIBER method

ASAquick method [52]; (3) evolutionary conservation (ECO) calculated from the outputs generated with a fast and sensitive HHblits tool [53]; (4) putative protein-binding disordered residues produced with computationally efficient ANCHOR tool [54]; (5) secondary structure (SS) predicted with a fast single-sequence version of the popular PSI-PRED method [55]; (6) the indices and physicochemical properties (PP) that quantify related properties of amino acids, such as hydrophobicity, polarity, and charge, which are estimated using the AAindex resource [56]; and (7) custom-designed features that combine information across multiple factors, such as presence of conserved amino acids on a putative protein surface defined by high RSA values or presence of protein-binding disordered amino acids that are conserved (*see* **Note 2**). Importantly, the entire profile is generated using fast tools, which result in a runtime-efficient implementation of SCRIBER. The profile is used as an input to implement five logistic regression-based predictors of binding residues that cover interactions with proteins, DNA, RNA, small ligands, and "other" molecules, where the latter represents all binding residues except for the protein-binding residues. The outputs of the first layer are the five propensities for protein binding, DNA binding, RNA binding, ligand binding, and other binding.

The second layer re-predicts the protein-binding residues that are predicted in the first layer with the objective to reduce cross-predictions. It relies on the logistic regression model that uses the outputs from the first layer in a sliding sequence window as its inputs. This approach exploits an observation that interacting amino acids typically cluster together in the protein sequence [2]. This means that amino acids predicted as protein binding by the first-layer regression that are nearby many other amino acids that are also predicted to interact with proteins and fewer residues predicted to bind other partner types (by the other first-layer regressions) are likely to in fact interact only with proteins. The logistic regression can detect such differences in the density of the predicted binding residues, and they can be used to make predictions that minimize the cross-predictions. The output of this regression quantifies the putative propensity for protein binding for each amino acid in the input protein sequence, which is the ultimate output generated by the SCRIBER tool.

3 Webserver

3.1 Architecture

SCRIBER is deployed as a convenient webserver that performs the entire prediction process on the server side, without the need to install any software on the user's computer end. The webserver is available at http://biomine.cs.vcu.edu/servers/SCRIBER/ and is free for academic use. The main webserver page is implemented

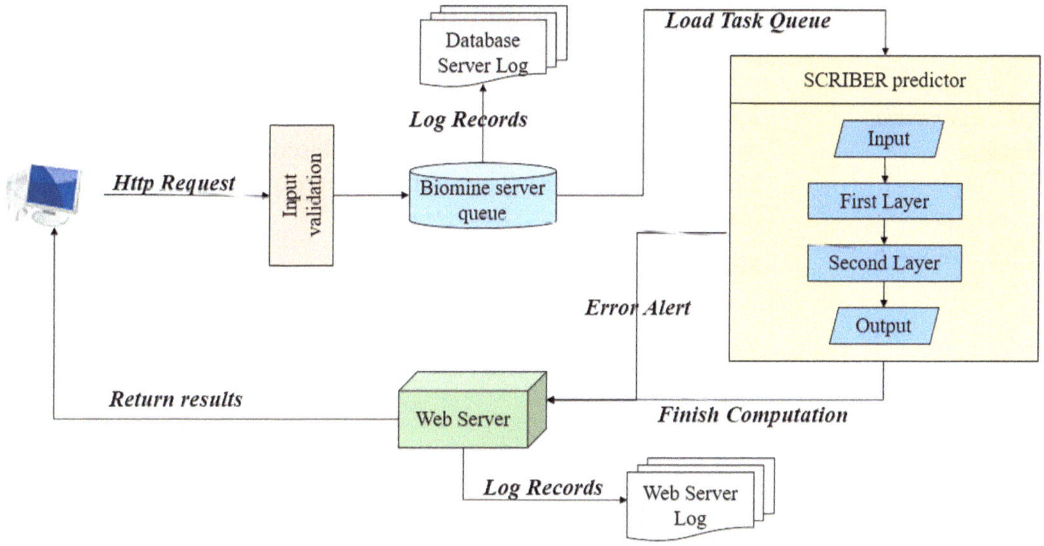

Fig. 2 Architecture of the SCRIBER webserver

with JSP and HTML, while the layers that compute the predictive model are written in Java, shell, Ruby, and Perl languages. The website can be accessed directly in the web browser window and supports multiple popular browsers including recent versions of Chrome, Firefox, and Internet Explorer.

Figure 2 provides a detailed flowchart of the prediction process for the SCRIBER webserver. Users submit sequence(s) of the query protein(s) via a browser window at http://biomine.cs.vcu.edu/servers/SCRIBER/. These inputs are checked for format and number, packaged together into a single job, and added to a queue of jobs that are processed by the biomine server, which is used to host the SCRIBER webserver. The biomine server incorporates over a dozen tools, such as disorder predictors, flDPnn [57] and MFDp [58, 59]; disorder function predictors, DEPICTER [60] and DisoRDPbind [61–63]; CONNECTOR that predicts protein–drug interactions [64]; predictors of nucleic acid binding, qNABpredict [65], DNAgenie [66], and DRNApred [48]; DisoLipPred tool that predicts protein–lipid interactions [67]; and a tool that predicts quality of crystal structures, XRRPred [68]. The queue is processed in the first-in-first-out manner that aims to ensure fair (balanced between users) and fast access to the biomine resource (*see* **Note 3**). The status of the queue refreshes automatically every 10 s. Once the user's job reaches the top of the server queue, it is passed to the SCRIBER predictor that performs the entire prediction process including calculations of results from the first and second layers and generation of the outputs (*see* **Note 4**). Finally, the results are sent back to the biomine server that generates a web page and an email with the results, completing the prediction process.

3.2 Submission Process

The inputs to the SCRIBER webserver are collected and submitted using the main server page at http://biomine.cs.vcu.edu/servers/SCRIBER/ (Fig. 3). The submission requires completion of three easy steps (Fig. 3):

Step 1. Copy FASTA-formatted protein sequence(s) into the white text entry field. An example of a properly formatted sequence can be obtained by clicking the "Example" button located below the text entry field. The "Reset sequence(s)" button

Fig. 3 Home page of the SCRIBER webserver. The web page is set up to make predictions for the PLK1 kinase protein (UniProt ID: Q29370)

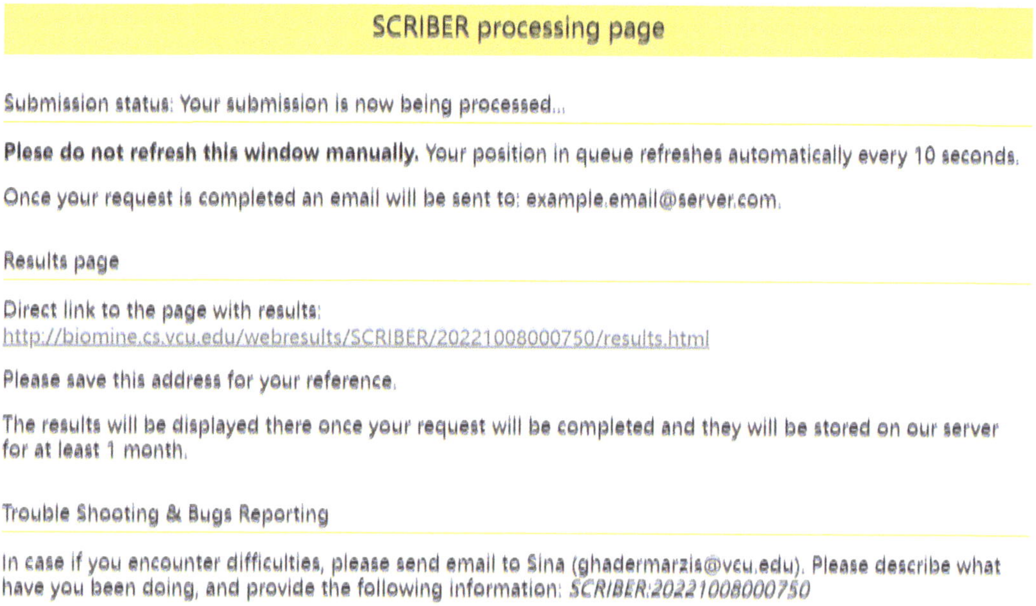

Fig. 4 Processing page generated by the SCRIBER webserver

can be used to clear the submission field. Users can submit up to ten protein sequences at the time (*see* **Note 5**).

Step 2. Provide an email address where a notification of completion for the submitted job will be delivered (*see* **Note 6**).

Step 3. Click the "Run SCRIBER" button to initiate the prediction. This triggers a check of the data submitted in the text entry field to verify that FASTA format is used and no more than ten proteins are entered. User is notified in case of a problem, and otherwise the prediction process is started, which we describe in Subheading 3.1 and Fig. 2. The submission page transitions into the processing page (Fig. 4), which guides the user through the prediction process. This page provides a unique identifier of the submitted job (bottom of Fig. 4), updates status of the server queue, and informs when the predictions are being processed. After predictions are completed, the processing page automatically redirects the user to the page with the results (*see* **Note 7**), and an email notification is sent to the email address provided by the user (*see* **Note 8**).

3.3 Outputs

The results are delivered via an email (*see* **Note 6**) and the results web page (Fig. 5), which shows predictions in a tabulated format and provides a hyperlink to the CSV-formatted text file with the predictions that are formatted in a similar way. The text file, which is the comma-separable format, can be easily parsed to extract predictions for individual residues. The results are shown for one protein at the time, with the data for the subsequent proteins given

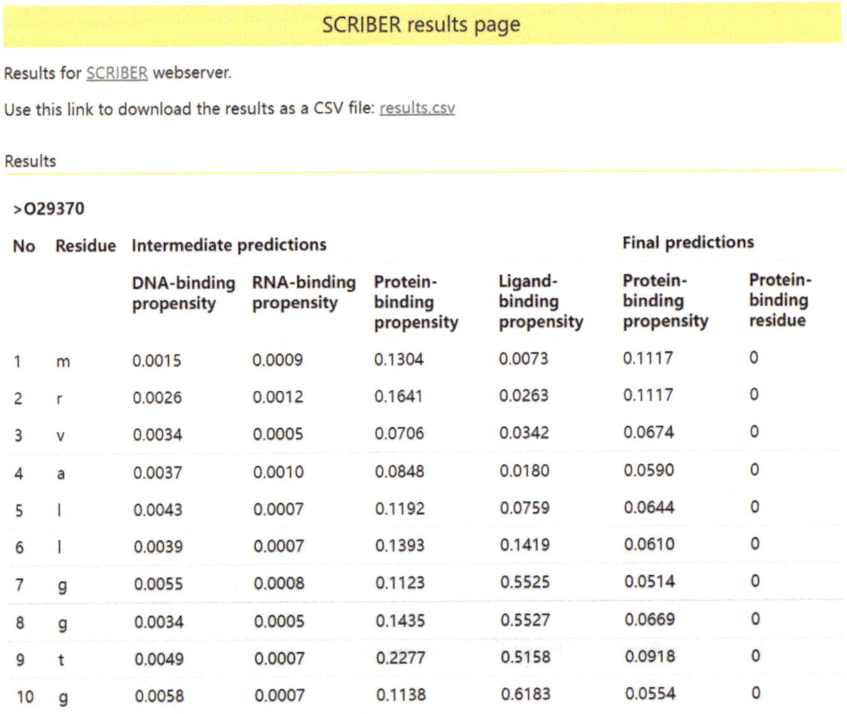
SCRIBER results page

Results for SCRIBER webserver.

Use this link to download the results as a CSV file: results.csv

Results

>O29370

No	Residue	Intermediate predictions				Final predictions	
		DNA-binding propensity	RNA-binding propensity	Protein-binding propensity	Ligand-binding propensity	Protein-binding propensity	Protein-binding residue
1	m	0.0015	0.0009	0.1304	0.0073	0.1117	0
2	r	0.0026	0.0012	0.1641	0.0263	0.1117	0
3	v	0.0034	0.0005	0.0706	0.0342	0.0674	0
4	a	0.0037	0.0010	0.0848	0.0180	0.0590	0
5	l	0.0043	0.0007	0.1192	0.0759	0.0644	0
6	l	0.0039	0.0007	0.1393	0.1419	0.0610	0
7	g	0.0055	0.0008	0.1123	0.5525	0.0514	0
8	g	0.0034	0.0005	0.1435	0.5527	0.0669	0
9	t	0.0049	0.0007	0.2277	0.5158	0.0918	0
10	g	0.0058	0.0007	0.1138	0.6183	0.0554	0

Fig. 5 Results page generated by the SCRIBER webserver

below. The data are organized into eight columns (Fig. 5). The first, left-most column lists residue position. The second column provides the corresponding amino acid types where capital letters indicate residues predicted as protein binding while lowercase is used for residues predicted as non-binding. Columns three to six give the intermediate predictions that are generated by the first predictive layer of the SCRIBER method, including the putative propensity for DNA, RNA, protein, and ligand binding. Columns seven and eight are the final predictions, including the numerical propensity and binary prediction (1 denotes protein binding vs. 0 for non-binding) of protein binding.

The binary prediction of protein binding is generated from the propensity where residues with propensities above a threshold are predicted as binding, while the remaining residues as non-binding. SCRIBER applied the threshold of 0.22. In general, predictions with high and low propensity values are more likely to provide correct predictions of protein-binding and non-binding residues, respectively, while predictions with propensity values that are near the threshold are less likely to be correct.

4 Case Study

We illustrate predictions from SCRIBER using an example protein, the NADPH oxidoreductase enzyme from *Archaeoglobus fulgidus* (UniProt ID: O29370). We pick this protein since it binds proteins (i.e., it forms a dimer) and several other ligands (e.g., sodium, magnesium, and NADP). This makes it possible for us to discuss predictions of protein-binding residues and the cross predictions for the other ligands. Also, structure of this protein in complex with the ligands is available (PDB ID: 1JAX [69]), and this allows us to visualize the sequence-based predictions in the context of their spatial proximity in the structure. Using the 212 amino-acid-long sequence of this protein as the sole input, SCRIBER predicts 19 protein binding amino acids that are mostly located at the C-terminus (19R, 139R, 186R, 187L, 191L, 192T, 195I-197N, 199M-202N, 204M, 206E, 207L, 210K-212L). Figure 6a compares native and putative protein-binding residues where green identifies true positives (correctly predicted protein-binding residues), red is for false negatives (native protein-binding residues incorrectly predicted as non-protein binding), and yellow is for false positives (non-protein-binding residues incorrectly predicted as protein binding). We find that SCRIBER correctly predicts 13 of the 29 native protein-binding residues (45%) and generates relatively few mispredictions. Figure 6b focuses on the

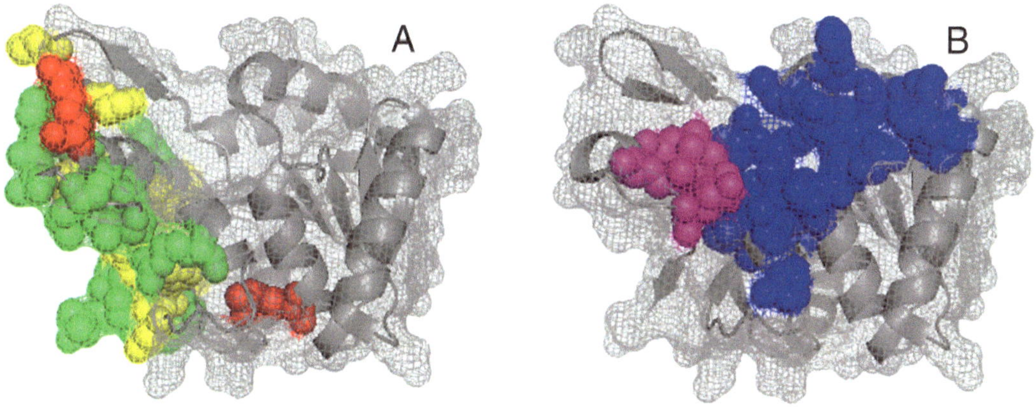

Fig. 6 Analysis of SCRIBER predictions that were generated from the sequence of the NADPH oxidoreductase enzyme (UniProt ID: 029370). The predictions are annotated onto the structure of this protein (PDB ID: 1JAX). Panel **a** compares native and putative protein-binding residues: green for true positives (correctly predicted protein-binding residues), red for false negatives (native protein-binding residues incorrectly predicted as non-protein binding), and yellow for false positives (non-protein-binding residues incorrectly predicted as protein binding). Panel **b** evaluates predictions for residues that interact with the other ligands and shows cross-predictions in pink (residues that interact with other/non-protein partners incorrectly predicted as interacting with proteins) and in blue (residues that interact with other/non-protein partners correctly predicted as not interacting with proteins

cross-predictions for the 27 amino acids that interact with the other ligands. SCRIBER cross-predicts only four of these residues as protein binding (15%) and correctly predicts that the remaining residues do not bind proteins. This means that SCRIBER selectively identifies residues that interact with proteins while excluding residues that interact with other partners, i.e., in other words, this tool produces relatively few cross-predictions. Interestingly, the predictions of the protein-binding residues are clustered in the same part of the protein surface, which partly explains the cross-predictions that are in proximity of the native interface of the dimer, i.e., residues that bind other ligands that are further away are correctly predicted as not protein binding (Fig. 6b).

5 Notes

1. The test dataset that was used to evaluate SCRIBER and other methods that predict protein-binding residues consist of 448 proteins. The sequences and annotations of binding amino acids for these proteins are available in the *Materials* section of the web page of the SCRIBER webserver at http://biomine.cs.vcu.edu/servers/SCRIBER/. This page also provides access to the training dataset and predictions generated with SCRIBER for the full human proteome.
2. The SCRIBER web page at http://biomine.cs.vcu.edu/servers/SCRIBER/ provides links to the websites where the code and/or webserver of the methods used to produce the profile are available. These details can be found in the *Acknowledgments* section.
3. Load-balancing between users of the biomine server is accomplished with two mechanisms. First, each user (IP address) is allowed to submit no more than two concurrent jobs. Second, the input size (i.e., the number of input protein sequences) is limited to a value (i.e., 10 proteins for the SCRIBER webserver) that ensures that the prediction can be completed in below 15 min.
4. The prediction performed by SCRIBER takes about 30 s for an average length sequence with about 200 amino acids and about 90 s for a longer sequence with around 500 residues. A more complete analysis of the runtime is provided in the original article that introduces SCRIBER [37].
5. Users who would like to predict a dataset with up to a few hundred proteins should split this dataset into ten-protein sized batches and run them using the SCRIBER webserver. At most two concurrent requests are allowed for the webserver. Users have two options to collect predictions for larger

requests that involve hundreds or thousands of proteins. First, they should consider the DescribePROT database that offers easy access to pre-computed predictions from 11 tools [70, 71], which include SCRIBER and 10 methods that predict other functional and structural aspects of proteins: ASAquick [52], DFLpred [72], DisoRDPbind [62], DRNApred [48], MMseqs2 [73], MoRFchibi [74], PSIPRED [55], SignalP [75], and VSL2B [76]. As of September 2024, DescribePROT covers 2.3 million proteins from 273 reference proteomes of popular and/or model organisms that include human, mouse, chicken, rat, dog, yeast, fruit fly, *A. thaliana*, *C. elegans*, herpes virus, papilloma virus, mumps virus, influenza A, and coronavirus. If the desired proteins are not available in the DescribePROT resources, then users should contact corresponding authors of SCRIBER [37] to request such predictions.

6. Users must provide the email address on the input webserver page. We encourage the users to double check correctness of the email address that they provide. This address is where the server sends an email notification when the results are ready. The email includes two unique hyperlinks to the web page with the results and the CSV-formatted file with the predictions and the job ID that can be used to identify the user's request. The email address is used solely for the purpose of delivering the results.
7. Users can gain access to the results in the same browser window where they submitted their request. However, they must ensure that this window is not closed for the entire prediction process. This could be a relatively long wait time when the queue of the biomine server is long.
8. The predictions of the query protein(s) are available via a unique hyperlink that is provided on the page with the results and in the email that is sent back to the user. These results are kept on the webserver for at least 1 month.

Acknowledgments

This work was supported in part by the Natural Science Foundation of Henan (242300421410) to J.Z., the Nanhu Scholars Program for Young Scholars of the Xinyang Normal University to J.Z., the Youth Research Fund Project of Xinyang Normal University (2023-QN-049) to F.Z., Robert J. Mattauch Endowments funds to L.K., and the National Science Foundation grants (IIS2125218 and DBI2146027) to L.K.

References

1. Chen K, Kurgan L (2009) Investigation of atomic level patterns in protein--small ligand interactions. PLoS One 4:e4473
2. Zhang J, Ma Z, Kurgan L (2019) Comprehensive review and empirical analysis of hallmarks of DNA-, RNA- and protein-binding residues in protein chains. Brief Bioinform 20:1250–1268
3. Cook KB, Hughes TR, Morris QD (2015) High-throughput characterization of protein-RNA interactions. Brief Funct Genomics 14: 74–89
4. Ellis JJ, Broom M, Jones S (2007) Protein-RNA interactions: structural analysis and functional classes. Proteins 66:903–911
5. Xie Z, Hu S, Qian J et al (2011) Systematic characterization of protein-DNA interactions. Cell Mol Life Sci 68:1657–1668
6. Jones S, Van Heyningen P, Berman HM et al (1999) Protein-DNA interactions: a structural analysis. J Mol Biol 287:877–896
7. Agasid MT, Robinson CV (2021) Probing membrane protein-lipid interactions. Curr Opin Struct Biol 69:78–85
8. Gerstberger S, Hafner M, Tuschl T (2014) A census of human RNA-binding proteins. Nat Rev Genet 15:829–845
9. Braun P, Gingras ACJP (2012) History of protein–protein interactions: From egg-white to complex networks. Proteomics 12:1478–1498
10. Cafarelli TM, Desbuleux A, Wang Y et al (2017) Mapping, modeling, and characterization of protein-protein interactions on a proteomic scale. Curr Opin Struct Biol 44:201–210
11. Bourret RB (2008) Signal transduction meets systems biology: deciphering specificity determinants for protein-protein interactions. Mol Microbiol 69:1336–1340
12. Athanasios A, Charalampos V, Vasileios TJCDM (2017) Protein-protein interaction (PPI) network: recent advances in drug discovery. Curr Drug Metabol 18:5–10
13. Kuzmanov U, Emili AJGM (2013) Protein-protein interaction networks: probing disease mechanisms using model systems. Genome Med 5:1–12
14. Uversky VN (2014) Wrecked regulation of intrinsically disordered proteins in diseases: pathogenicity of deregulated regulators. Front Mol Biosci 1:6
15. Vakser IA (2014) Protein-protein docking: from interaction to interactome. Biophys J 107:1785–1793
16. Scott DE, Bayly AR, Abell C et al (2016) Small molecules, big targets: drug discovery faces the protein-protein interaction challenge. Nat Rev Drug Discov 15:533–550
17. Petta I, Lievens S, Libert C et al (2016) Modulation of protein-protein interactions for the development of novel therapeutics. Mol Ther 24:707–718
18. Andrei SA, Sijbesma E, Hann M et al (2017) Stabilization of protein-protein interactions in drug discovery. Expert Opin Drug Discov 12: 925–940
19. Lu H, Zhou Q, He J et al (2020) Recent advances in the development of protein-protein interactions modulators: mechanisms and clinical trials. Signal Transduct Target Ther 5:213
20. Zhang J, Chai H (2019) Recent in Silico research in high-throughput drug discovery and molecular biochemistry. Curr Top Med Chem 19:103–104
21. Wwpdb Consortium (2019) Protein Data Bank: the single global archive for 3D macromolecular structure data. Nucleic Acids Res 47:D520–D528
22. Hatos A, Hajdu-Soltesz B, Monzon AM et al (2020) DisProt: intrinsic protein disorder annotation in 2020. Nucleic Acids Res 48: D269–D276
23. Yang J, Roy A, Zhang Y (2013) BioLiP: a semi-manually curated database for biologically relevant ligand-protein interactions. Nucleic Acids Res 41:D1096–D1103
24. Szklarczyk D, Gable AL, Nastou KC et al (2021) The STRING database in 2021: customizable protein-protein networks, and functional characterization of user-uploaded gene/measurement sets. Nucleic Acids Res 49: D605–D612
25. Calderone A, Castagnoli L, Cesareni G (2013) mentha: a resource for browsing integrated protein-interaction networks. Nat Methods 10:690–691
26. Oughtred R, Rust J, Chang C et al (2021) The BioGRID database: a comprehensive biomedical resource of curated protein, genetic, and chemical interactions. Protein Sci 30:187–200
27. Zhang J, Kurgan L (2018) Review and comparative assessment of sequence-based predictors of protein-binding residues. Brief Bioinform 19:821–837

28. Chen H, Li F, Wang L et al (2020) Systematic evaluation of machine learning methods for identifying human-pathogen protein-protein interactions. Brief Bioinform 22
29. Kotlyar M, Rossos AEM, Jurisica I (2017) Prediction of protein-protein interactions. Curr Protoc Bioinform 60:8 2 1–8 2 14
30. Murakami Y, Mizuguchi K (2022) Recent developments of sequence-based prediction of protein-protein interactions. Biophys Rev 14: 1393–1411
31. Biró B, Zhao B, Kurgan L (2022) Complementarity of the residue-level protein function and structure predictions in human proteins. Comput Struct Biotechnol J 20:2223–2234
32. Wang L, Wang HF, Liu SR et al (2019) Predicting protein-protein interactions from matrix-based protein sequence using convolution neural network and feature-selective rotation forest. Sci Rep 9:9848
33. Zhao N, Zhuo MJ, Tian K et al (2022) Protein-protein interaction and non-interaction predictions using gene sequence natural vector. Commun Biol 5:652
34. Canzler S, Fischer M, Ulbricht D et al (2022) ProteinPrompt: a webserver for predicting protein-protein interactions. Bioinform Adv 2:vbac059
35. Albu AI, Bocicor MI, Czibula G (2023) MM-StackEns: a new deep multimodal stacked generalization approach for protein-protein interaction prediction. Comput Biol Med 153:106526
36. Li X, Han P, Chen W et al (2023) MARPPI: boosting prediction of protein-protein interactions with multi-scale architecture residual network. Brief Bioinform 24
37. Zhang J, Kurgan L (2019) SCRIBER: accurate and partner type-specific prediction of protein-binding residues from proteins sequences. Bioinformatics 35:i343–i353
38. Zeng M, Zhang F, Wu FX et al (2020) Protein-protein interaction site prediction through combining local and global features with deep neural networks. Bioinformatics 36:1114–1120
39. Zhang F, Shi W, Zhang J et al (2020) PROBselect: accurate prediction of protein-binding residues from proteins sequences via dynamic predictor selection. Bioinformatics 36:i735–i744
40. Qiu J, Bernhofer M, Heinzinger M et al (2020) ProNA2020 predicts protein-DNA, protein-RNA, and protein-protein binding proteins and residues from sequence. J Mol Biol 432: 2428–2443
41. Zhang J, Ghadermarzi S, Kurgan L (2020) Prediction of protein-binding residues: dichotomy of sequence-based methods developed using structured complexes versus disordered proteins. Bioinformatics 36:4729–4738
42. Li Y, Golding GB, Ilie L (2021) DELPHI: accurate deep ensemble model for protein interaction sites prediction. Bioinformatics 37: 896–904
43. Khan SH, Tayara H, Chong KT (2022) ProB-site: protein binding site prediction using local features. Cells 11
44. Hosseini S, Ilie L (2022) PITHIA: protein interaction site prediction using multiple sequence alignments and attention. Int J Mol Sci 23
45. Zhang F, Zhao B, Shi W et al (2022) DeepDISOBind: accurate prediction of RNA-, DNA- and protein-binding intrinsically disordered residues with deep multi-task learning. Brief Bioinform 23
46. Zhang F, Li M, Zhang J et al (2023) DeepPRObind: modular deep learner that accurately predicts structure and disorder-annotated protein binding residues. J Mol Biol:167945
47. Zhang J, Chen Q, Liu B (2021) NCBRPred: predicting nucleic acid binding residues in proteins based on multilabel learning. Brief Bioinform 22
48. Yan J, Kurgan L (2017) DRNApred, fast sequence-based method that accurately predicts and discriminates DNA- and RNA-binding residues. Nucleic Acids Res 45: e84
49. Su H, Liu M, Sun S et al (2019) Improving the prediction of protein-nucleic acids binding residues via multiple sequence profiles and the consensus of complementary methods. Bioinformatics 35:930–936
50. Wei Z-S, Han K, Yang J-Y et al (2016) Protein–protein interaction sites prediction by ensembling SVM and sample-weighted random forests. Neurocomputing 193:201–212
51. Zhang J, Ma Z, Kurgan L (2017) Comprehensive review and empirical analysis of hallmarks of DNA-, RNA- and protein-binding residues in protein chains. Brief Bioinform 20:1250
52. Faraggi E, Zhou YQ, Kloczkowski A (2014) Accurate single-sequence prediction of solvent accessible surface area using local and global features. Proteins 82:3170–3176
53. Remmert M, Biegert A, Hauser A et al (2012) HHblits: lightning-fast iterative protein sequence searching by HMM-HMM alignment. Nat Methods 9:173–175

54. Dosztanyi Z, Meszaros B, Simon I (2009) ANCHOR: web server for predicting protein binding regions in disordered proteins. Bioinformatics 25:2745–2746
55. Buchan DWA, Minneci F, Nugent TCO et al (2013) Scalable web services for the PSIPRED protein analysis workbench. Nucleic Acids Res 41:W349–W357
56. Kawashima S, Pokarowski P, Pokarowska M et al (2008) AAindex: amino acid index database, progress report 2008. Nucleic Acids Res 36:D202–D205
57. Hu G, Katuwawala A, Wang K et al (2021) flDPnn: accurate intrinsic disorder prediction with putative propensities of disorder functions. Nat Commun 12:4438
58. Mizianty MJ, Stach W, Chen K et al (2010) Improved sequence-based prediction of disordered regions with multilayer fusion of multiple information sources. Bioinformatics 26: i489–i496
59. Mizianty MJ, Uversky V, Kurgan L (2014) Prediction of intrinsic disorder in proteins using MFDp2. Methods Mol Biol 1137:147–162
60. Barik A, Katuwawala A, Hanson J et al (2020) DEPICTER: intrinsic disorder and disorder function prediction server. J Mol Biol 432: 3379–3387
61. Oldfield CJ, Peng Z, Kurgan L (2020) Disordered RNA-binding region prediction with DisoRDPbind. Methods Mol Biol 2106:225–239
62. Peng Z, Kurgan L (2015) High-throughput prediction of RNA, DNA and protein binding regions mediated by intrinsic disorder. Nucleic Acids Res 43:e121
63. Peng Z, Wang C, Uversky VN et al (2017) Prediction of disordered RNA, DNA, and protein binding regions using DisoRDPbind. Methods Mol Biol 1484:187–203
64. Wang C, Kurgan L (2019) Review and comparative assessment of similarity-based methods for prediction of drug-protein interactions in the druggable human proteome. Brief Bioinform 20:2066–2087
65. Wu Z, Basu S, Wu X et al (2023) qNABpredict: quick, accurate, and taxonomy-aware sequence-based prediction of content of nucleic acid binding amino acids. Protein Sci 32:e4544
66. Zhang J, Ghadermarzi S, Katuwawala A et al (2021) DNAgenie: accurate prediction of DNA-type-specific binding residues in protein sequences. Brief Bioinform 22
67. Katuwawala A, Zhao B, Kurgan L (2021) DisoLipPred: accurate prediction of disordered lipid binding residues in protein sequences with deep recurrent networks and transfer learning. Bioinformatics
68. Ghadermarzi S, Krawczyk B, Song J et al (2021) XRRpred: accurate predictor of crystal structure quality from protein sequence. Bioinformatics 37:4366
69. Warkentin E, Mamat B, Sordel-Klippert M et al. (2001) Structures of F420H2 : NADP (+) oxidoreductase with and without its substrates bound. EMBO J 20:6561–6569
70. Zhao B, Kurgan L (2023) Databases of protein structure and function predictions at the amino acid level. In: Machine learning in bioinformatics of protein sequences, pp 329–353
71. Zhao B, Katuwawala A, Oldfield CJ et al (2021) DescribePROT: database of amino acid-level protein structure and function predictions. Nucleic Acids Res 49:D298–D308
72. Meng F, Kurgan L (2016) DFLpred: high-throughput prediction of disordered flexible linker regions in protein sequences. Bioinformatics 32:i341–i350
73. Mirdita M, Steinegger M, Soding J (2019) MMseqs2 desktop and local web server app for fast, interactive sequence searches. Bioinformatics 35:2856–2858
74. Malhis N, Jacobson M, Gsponer J (2016) MoRFchibi SYSTEM: software tools for the identification of MoRFs in protein sequences. Nucleic Acids Res 44:W488
75. Teufel F, Almagro Armenteros JJ, Johansen AR et al (2022) SignalP 6.0 predicts all five types of signal peptides using protein language models, vol 40. Nat Biotechnol, p 1023
76. Obradovic Z, Peng K, Vucetic S et al (2005) Exploiting heterogeneous sequence properties improves prediction of protein disorder. Proteins 61(Suppl 7):176–182

Chapter 16

LMPTMSite: A Platform for PTM Site Prediction in Proteins Leveraging Transformer-Based Protein Language Models

Pawel Pratyush, Suresh Pokharel, Hamid D. Ismail, Soufia Bahmani, and Dukka B. KC

Abstract

Protein post-translational modifications (PTMs) introduce new functionalities and play a critical role in the regulation of protein functions. Characterizing these modifications, especially PTM sites, is essential for unraveling complex biological systems. However, traditional experimental approaches, such as mass spectrometry, are time-consuming and expensive. Machine learning and deep learning techniques offer promising alternatives for predicting PTM sites. In this chapter, we introduce our LMPTMSite (language model-based post-translational modification site predictor) platform, which emphasizes two transformer-based protein language model (pLM) approaches: pLMSNOSite and LMSuccSite, for the prediction of S-nitrosylation sites and succinylation sites in proteins, respectively. We highlight the various methods of using pLM-based sequence encoding, explain the underlying deep learning architectures, and discuss the superior efficacy of these tools compared to other state-of-the-art tools. Subsequently, we present an analysis of runtime and memory usage for pLMSNOSite, with a focus on CPU and RAM usage as the input sequence length is scaled up. Finally, we showcase a case study predicting succinylation sites in proteins active within the tricarboxylic acid (TCA) cycle pathway using LMSuccSite, demonstrating its potential utility and efficiency in real-world biological contexts. The LMPTMSite platform, inclusive of pLMSNOSite and LMSuccSite, is freely available both as a web server (http://kcdukkalab.org/pLMSNOSite/ and http://kcdukkalab.org/LMSuccSite/) and as standalone packages (https://github.com/KCLabMTU/pLMSNOSite and https://github.com/KCLabMTU/LMSuccSite), providing valuable tools for researchers in the field.

Key words Post-translational modification (PTM), Protein language models, Transformer, Natural language processing, Deep learning, CPU runtime, RAM usage, Succinylation, S-nitrosylation, Tricarboxylic acid (TCA) cycle

1 Introduction

Proteins, one of the most critical organic compounds produced within living organisms' cells, are manufactured when specific genes are first transcribed into mRNA and then later translated into chains of amino acids through the process called as translation.

Andrzej Kloczkowski et al. (eds.), *Prediction of Protein Secondary Structure*, Methods in Molecular Biology, vol. 2867, https://doi.org/10.1007/978-1-0716-4196-5_16,

The translation process occurs within ribosomes and results in the assembly of proteins from amino acids. Post these stages, proteins may undergo various biochemical modifications to expand their functional roles within biological activities. These modifications, known as post-translational modifications (PTMs), involve the severance of bonds between specific amino acids in proteins and the subsequent addition of functional biochemical groups such as methyl, phosphoryl, acetyl, glycosyl, nitrosyl, succinyl, or ubiquitin. Once synthesized, most proteins are transported to the endoplasmic reticulum (ER), where PTMs activate them. These well-folded and functional proteins are then moved to the Golgi body and eventually secreted through exocytosis, performing critical roles as secretory or plasma membrane proteins. Other proteins, such as histones found in the cytosol and nucleus, may also undergo PTMs to maintain cellular functions.

PTMs were initially studied for their role in the enzymatic activity of kinases. However, they are now known to be involved in nearly all cellular activities, including gene regulation, localization, signal transduction, and interaction with other biochemical molecules such as proteins, nucleic acids, and lipids. Consequently, PTMs are implicated in the pathogenesis of numerous diseases, including cancers and cardiovascular diseases, and in a wide range of cellular-level biological activities. Research into PTMs and the identification of modification sites is a burgeoning field, with both biologists and computational scientists working in tandem to identify these sites. Although around 400 PTMs exist [1]; some of the well-studied PTMs include phosphorylation, succinylation, S-nitrosylation, acetylation, ubiquitination, methylation, glycosylation, and SUMOylation. Owing to their roles in virtually all cellular biological activities, both normal and abnormal, PTMs are a heavily researched field in biology. The advent of immunoprecipitation, mass spectrometry, and high-throughput sequencing technologies has facilitated large-scale detection of PTMs and data collection. This data is stored in databases for experimentally validated PTM sites, such as Swiss-Prot [2], PhosphoELM [3], O-GLYCBASE [4], dbPTM [5], PTMcode [6], PTMCuration [1], PhosphoSitePlus [7], etc. The dbPTM [5] database alone has amassed more than two million experimentally validated PTM substrate sites.

This ever-increasing accumulation of PTM data has spurred the development of computational methods, both in terms of algorithms and tools, to handle and analyze this wealth of information [8]. Laboratory detection of PTM sites via mass spectrometry is not only costly but also labor-intensive and time-consuming. Computational methods have consistently proven their efficacy in detecting various types of PTM sites and their functional significance, enabling researchers to prioritize targets for functional validation. Machine learning tools for PTM site detection rely on features extracted from the targeted protein sequences. The first generation of computational tools in this domain focused on deriving physicochemical and structural properties of amino acids from protein sequences and applying shallow machine learning algorithms such as support vector machines (SVM) and random forest [9]. These properties may encompass a broad spectrum of information, including physicochemical, evolutionary, structural, autocorrelational, and compositional data, all contributing to patterns that distinguish between positive sites (where modifications occur) and negative sites (where modifications do not occur). These features may include amino acid composition (AAC); pseudo-amino acid composition (PseAAC); composition, transition, and distribution (CTD); solvent accessibility of amino acids; evolutionary features based on entropy and position-specific scoring matrix (PSSM); and more.

One of the significant challenges in PTM site prediction is the site-specific nature of modifications. For instance, phosphorylation only affects serine (S), threonine (T), and tyrosine (Y). However, not all occurrences of these three amino acid residues in a protein sequence are always affected, but only at specific positions, which complicates prediction. This challenge also extends to other PTM sites, making their prediction difficult. Machine learning-based approaches typically face numerous challenges centered around feature extraction that provides sufficient discriminatory information and the algorithms that synthesize this information for knowledge learning and prediction. The probabilistic nature of machine learning prediction sparks debate regarding the efficacy of the prediction method. Nevertheless, researchers strive to minimize false discovery errors by using a large number of experimentally verified instances, eliminating biases, applying validation metrics of varying interpretations, and leveraging powerful algorithms.

Recently, the field of PTM site prediction has evolved with the advent of emerging variants of deep learning and language models [8], including transformer-based models. Additionally, the field of natural language processing (NLP) has seen remarkable advancements with sequential models such as recurrent neural network (RNN) [10], long short-term memory (LSTM) [11], and convolutional neural network (CNN) [12]. These advanced algorithms have revolutionized the ability of machines to understand and generate human language. RNNs and LSTMs are adept at capturing temporal dependencies, enabling tasks like language modeling and sentiment analysis. CNNs, originally designed for computer vision, have demonstrated success in analyzing sequential data, particularly for text classification and sentiment analysis. Transformers, with their attention mechanisms, have reshaped the NLP landscape, bringing about breakthroughs in language understanding, question answering, and text classification.

The application of natural language processing (NLP) to the life sciences has garnered significant attention and recognition within the scientific community. Given the sequential nature common to nucleotide sequences and natural languages, extensive efforts have been put into developing frameworks that utilize shared mechanisms between these domains. As a result, substantial advancements have been made in developing NLP-based encoding techniques for representing protein sequences. The use of sequence encoding techniques derived from natural language processing, such as one-hot encoding, embedding (including Word2Vec [13] and GloVe [14], BERT [15]), and other generative models, has increased for protein sequences. These methods offer potential solutions for a wide range of downstream protein tasks by capturing semantic relationships and contextual information. Whether representing amino acids as binary vectors, mapping them to continuous vector spaces, or leveraging transformer architectures, these techniques enable protein analysis and understanding, with implications for drug discovery, protein engineering, and systems biology research. Please refer to Pokharel et al. [16] for the in-depth details of various NLP-based encoding techniques of protein sequences for the PTM site prediction.

The applications of these emerging NLP technologies for PTM site prediction are elevated by the resemblance of biological sequences to text. Since the first transformer model [17], various families of transformer-based model have been developed. More recently, language models like GPT (Generative Pre-trained Transformer) [18], BERT (Bidirectional Encoder Representations from Transformers [15], and T5 (Text-to-Text Transfer Transformer) [19] are examples of transformer-based models that have attained state-of-the-art performance in various NLP tasks. These models are pre-trained on large text corpora, allowing them to learn general language representations, which can then be fine-tuned for specific tasks. The principles behind attention mechanisms and transformers have been applied to the field of protein language modeling to predict and analyze protein sequences, structures, and functions. Analogous to how natural language models learn the grammar and context of human language, protein language models aim to learn the "grammar" of amino acid sequences that form proteins, capturing patterns and relationships within these sequences. These models work by applying techniques from natural language processing and deep learning to learn the underlying patterns and relationships within protein sequences. These models are trained on large databases of protein sequences, treating amino acids as analogous to words in human language. By employing techniques such as positional embeddings, multi-head attention mechanisms, and masked language modelling (MLM), protein language models can effectively capture both local and global sequence features that contribute to the structure and function of proteins.

Building on this foundation, pre-trained protein language models can be fine-tuned to predict post-translational modifications (PTMs) in protein sequences by leveraging transfer learning. The general knowledge of protein sequences acquired during pre-training enables these models to capture the context and patterns of amino acid sequences surrounding the modification sites. This ability to recognize local and global sequence features, along with the use of task-specific loss functions, aids in the identification of PTM sites with higher accuracy.

The recent emergence of these large protein language models and the potential of these models to distill information from the sequences unleash various opportunities to use them as feature extractors in subsequent downstream prediction tasks. Motivated by the development of these language models and their ability to distill global contextual information, we developed various tools [20–23] for predicting different types of PTM sites. We refer to this platform of various PTM site prediction tools as LMPTMsite (language model-based PTM site predictor). This chapter primarily focuses on two tools within the LMPTMSite platform: pLMSNO-Site [20] and LMSuccSite [21], tools that have demonstrated superior performance in predicting S-nitrosylation and succinylation sites, respectively. We discuss these tools in depth, focusing on the following aspects: the benchmark datasets used (Subheading 2.2), sequence encoding based on language models (Subheading 2.3), the underlying deep learning (DL) architectures (Subheading 2.4), and the protocols for model evaluation and training (Subheadings 2.5 and 2.6). Subsequently, we recap the comparative performance results from the tools' literature, emphasizing the comparison of different DL architectures, various pLMs, and other widely available tools (Subheading 3.1). We also provide comprehensive instructions for using the web servers and standalone programs, as well as interpreting their outputs (Subheadings 3.2 and 3.3). In addition, we conduct a new study in this chapter, reporting on the tools' resource utilization, including CPU and RAM usage as a function of protein sequence length (Subheading 3.4). We conclude with a case study where LMSuccSite is used to predict succinylation sites in the tricarboxylic acid (TCA) cycle pathway (Subheading 3.5) and a discussion on the limitations of our study and potential future work (Subheading 3.6). Accompanying the detailed methodology, we provide helpful notes to guide readers in developing their own machine learning-based approaches for bioinformatics problems.

2 Materials and Methods

In this section, we describe our recently developed two tools within the LMPTMSite platform: pLMSNOSite [20] and LMSuccSite [21], designed for the prediction of S-nitrosylation and succinylation sites, respectively. We delve into the particulars of the benchmark dataset used, the process of sequence encoding (also referred to as feature extraction), the underlying deep learning architectures, the intricacies of model training, as well as the evaluation strategy and metrics employed.

2.1 S-nitrosylation and Succinylation

S-nitrosylation (SNO) and succinylation are some of the vital protein post-translational modifications (PTMs) that have profound regulatory effects on cellular functions. SNO is characterized by the covalent attachment of a nitric oxide (NO) group to the thiol side chain of cysteine (C) residues within proteins (Fig. 1a). This PTM is of great biological significance as it modulates protein function and plays a significant role in cellular signal transduction. Recently, it has been discovered that the protein SNO can contribute to protein misfolding events that are typically observed in neurodegenerative diseases. Consequently, abnormality in SNO has been linked to numerous neurodegenerative diseases, including Alzheimer's disease and Parkinson's disease, where abnormal protein folding patterns are a common hallmark. This highlights the critical need for continued research into SNO, as it can not only enhance our understanding of the molecular mechanisms underlying these diseases but also aid in the development of potential therapeutic strategies [24].

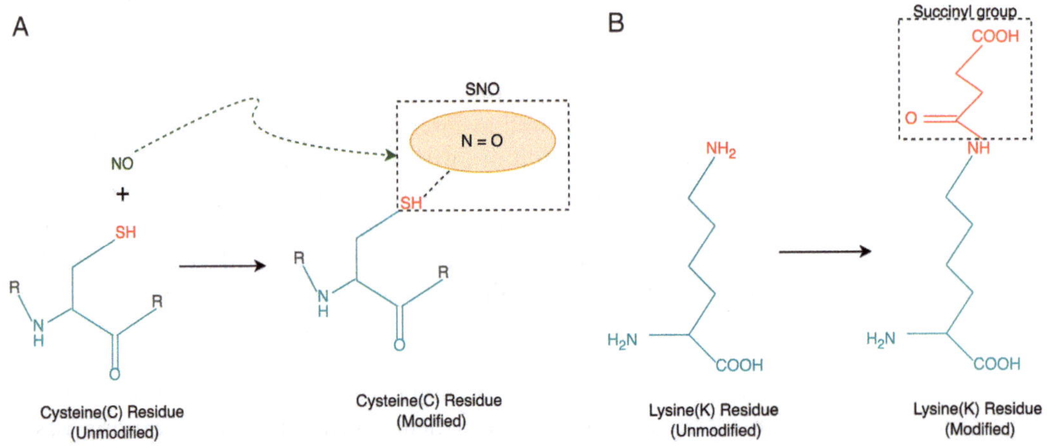

Fig. 1 (**a**) Illustration depicting S-nitrosylation modification on a cysteine (C) residue, (**b**) illustration depicting succinylation modification on a lysine (K) residue

On the other hand, succinylation involves the modification of lysine (K) residues in a protein by attaching a succinyl group (-CO-CH_2-CH_2-CO_2H) (Fig. 1b). This PTM has garnered significant attention due to its integral roles in enzyme function and metabolic pathways. Recent research has illustrated that succinylation can fundamentally alter the enzymatic rates of proteins and influence the pathways they operate in. Predominantly, succinylation targets the mitochondrial metabolic pathways, making it an essential mechanism for the regulation of metabolism and signaling pathways. Additionally, these alterations have implications that extend beyond normal cellular function. Dysregulation in succinylation has been associated with a variety of health conditions and diseases, including, but not limited to, tumors, cardiometabolic diseases, liver metabolic disorders, and various nervous system diseases. Therefore, understanding succinylation and its implications can aid in developing therapeutic strategies for these diseases [25].

Given the crucial role these PTMs play in cellular function and disease, predicting SNO and succinylation sites using computational methods can be immensely beneficial. Such prediction tools can expedite the identification of modification sites, thus accelerating our understanding of their functional implications. Moreover, computational prediction can streamline experimental designs and help prioritize targets for experimental validation, making it a cost-effective strategy. These advantages have the potential to significantly speed up both basic research and the development of novel therapeutic strategies targeting aberrant PTMs in a range of diseases.

Several computational tools have been developed for predicting these sites (*see* **Note 1**). Below, we provide a brief summary of these existing approaches, with a focus on the feature encoding approach and computational method utilized in each case (see Table 1).

2.2 Creation of Benchmark Datasets

The quality and diversity of datasets are of paramount importance in ensuring the reliable performance of machine learning and deep learning models. Thus, it is crucial to have high-quality datasets that accurately represent the problem space, encompassing a wide array of patterns and variations within the dataset. For the creation of a robust dataset geared toward PTM site prediction, robust preprocessing steps should be executed before proceeding to the modeling stage.

The datasets used for building our tools are carefully curated from sources that have been used by several state-of-the-art tools. These datasets have undergone rigorous preprocessing, following the steps outlined in the notes (*see* **Notes 2–5**), to ensure their quality and reliability. Subsequently, homologous sequences have been removed using the CD-HIT algorithm [26], which is a greedy incremental clustering algorithm that groups sequences based on a predetermined sequence identity threshold. This helps to reduce

Table 1
Summary of existing tools for the prediction of S-nitrosylation and succinylation sites

Existing tools	PTM type	Published year	Encoding method	Computational method
GPS-SNO [40]	S-nitrosylation (SNO)	2010	BLOSUM62	GPS2.0 algorithm
iSNO-PseAAC [41]		2013	PseAAC	Conditional random field (CRF) algorithm
SNOSite [42]		2011	Physicochemical properties, positional weighted matrix	SVM
DeepNitro [28]		2018	One-hot encoding, PSSM, SAC, PFR	ANN
PreSNO [27]		2019	CPA, SAC, PSSM, physicochemical properties	SVM
DeepSuccinylSite [43]	Succinylation	2018	Word embedding	CNN
GPSuc [35]		2018	AAindex, AAC, PSSM, pCKSAAP	RF, logistic regression
SuccineSite2.0 [29]		2017	pbCKSAAP, orthogonal binary features	SVM
pSuc-FFSEA [44]		2022	EBGW, One-Hot, CBOW, CGR, and AAF_DWT	SVM, BLS, LightGBM, LR

biases during model training. Duplicate or contradicting sites and overlapping sequences between the training and independent testing sets have been removed to maintain the distinctness of each set. Lastly, to handle data imbalance, random undersampling (RUS) of the training set is performed, which helps preserve the distribution between the training and testing sets. It is important to note that pLMSNOSite [20] and LMSuccSite [21] also underwent imbalance learning via cost-sensitive (CS) learning (*see* **Note 6**). However, the results were suboptimal, potentially attributable to the presence of false negatives when considering the entire negative set. Please refer to the supplementary materials of the literature of the respective tools for details about the cost-sensitive learning.

2.2.1 S-nitrosylation Dataset

The SNO dataset has been sourced from PreSNO [27], which is derived from the original DeepNitro [28] dataset, a dataset compiled through a rigorous literature search for experimentally validated SNO sites. This includes 4762 sites that have been confirmed experimentally across 3113 protein sequences. The CD-HIT algorithm [26] was employed to eliminate homology, with a similarity cut-off of 30%, resulting in 3734 positive sites. The remaining cysteine (C) residues from the same protein sequences not annotated as positive were designated as negative SNO sites, with a total of 20,548 negative sites (*see* **Note 7**). A further refinement was conducted by removing any negative sites that matched a window sequence in the positive set, resulting in 20,333 negative sites. An independent test set was then created by randomly selecting 20% of the total sites, with the remaining data utilized to create the training set. The training set consisted of 3383 SNO sites and 17,165 non-SNO sites, while the independent test set consisted of 351 SNO sites and 3168 non-SNO sites. Given the imbalance between positive and negative sites in the training set, random undersampling (RUS) of non-redundant negative sites was employed to mitigate modeling bias, while the independent test was left as it is to mimic the distribution of positive and negative sites in the real-world scenario.

2.2.2 Succinylation Dataset

This dataset was constructed from experimentally verified succinylation sites provided by Hasan et al. [29], which were acquired from the UniProtKB/Swiss-Prot database [30] and the NCBI protein sequence database. Initially, the sequences were subjected to homology removal using a CD-hit algorithm [26] with a similarity cut-off of 30%, resulting in 5009 succinylated sites from 2322 protein sequences. Thereafter, the sequences were randomly partitioned into a training set (2192 protein sequences) and a testing set (124 proteins), containing 4755 and 254 succinylation sites, respectively. Negative sites were subsequently derived from the same protein sequences, with all lysine residues (K) from the same protein sequences that were not annotated as succinylated sites (*see* **Note 7**). This resulted in 50,565 negative succinylation sites in the training set and 2977 negative succinylation sites in the test set. To address the imbalanced training set, Random Under Sampling (RUS) was used on the negative training set to obtain an equal number of negative sites as the positive sites in the training set. To evaluate the performance of the model in a real-world scenario, the independent test set was left as it is obtained from 124 proteins resulting imbalance in the number of positive and negative sites. The two datasets are summarized in Table 2.

2.3 Sequence Encoding aka Feature Extraction

The process of converting amino acid sequences into numerical space representations presents a significant challenge when developing deep learning-based predictors for PTM sites. An effective vector space encoding should capture crucial sequence features while minimizing noise and redundancy. The predictive models' performance and generalizability are directly influenced by the quality of encoding, particularly its capacity to handle variations in sequence length and context. Various techniques for encoding protein sequences have been devised and employed in numerous downstream prediction tasks, including PTM prediction. As discussed in the introduction, traditional manual feature extraction methods, typically based on physicochemical properties, require a significant amount of manual labor and expertise. It is a time-consuming and complex task to select an optimal set of

Table 2
Description of the training and independent test datasets

PTM	Dataset type	Site status	#Proteins	#Sites	Total sites
S-nitrosylation (SNO)	Train	Positive (SNO)	1962	3383	20546 (after balancing:6766)
		Negative (non-SNO)	340	17165 (after balancing: 3383)	
	Independent test	Positive (SNO)	267	351	3519
		Negative (non-SNO)	231	3168	
Succinylation	Train	Positive (succinylated)	2192	4750	55315 (after balancing: 9500)
		Negative (non-Succinylated)		50565 (after balancing: 4750)	
	Independent test	Positive (succinylated)	124	254	3231
		Negative (non-succinylated)		2977	

physicochemical features that can effectively represent the properties and behaviors of amino acids in protein sequences. Additionally, existing tools have relied mainly on peptide sequences, which restrict the feature extraction to only local information, hence potentially missing out on crucial global information that might be embedded in the larger protein sequence context. To overcome these limitations, we adopted two types of embedding techniques for our tools, each addressing different levels of sequence context: the first technique involves employing unsupervised transformer-based protein language models (pLMs) to embed dynamic global contextual information influencing the site of interest, while on the other hand, we utilize a supervised word embedding layer trained on peptide sequences to emphasize the local context of the site of interest. By combining these two encoding techniques, we can capture a wide range of sequence dependencies, from local to global, without the need for manual feature extraction. This hybrid approach enhances the representation of the protein sequences, leading to improved prediction performance as evidenced by the results achieved by our tools, pLMSNOSite and LMSuccSite. The details of these encoding techniques are described below.

2.3.1 Embeddings from Protein Language Models (pLMs)

As discussed in the Introduction section, language models (LMs) have transformed the landscape of natural language processing (NLP) by learning embeddings directly from expansive, unlabeled datasets of natural language. Unlike traditional uncontextualized word embeddings, which allocate a fixed embedding for a word regardless of its surrounding context, embeddings generated by LMs offer contextualized understanding, adjusting based on the words adjacent to them. This breakthrough in NLP has now found its way into the realm of proteins through the development of protein language models (pLMs).

In pLMs, each amino acid is mapped to a vector of fixed length via an embedding layer. The model also applies position embedding to encode the relative positions of each amino acid within its respective protein sequence. A third type of embedding, segment embedding, is used to differentiate between distinct protein sequences. The integrated use of token (amino acid), position, and segment embeddings allows pLMs to provide more than just a non-contextual mapping of amino acids to the vector space. It also captures the dependencies of amino acids within each protein sequence and the contextual relationships between different protein sequences.

The pLMs have demonstrated remarkable success in various downstream bioinformatics prediction tasks. The wealth of information learned by these models can be transferred to other tasks by generating embeddings from them and subsequently using them as inputs to predict other properties of proteins. This has opened up a whole new avenue of research in the field of bioinformatics, as researchers can now leverage the power of pLMs for a wide range of applications. In our work, we employed three prominent protein language models (pLMs)—ProtT5, ESM-1b, and ProtBert. These pLMs have demonstrated significant success in various downstream tasks, such as protein–protein interaction prediction, protein function prediction, and subcellular localization prediction. The following subsections provide a brief introduction to each of these pLMs.

(a) ProtT5-XL-UniRef50

ProtT5-XL-UniRef50 (also ProT5-XL-U50) [31] (referred to as ProtT5 onwards) has been pre-trained by Rostlab in a self-supervised fashion, utilizing Google's T5 (t5-3b version) transformer architecture. Its operational premise involves the prediction of masked or absent amino acids using a vast collection of protein sequences for training. The Uniref50 dataset, employed for training ProtT5, encompasses 45 million protein sequences, which are composed of 15 billion amino acids. This enormous corpus allows the model to implicitly grasp the structural and functional interconnections among various protein types.

The attention mechanism of ProtT5 is built on a stack of 24 transformer layers. Each of these layer houses 32 attention heads and has a final hidden layer size of 1024 units (thus producing feature vector of size 1024) (*see* **Note 8**). This stacked arrangement allows for each layer to operate sequentially on the output generated by the layer that precedes it. By iteratively blending word embeddings in this way, ProtT5 succeeds in generating rich representations as the input travels through the model's deepest layers.

(b) ESM-1b

Evolutionary Scale Modeling (ESM) [32] is another example of a pre-trained language model for proteins, released by Meta. Among its different versions and variants, ESM-1b utilizes a transformer architecture with 33 layers, featuring 650 million trainable parameters, underscoring its potential for complex pattern recognition. Like ProtT5, ESM-1b is also trained using the UniRef50 dataset, thus inheriting the wide scope of 45 million protein sequences represented therein. Operating on an input of protein sequences, ESM-1b generates feature vectors of 1280 dimensions for each individual amino acid (*see* **Note 8**).

(c) ProtBERT-UniRef100

ProtBERT-UniRef100 [31] is a protein language model based on the BERT architecture, trained by Rostlab using a self-supervised approach on the expansive Uniref100 dataset, which includes 217 million protein sequences. Its architecture comprises 24 attention layers that generate an output embedding of 1024 dimensions from the final layer (*see* **Note 8**).

We summarize these three pLMs based on the architecture, number of layers, number of attention heads, training strategy, etc. in Table 3.

2.3.2 Various Approaches for Extracting Embeddings from pLMs

As mentioned earlier, these language models take a protein sequence (or peptide) as input and provide contextual per residue embeddings. Depending upon the input to these language models and/or whether only the embeddings of the particular residue or windows of residues is considered, the approaches can be divided into various categories. Based on the input to the pLM, the approaches for extracting embedding can be broadly classified into three categories (Fig. 2): (a) full-sequence-based, (b) window sequence-based, and (c) multiple windows of varying length.

(a) Full Sequence-Based Embeddings

This class of approaches leverages the entire protein sequence as input to the pLM to extract features. These approaches can be further divided into two types based on whether the output features are generated for the site of interest only or the window around the site of interest.

(i) *Full Sequence-Based Per-Residue Embeddings (FSPE)*

In this method, for each amino acid residue, the input to the pLM is the entire protein sequence. Subsequently, the feature vector corresponding to only the residue of interest (target residue) is extracted for constructing classifiers. If the feature vector generated from pLM has a size of L, then the feature-length for each instance (or site) will be represented as $1 \times L$. This approach incorporates global

Table 3
Summary of the three protein language models (pLMs) utilized in this study

pLM	Based transformer architecture	#Parameters	Dataset for pre-training	#Layers	#Attention heads	Training strategy	Embedding dimension per residue *(1×L)*
ProtT5 [31]	T5	3 B	Uniref50	24	32	Masked Language Modelling (MLM) Objective	1 × 1024
ESM-1b [32]	RoBERTa	650 M	Uniref50	33	12	MLM	1 × 1280
ProtBERT [31]	BERT	420 M	Uniref100	30	16	MLM	1 × 1024

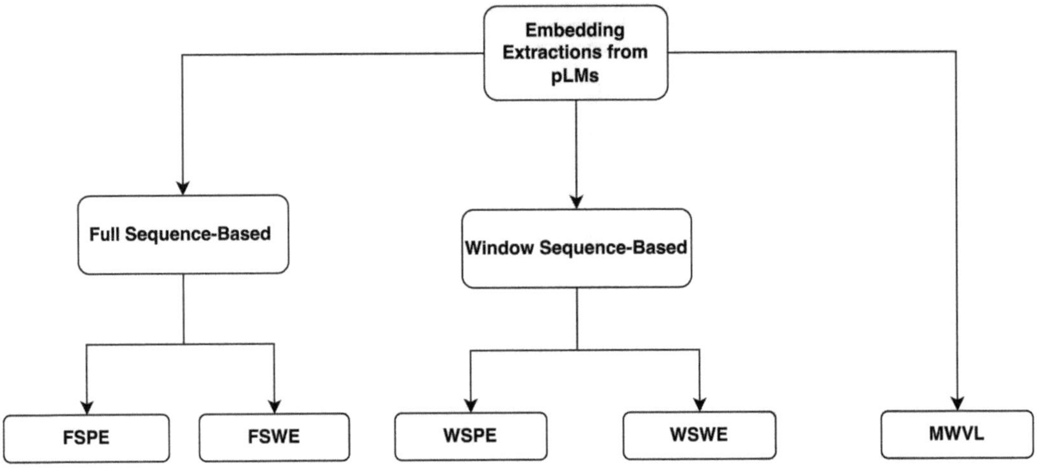

Fig. 2 Various approaches for extracting features from pLMs

contextual information from the protein sequence while focusing on specific sites of interest.

(ii) *Full Sequence-Based Window Embeddings (FSWE)*

This approach bears similarity to FSPE; however, it incorporates a key distinction. Instead of focusing solely on the embedding of the site of interest, a fixed-length window centered around this site is employed. The feature vectors of all residues within this window are extracted (either by concatenation or averaging) for classifier construction. Consequently, the feature-length for each instance (or site) is represented as $W \times L$, where W denotes the window size. The integration of local context information around the site of interest potentially improves the model's performance in capturing residue interactions.

(b) Window Sequence-Based Embedding

This class of approaches diverges from considering the entire protein sequence by focusing on the window sequence fragment centered around the site of interest as input to the pLM.

(i) *Window Sequence-Based Per Residue Embeddings (WSPE)*

Once the fragment is isolated by taking a fixed-size window around the target residue, feature vectors are extracted for each of the residues within it. However, only the feature vector of the site of interest (the central residue) is used for modeling purposes. As a result, the feature-length for each instance remains $1 \times L$. This approach emphasizes local context information, which can be beneficial in capturing residue interactions within the immediate vicinity of the site of interest.

(ii) *Window Sequence-Based Window Embeddings (WSWE)*

In this approach, the focus is on the window sequence fragment centered around the site of interest, similar to WSPE. However, instead of only considering the feature vector of the central residue, the feature vectors for all residues within the fragment are taken into account. These feature vectors collectively represent the local context and interactions around the site of interest. Consequently, the feature length for each instance (or site) becomes W×L, where W denotes the window size.

(c) Multiple Windows of Varying Length (MWVL)

The multiple windows of varying length (MWVL) technique provides a versatile method for PTM prediction by accommodating diverse protein sequences and residue interactions through its adaptability to different scales. Essentially, this approach employs multiple windows of varying lengths centered around the site of interest to capture information at different scales. For each window length, embeddings for all residues within the window are generated and aggregated to create a single representation. The aggregated embeddings from all window sizes are then concatenated to form the final feature vector for the site of interest. The feature-length of each instance (or site) is represented as the total aggregated size of all window embeddings (e.g., $W_1 \times L + W_2 \times L + \cdots + W_n \times L$, where W_i is the size of the ith window and L is the size of the individual residue embeddings).

In both pLMNOSite and LMSuccSite, we utilize the encoder part of the pLMs to extract FSPE-based features (use the entire protein sequence as an input and extract the embeddings only for the site of interest). As a result, each site of interest, whether it is a cysteine (C) in case of SNO or lysine (K) in case of succinylation, is represented using a feature vector of dimension $1 \times L$ (ProtT5, $L = 1024$; ESM-1, $L = 1280$; ProtBERT, $L = 1024$) where L is the length of embedding per residue for each of these pLMs.

2.3.3 Word Embedding

In the architecture of various deep learning models, particularly when dealing with sequential data such as text or, in our case, protein sequences, the Keras embedding layer can be utilized to extract word embedding. In the context of PTM prediction tasks, n-mer window sequences are extracted centered around the site of interest, ensuring an equal number of flanking residues exist on both sides. If there are insufficient residues to form a window, virtual amino acids (X) are padded to maintain sequence length.

Each unique amino acid in these n-mer window sequences is represented by a unique integer in a process known as integer encoding. This integer-encoded representation then serves as the input to the embedding layer. The crux of the embedding layer's functionality lies in its ability to transform these positive integers (indexes) into fixed-size dense vectors. Each unique index is thereby mapped to a distinct vector, facilitating the encoding of amino acids into a format suitable for ML/DL models.

The initial state of the embedding layer consists of random weights. During the training process, the network adjusts these weights through backpropagation, which aims to minimize the loss function. The outcome is that semantically similar inputs (amino acids in our case) are mapped to proximal points in the high-dimensional vector space. Over time, this process results in a learned dense representation of the input data, where the semantic relationships between different amino acids are encoded in the spatial relationships between their corresponding vectors.

The layer is defined by three salient parameters: input_dim, output_dim, and input_length.

- *input_dim:* This is the size of the vocabulary, essentially the number of unique amino acids (including canonical, non-canonical, and virtual amino acids) in the input dataset.
- *output_dim:* This is the size of the dense vector space into which the inputs are embedded. The optimal value for output_dim can vary and is typically determined through methods like *k*-fold cross-validation. Larger values can capture more detailed representations at the cost of increased computational complexity.
- *input_length:* This is the maximum permissible length of input sequences, in our case the n-mer window size centered around the site of interest in the protein sequence. Again, the optimal value for input_length is determined empirically.

The output of the embedding layer is a 2D vector of dimension $W \times D$, where W represents the sequence (window size) and the second dimension D is the learned dense vector (embedding dimension) representing each unique amino acid for the particular task (*see* **Note 9**). The optimal values of vocabulary size (input_-dim), embedding dimension (output_dim), and Window size (input_length) obtained from *k*-fold cross-validation (coupled with exhaustive grid search) in pLMNOSite and LMSuccSite are provided below in Table 4.

2.4 Deep Learning Architectures

Given that only window sequences are fed into the word embedding layer, it may primarily capture the local contextual information surrounding the site of interest. Leveraging the representation obtained by pLMs seems logical as they consider the entire protein sequence to generate embeddings, thereby possessing the capacity to convey the global contextual information that influences the function of the sites of interest. Consequently, the proposed architecture of our tools (pLMNSNOSite and LMSuccSite) is rooted in

Table 4
Various parameters of the word embedding layer selected for pLMSNOSite and LMSuccSite

Tool	Embedding module architecture	Vocabulary size (V)	Embedding dimension (D)	N-mer window sequence length (W)	Output dimension of the embedding layer (W × D)
pLMSNOSite	2D-CNN	23	4	37	37 × 4
LMSuccSite	2D-CNN	21	21	33	33 × 21

an ensemble approach (via stacking). It is designed to integrate the representational capabilities of the protein language model pLM and the supervised word embedding layer, thereby merging both local and global contextual information related to the site of interest.

The overall architecture can be divided into three modules, with each module aiming to learn a specific representation. The first module, known as the *Embedding Layer Module (ELM)*, takes in an n-mer window sequence around the site of interest (followed by integer encoding of the window sequence). The deep learning model underlying this module learns the embeddings each of size $W x D$, where W is the size of the window and D is the embedding dimension, generated by the supervised word embedding layer, the details of which are discussed in Subheading 2.3.3. Since we are considering a window sequence, we establish a spatial correlation between proximal amino acids surrounding the site of interest within the window. To capture this local interaction of amino acids (spatial correlation), we utilize a two-dimensional convolutional neural network (2D-CNN). There are two key reasons for choosing the CNN architecture over other sequence-based models such as RNN, LSTM, BiLSTM (bidirectional LSTM), etc. Firstly, the computational efficiency of CNN models surpasses that of sequence-based models. Secondly, experiments from *k*-fold cross-validation showed superior performance of the CNN model compared to ANN, LSTM, BiLSTM, and ConvLSTM (convolutional LSTM). The convolution layer in the CNN extracts feature maps from the spatial interaction of the amino acids, while the max-pooling layer captures the most prominent features from these maps, reducing their dimensionality with minimal information loss. Finally, the fully connected (FC) layer employs a feed-forward network to produce a classification inference for the site of interest.

The second module, known as the *pLM Module (PLM)*, processes FSPE-based feature vectors of the site of interest ("C" in the case of pLMSNOSite and "K" in the case of LMSuccSite) obtained from the encoder side of the pre-trained ProtT5 in half-precision mode (*see* **Note 10**). These features have a dimension of $1{\times}L$, where L is 1024 (the length of the feature vector per residue produced by ProtT5). As only the embeddings of the site of interest are considered, a simple artificial neural network (ANN) is utilized to learn these contextualized embeddings. This approach aligns with the findings presented by Villegas-Morcillo et al. [33], which suggest that learning embeddings from pLMs may not necessitate complex architectural designs.

The third and final module, known as the *Stacked Generalization Module (SGM)*, consists of a meta-classifier that performs a stacking ensemble of the embedding layer and ProtT5 modules. A stacking ensemble can be implemented either through intermediate fusion or decision-level fusion (also known as late fusion). In our tools, we utilize an intermediate fusion-based stacking to learn the marginal (or shared) representation of the other two modules. Detailed discussions on various fusion strategies and their drawbacks are described in the Notes (*see* **Notes 11–14**). To train the meta-classifier, designed on an ANN architecture, we freeze the 2D-CNN model within the ELM and the ANN model in the PLM. This effectively locks the internal parameters of these models and prevents them from updating during the training of the meta-classifier. Subsequently, predictions are extracted from the final hidden layer of each model. In the case of pLMSNOSite, this amounts to a 4×1 feature vector from the PLM and a 16×1 feature vector from the ELM. Similarly, for LMSuccSite, the corresponding feature vectors are of length 128×1 from the PLM and 16×1 from the ELM. These extracted predictions are then concatenated together to form a unified feature representation. These intermediate predictions, or what we refer to as learned representations, are then utilized to train the meta-classifier. Therefore, the input to the meta-classifier is of length 20×1 in pLMSNOSite and 144×1 in LMSuccSite.

The determination of the optimal architecture for each module was achieved through the utilization of k-fold cross-validation in conjunction with grid search. For detailed information regarding the hyperparameter space and the optimal architecture of various models, please refer to the supplementary materials provided in the individual literature of pLMSNOSite and LMSuccSite. Table 5 provides a comprehensive breakdown of the final architecture specifics, including the type of layers, dimensions of each layer, and the employed activation functions in ELM, PLM, and SGM modules of pLMSNOSite and LMSuccSite.

It should be noted that pLMSNOSite and LMSuccSite utilize intermediate fusion for two primary reasons: (i) we are employing two distinct representations, namely, ProtT5 pLM, which utilizes the entire sequence context (global sequence context), and the embedding layer, which operates on the n-mer window sequence (local sequence context). Early fusion might struggle to capture the differing characteristics of these two modalities. (ii) Merging at the score level may be insufficient in effectively capturing the correlation between these two representations.

Table 5
Architectural details of different modules in pLMSNOSite and LMSuccSite

Module	Architecture	Tool	Input shape	Description	Activation function
ELM	2D-CNN	pLMSNOSite	37×4	Embedding (37×4) - Convolution (64×19×1) - Dropout (0.3) - Max Pooling (5×2) - Flatten (384) - Dense (16) - Output (1)	Rectified Linear Unit (ReLU) activation function across all hidden layers and sigmoid activation in output layers
		LMSuccSite	33×21	Embedding (33×21) - Convolution (32×17×3) - Dropout (0.2) - Max Pooling (2×2) - Flatten (2304) - Dense (16) - Dropout (0.2) - Output (1)	
PLM	ANN	pLMSNOSite	1024×1	Dense (128) - Dropout (0.4) - Dense (16) - Dropout (0.2) - Dense (4) - Output (1)	
		LMSuccSite		Dense (256) - Dropout (0.2) - Dense (128) - Dropout (0.2) - Output (1)	
SGM	ANN	pLMSNOSite	20×1	Dense (8) - Output (1)	
		LMSuccSite	144×1	Dense (16) - Dense (4) - Output (1)	

The symbol "-" denotes the sequence of layers or operations in the model, and values inside the brackets represent the dimensions or shape of specific layers

2.5 Model Evaluation and Performance Measures

A stratified k-fold cross-validation (pLMSNOSite $k = 5$ and LMSuccSite $k = 10$) coupled with a grid-search strategy was implemented for the optimal selection of models and hyperparameters (*see* **Note 15**). Additionally, independent testing was conducted to assess the models on unseen data and benchmark them against other existing methodologies. Performance metrics used for evaluating the accuracy of model predictions against ground truth labels are outlined in Table 6.

2.5.1 Dealing with Class Imbalance

The data distribution in PTM prediction tasks is often skewed toward the negative class, potentially biasing performance estimates toward the negative set. As such, metrics like ACC, SN, FPR, FNR, and AUROC may be significantly impacted and should not be solely relied upon. In our case, the training sets in pLMSNOSite

Table 6
Various performance metrics used for evaluating pLMSNOSite and LMSuccSite

Metric name	Definition	Range[a]
True positive (TP)	Count of correctly predicted PTM sites	[0, +∞)
True negative (TN)	Count of correctly predicted non-PTM sites	
False positive (FP)	Count of incorrectly predicted PTM sites	
False negative (FN)	Count of incorrectly predicted non-PTM sites	
Accuracy (ACC)	$ACC = \frac{TP+TN}{TP+FN+FP+FN}$	[0, 1]
Sensitivity (SN)/recall/true positive rate (TPR)	$TPR = \frac{TP}{TP+FN}$	
Specificity (SP)/true negative rate (TNR)	$TNR = \frac{TN}{FP+TN}$	
Fall-out/false alarm rate/type I error/false positive rate (FPR)	$FPR = \frac{FP}{FP+TN}$	
Miss rate/type II error/false negative rate (FNR)	$FNR = \frac{FN}{TP+FN}$	
g-mean	$gmean = \sqrt[2]{SN \times SP}$	
Matthews correlation coefficient (MCC)	$MCC = \frac{(TP \times TN) - (FP \times FN)}{\sqrt[2]{(TP+FP)(TP+FN)(TN+FP)(TP+FN)}}$	[−1, 1]
Area under receiver operating characteristic curve (AUROC)	Area under the curve of sensitivity (TPR) plotted against 1 - specificity (FPR) at various decision threshold cut-offs (range, 0.0–1.0)	[0, 1]
Area under precision–recall curve (AUPR/PrAUC)	Area under the curve of precision plotted against recall (SN) at various decision thresholds cut-offs (range, 0.0–1.0)	

[a]Note: The ranges provided within brackets are general guidelines, but they may vary depending on the specific implementation and context of the metrics

and LMSuccSite were balanced using random undersampling, mitigating the potential effects of class imbalance on cross-validation evaluations. However, the independent test datasets employed in both these tasks were considerably imbalanced (majority-to-minority ratios, ~9 for pLMSNOSite, ~12 for LMSuccSite). Consequently, we focused on performance measures less affected by class imbalance, such as MCC, g-mean, and AUPR. We considered all these metrics in conjunction to gain a comprehensive view of our model's performance on the independent test set.

2.5.2 Dealing with Varying Decision Threshold

In binary classification tasks, a critical process involves determining an appropriate decision threshold cut-off. This threshold transforms the model's raw probability outputs into definitive categorical labels. Depending on the selected threshold cut-off, the value of computed metrics can vary significantly. The decision of selecting an optimal cut-off value typically relies on the specific use case. In the context of pLMSNOSite and LMSuccSite, we consistently utilized a widely adopted cut-off value of 0.5. However, we emphasized on threshold-independent metrics like AUROC and AUPR when evaluating our models. These metrics are particularly handy in situations where the optimal cut-off point is not known a priori. In addition, our web servers give users the flexibility to select a desired cut-off value within a range of 0.1–0.9 (refer to Subheading 3.3). Similarly, in our standalone version, we have facilitated an easy process to modify the cut-off value prior to running the prediction (see Subheading 3.2).

2.5.3 Dealing with Data Leakage

In order to address the issue of potential target information leakage from base models to the meta-classifier in stacked generalization (*see* **Note 16**), which can result in overestimated cross-validation performance, we have implemented Wolpert's stacking algorithm with *k*-fold cross-validation [34]. This strategy starts with random partitioning of the total training data into *k* folds. Then the base models are trained on *k*-1 of these folds and validated on the remaining one fold. This procedure is performed *k* times, rotating the validation fold each time, and the predictions from base models are gathered as new features. The gathered predictions from each base model train the higher-level meta-classifier, ensuring it is trained on a non-overlapping dataset, thus effectively preventing data leakage and preserving the model selection process' integrity.

2.6 Deep Learning Model Training

All deep learning models in pLMSNOSite and LMSuccSite were trained with the objective of minimizing the binary cross-entropy loss, also known as log loss, represented by the following equation:

$$-\frac{1}{N}\sum_{i=1}^{N}[y_i \log y_i' + (1 - y_i)\log(1 - y_i')]] \qquad (1)$$

In Eq. (1), y_i and y_i' denote the actual and predicted probabilities for the *i*th instance out of *N* instances, respectively.

To minimize this loss function, the Adam stochastic optimization method was utilized. In addition, an early stopping mechanism was implemented with a patience value of P_L (range, 0–total number of epochs) halting the training process if there's no loss improvement detected after P_L consecutive epochs. This helps avoid unnecessary computational expenditure and overfitting. The progress of training was continuously scrutinized by monitoring the accuracy/loss curves to further prevent overfitting. The values of major parameters chosen for the Adam optimizer in pLMSNOSite and LMSuccSite are detailed in Table 7 below. This includes learning rate, decay rates, number of epochs, batch size, and patience (P_L) for early stopping, all optimized via a *k*-fold cross-validation strategy with grid search for maximum model performance.

Table 7
Various model training parameters chosen for pLMSNOSite and LMSuccSite

Tool	AMSgrad variant?	Adaptive learning rate	Decay rate		Batch size	No. of epochs	Patience (P_L)
			First moment	Second moment			
pLMSNOSite	Yes	0.001	0.900	0.999	128	200	5
LMSuccSite	No	0.001	0.900	0.999	256	100	7

3 Results

In the subsequent sections, we present the results for our recently developed SNO site predictor, pLMSNOSite [20], and succinylation site predictor, LMSuccSite [21]. In Subheading 3.1, we summarize the *k*-fold cross-validation and independent testing results from the published works [20] [21] on these predictors. In Subheadings 3.2 and 3.3, we provide a comprehensive guide to using the standalone and web server versions of the tools, respectively. In Subheading 3.4, we describe an additional experiments performed in this chapter, where we evaluated the runtime and memory usage of pLMSNOSite as a function of the input sequence length. Finally, in Subheading 3.5, we outline a case study showcasing the application of LMSuccSite in identifying succinylated sites within proteins participating in the TCA pathway.

3.1 k-Fold Cross-validation and Independent Testing Summary

The predictive performance of our approach was assessed in greater details in the original manuscripts [20, 21]. Here, we briefly summarize the results of the k-fold cross-validation experiment, as detailed in the published literature for pLMSNOSite [20] and LMSuccSite [21]. This encompasses all modules (ELM, PLM, and SGM) for both tools, with $k = 5$ for pLMSNOSite and $k = 10$ for LMSuccSite, as shown in Table 8. In the ELM module of both tools, the 2D-CNN emerged as the optimal model, providing a mean MCC of 0.382 ± 0.03 for pLMSNOSite and 0.43 ± 0.02 for LMSuccSite, outshining other architectures such as ANN, LSTM, BiLSTM, and ConvLSTM. In the PLM module, the ANN proved to be the optimal model for both tools, delivering a mean MCC of 0.42 ± 0.03 for pLMSNOSite and 0.47 ± 0.02 for LMSuccSite, outperforming other architectures like 1D-CNN, SVM (support vector machine), RF (random forest), XGBoost (extreme gradient boosting), and AdaBoost (adaptive boosting). As for the SGM module, ANN again surfaced as the optimal model, demonstrating a mean MCC of 0.46 ± 0.03 for pLMSNOSite and 0.56 ± 0.02 for LMSuccSite, superior to architectures like LR (logistic regression), SVM, RF, and XGBoost.

Table 8
***k*-fold cross-validation comparison of various machine learning/deep learning architectures in the modules of the architecture**

Module	Tool	Optimal architecture	ACC	MCC	SN	SP	ROA UC	Compared against[a]
Embedding Layer (ELM)	pLMSNOSite LMSuccSite	2D-CNN	0.69 0.73	0.38 0.47	0.76 0.76	0.62 0.70	0.73 0.79	ANN, LSTM, BiLSTM, ConvLST
ProtT5 (PLM)	pLMSNOSite LMSuccSite	ANN	0.71 0.74	0.42 0.47	0.75 0.76	0.67 0.71	0.73 0.80	1D-CNN, SVM, RF, XGBoost, AdaBoost
Stacked Generalization (SGM)	pLMSNOSite LMSuccSite	ANN	0.73 0.77	0.46 0.56	0.77 0.80	0.69 0.76	0.80 0.85	LR, SVM, RF, XGBoost

Adopted from pLMSNOSite [20] and LMSuccSite [21]
[a]For detailed architectural specifics of each model, please refer to the supplementary materials in the literature corresponding to each respective tool

Table 9
Independent test comparison of various protein language models (pLMs) on the pLMSNOSite dataset

pLM	MCC	Sensitivity	Specificity	g-mean	ROAUC
ProtBERT[31]	0.20	0.48	0.80	0.62	0.64
ESM-1 [33]	0.25	0.56	0.79	0.67	0.69
ProtT5[31]	**0.29**	**0.60**	**0.81**	**0.70**	**0.71**

Adopted from pLMSNOSite [20]

Next, we summarize the comparative study from the original manuscripts [20, 21] where three protein language models (pLMs), ProtBERT, ESM-1, and ProtT5 were used to extract the features. These findings indicated that ProtT5 consistently outperformed the other two pLMs in both pLMSNOSite and LMSuccSite, leading to its choice as a feature extractor in both instances. Table 9 demonstrates the performance differences among these three pLMs in pLMSNOSite on the independent test set. A similar trend was observed in the case of LMSuccSite, with ProtT5 surpassing the other pLMs in performance.

Additionally, we also summarize from the original manuscripts [20, 21] the comparison of pLMSNOSite and LMSuccSite against other existing tools, and these results are presented in Table 10. It can be seen from the table that both these tools perform quite well in comparison to current state-of-the-art tools. Notably, pLMSNOSite demonstrated an impressive 35.0% increase in MCC and a 10.6% improvement in g-mean compared to its closest

Table 10
Comparison of pLMSNOSite and LMSuccSite against other existing tools

PTM	Predictor	ACC	MCC	SN	SP	g-mean
S-nitrosylation	GPS-SNO [40]	0.693	0.014	0.281	0.739	0.455
	iSNO-PseAAC [41]	0.710	0.031	0.287	0.757	0.466
	SNOSite [42]	0.469	0.069	0.668	0.447	0.546
	DeepNitro [28]	0.737	0.222	0.578	0.737	0.653
	pLMSNOSite [20]	0.769	0.340	0.735	0.773	0.754
Succinylation	SuccineSite2.0 [29]	0.85	0.26	0.40	0.88	0.63
	GPSuc [35]	0.85	0.30	0.50	0.88	0.66
	PSuccE [45]	0.85	0.20	0.38	0.89	0.58
	DeepSuccinylSite [43]	0.70	0.27	0.79	0.69	0.74
	LMSuccSite [21]	0.79	0.36	0.79	0.79	0.79

Adopted from pLMSNOSite [20] and LMSuccSite [21]

competitor, PreSNO [27]. Importantly, the pLM-based model alone was able to outperform PreSNO, showing a significant 16.3% increase in MCC. It is worth highlighting that the pLM-based model alone was able to outperform PreSNO with an approximately 16.3% increment in MCC. This underlines the capability of pLM-based embeddings in predicting PTM sites. In a similar vein, LMSuccSite exhibited a 20.0% improvement in MCC and approximately a 19.7% increase in g-mean when compared to GPSuc [35], the next best predictor. These results affirm that our novel approach, which combines global contextual information using pLM and local contextual information using a word embedding layer, is indeed a robust predictor of PTM sites such as S-nitrosylation and succinylation.

3.2 pLMSNOSite and LMSuccSite: Standalone Version

The standalone version of pLMSNOSite is provided on GitHub and can be accessed freely via this link: http://kcdukkalab.org/pLMSNOSite/. The source code and all its dependencies fall under the open-source Apache-2.0 license, providing the freedom to modify and distribute the software.

In the repository, a *requirements.txt* file is provided which details all the necessary libraries with their appropriate versions required to run pLMSNOSite (*see* **Notes 17** and **18**). The instructions for executing pLMSNOSite are outlined below.

Use the following commands to install the libraries and dependencies:

1. *pip install -r requirements.txt*
2. *pip install -q SentencePiece transformers*

After installing all dependencies, one can evaluate the proposed model on the independent test set or run your own predictions. To evaluate the model on the independent test set:

1. Navigate to the *data/test/folder* where the test sequences and corresponding ProtT5 features are placed.
2. Run the command *python3 evaluate_model.py.*

To predict using your own sequence:

1. Place the FASTA file, which should contain the protein sequence data, in the directory *input/sequence.fasta.*
2. Run the script for generating predictions with the command *python3 predict.py.*
3. Once the prediction process is complete, the results will be stored in the *output/ folder.* The output file will provide predictions for each cysteine (C) residues present in the input sequence.

The screenshot of pLMSNOSite repository on GitHub with its command line interface (CLI) is shown in Fig. 3. LMSuccSite's setup process on a local system shares a similar set of procedures to those necessary for establishing pLMSNOSite. To begin with, one should clone the LMSuccSite repository from GitHub (https://github.com/KCLabMTU/LMSuccSite) and ensure that one is covered under the Apache-2.0 license. Then, one needs to adhere to the instructions similarly outlined in the pLMSNOSite setup and its implementation steps.

It is important to highlight that pLMSNOSite and LMSuccSite were developed using Python 3.9.7, TensorFlow 2.9.1, Keras 2.9.0, PyTorch 1.11.0, and Transformers 4.18.0. Consequently, to ensure seamless operation, your system should be equipped with corresponding compatible versions (*see* **Note 18**).

3.3 pLMSNOSite and LMSuccSite: Web Server

The pLMSNOSite and LMSuccSite web servers within the LMPTMSite platform were developed using the Python framework and served through the Apache web service. These tools operate entirely on the server side, with client queries sent via an Internet browser. The web user interfaces for pLMSNOSite and LMSuccSite are accessible at http://kcdukkalab.org/pLMSNOSite/ and http://kcdukkalab.org/LMSuccSite/, respectively. The screenshot of web server of pLMSNOSite is shown in Fig. 4.

Users have the option to upload a FASTA file containing protein sequences along with their respective accession IDs. An example FASTA file is readily available for download via a link on the

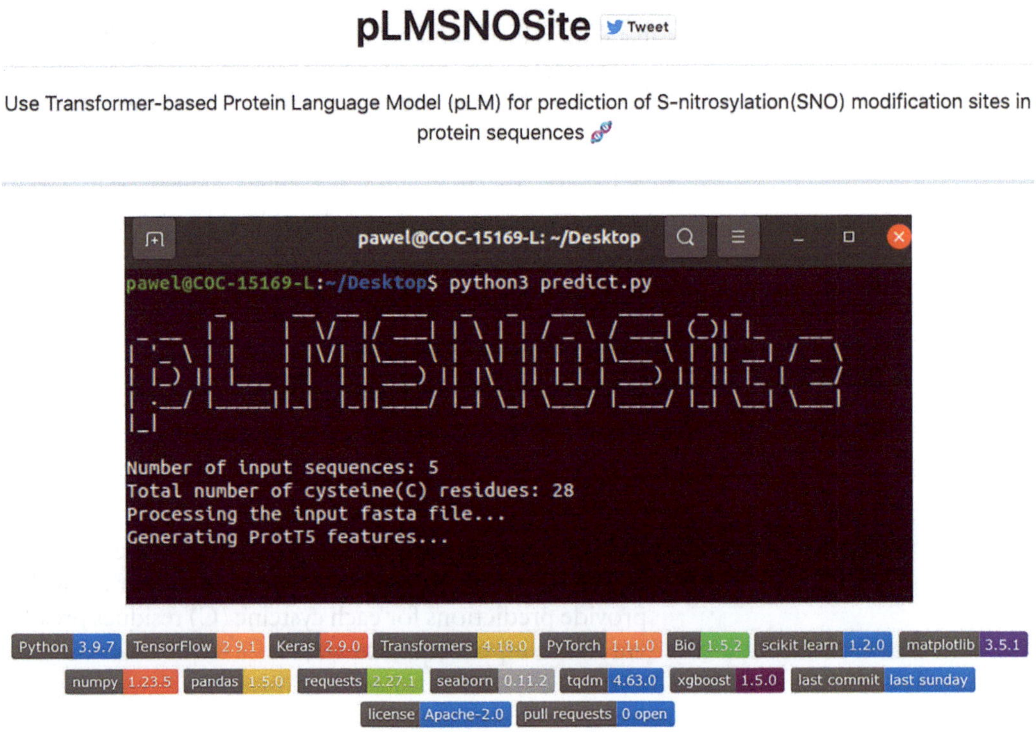

Fig. 3 Screenshot of standalone version of pLMSNOSite along with its CLI hosted at GitHub

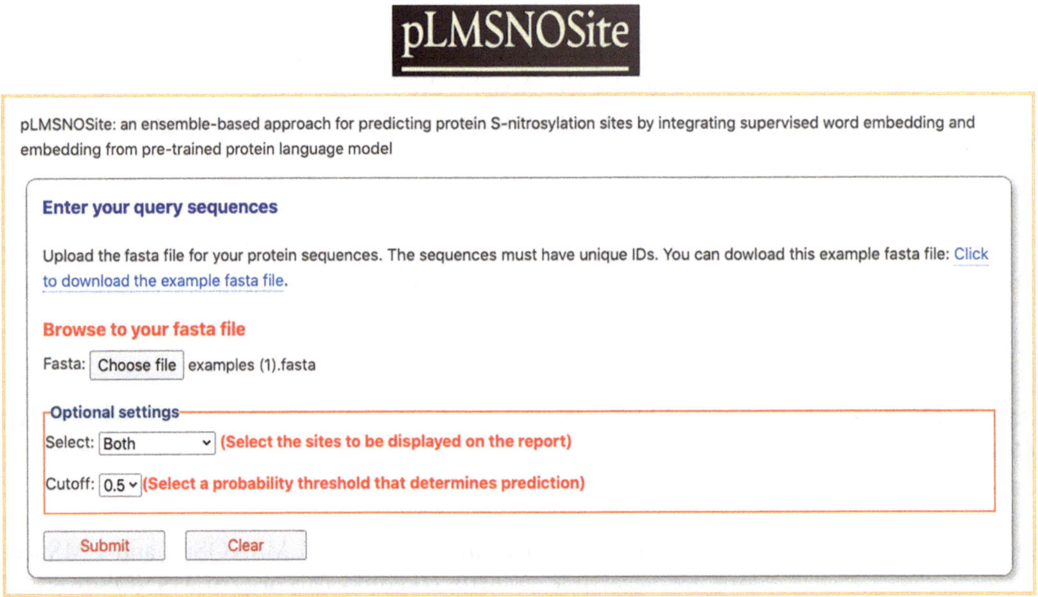

Fig. 4 Screenshot displaying the homepage of the pLMSNOSite web server

webpage (*Click to download the example fasta file*). Additionally, using optional settings, users can further customize their output by selecting whether they prefer to display the results of only positive sites or only negative sites or both. Along with it, users can set their desired decision threshold cut-off, ranging from 0.1 to 0.9, to convert raw probability scores into crisp labels (positive or negative). Please note that the default cut-off for the tools is set at 0.5, as this was the value utilized during the analytical phase of these tools.

Input Once a valid FASTA file is uploaded and the *Submit* button is pressed, the user's request will be sent to the server using an encrypted POST method. The processing time required to display the output will depend on the sequence length, the size of the FASTA file, and the current server load. The uploaded FASTA file will undergo a validity check; if the file is valid, the sequences will be pre-processed and fed into the model for feature extraction and site prediction, as detailed in the Materials and Methods section. An example output from the pLMSNOSite web server is shown in Fig. 5.

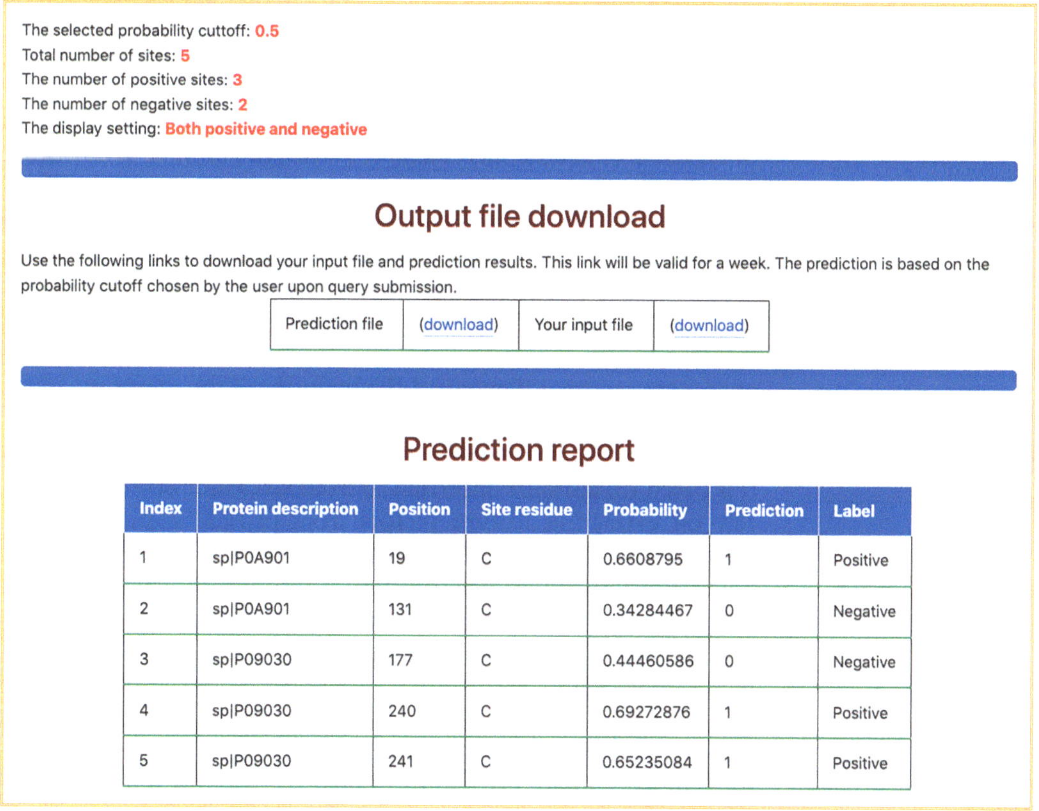

Fig. 5 Screenshot displaying an example output page of the pLMSNOSite web server

Output Upon successful processing of the input FASTA file, users will be directed to the results page, which displays various details such as the selected cut-off threshold, the total count of identified potential sites, and the chosen display setting (whether it is positive only, negative only, or both). The "Download" section provides links to export the results as a *csv* (comma-separated values) file. In addition to this, a "Prediction Report" section displays the prediction results, which include the probability scores and the predicted labels for each site. The results will be retained on the server for 1 week before deletion.

3.4 Runtime and Memory Usage Analysis

Additionally, for this book chapter, we conducted an analysis of CPU runtime and RAM memory usage by computing the time (in seconds) and memory (in megabytes) required for the pLMSNOSite tool to generate predictions across the input sequence (*see* **Note 19**). This process was examined as the length of the sequence was progressively scaled up, ranging from 100 to 4000. To account for variability, these experiments were repeated for ten runs for each length scale, and the results were averaged out. Furthermore, we separately analyzed the average time and average memory usage for the following components: *Environment setup* (which involves importing transformer and other libraries, loading models, defining devices (CPU or GPU), etc.), *ProtT5 encoding (Embedding Generation)*, *Prediction*, and *Encoding+Prediction*.

As the length of the sequence was scaled up (see Fig. 6a), we observed a quadratic increase in the average runtime of Encoding +Prediction (blue curve), which is mainly due to the quadratic nature of generating embeddings from the ProtT5 encoder (green curve). In terms of memory usage (see Fig. 6b), a loosely linear trend was observed for Encoding+Prediction (blue curve). Interestingly, ProtT5 encoding (green curve) did not contribute significantly to this trend—it was primarily the Prediction (red curve) process that demanded the majority of the memory.

It is important to highlight that LMSuccSite is expected to show similar trends regarding time and memory usage, considering it is based on a similar approach to pLMSNOSite. Both these tools rely on extracting feature vectors from protein sequences using ProtT5 and the same underlying architecture for predictions; hence their computational resource consumption would follow a similar pattern. This investigation provides valuable insights into the computational efficiency of our tools and highlights the balance between predictive performance and computational resources.

3.5 Investigating TCA Pathway Proteins: A Case Study

We also performed a case study to demonstrate the utility of our tools pLMSNOSite [20] and LMSuccSite [21]. In this case study, we selected five functional protein sequences originating from the *Homo sapiens*, all significantly involved in the citric acid cycle (also

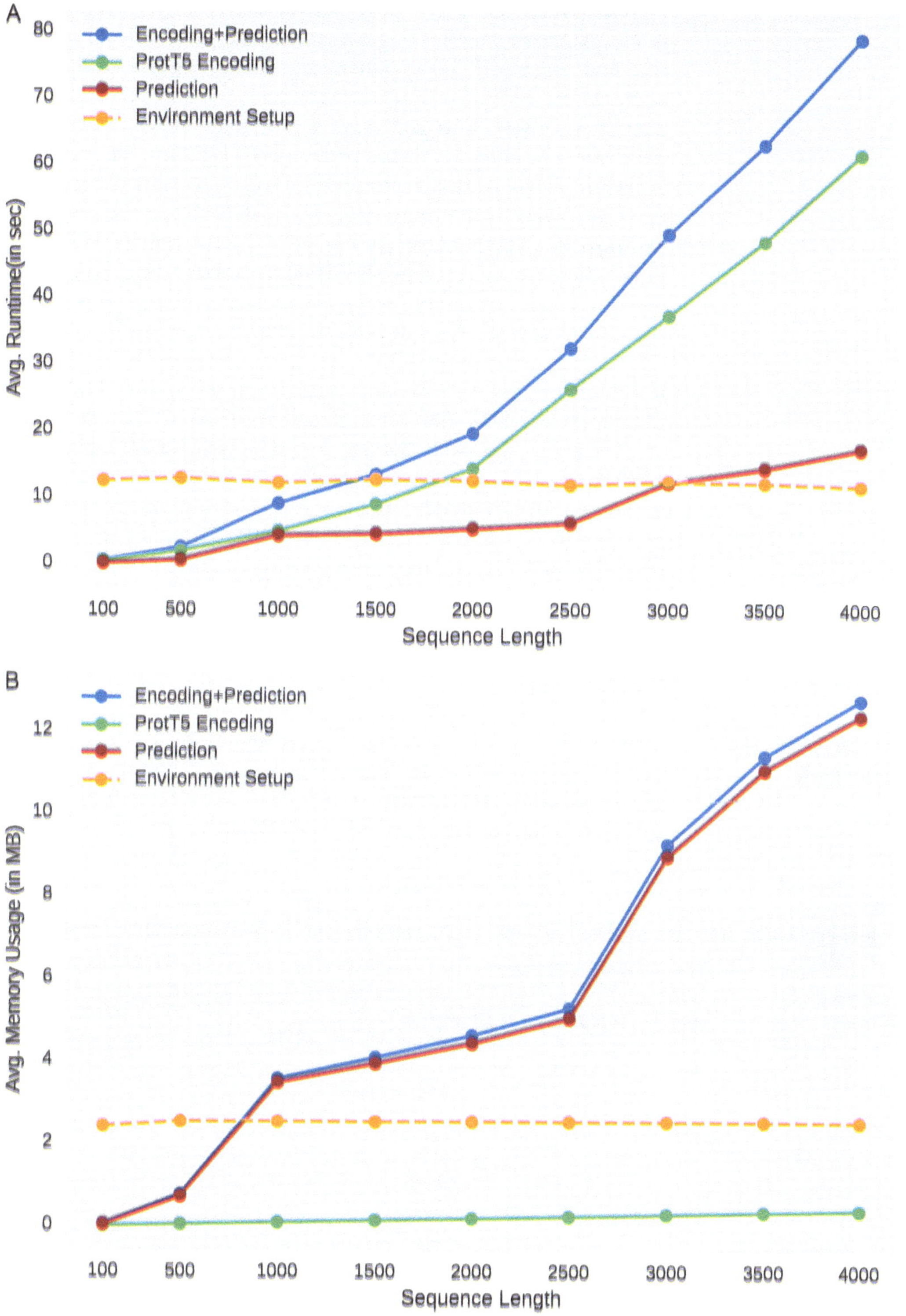

Fig. 6 (**a**) Relationship between average runtime (in seconds/sec) and sequence length and (**b**) average memory usage (in megabytes/MB) and sequence length for pLMSNOSite

known as the tricarboxylic acid cycle or TCA cycle). The TCA cycle is a sequence of chemical reactions that take place in both prokaryotic and eukaryotic cells, generating energy via the oxidation of acetyl-CoA, derived from carbohydrates, fats, and proteins. It is a central metabolic hub and plays a fundamental role in both the generation of energy and the provision of building blocks for the biosynthesis of complex molecules. The regulation of the TCA cycle is driven by a variety of genes that facilitate the process of carbohydrate metabolism. It has been found that proteins produced by these TCA pathway genes are primarily affected by succinylation [36]. For the purposes of our case study, we downloaded the canonical FASTA sequences of the proteins generated by these TCA cycle genes from KEGG pathway database [37]. Detailed information about these sequences along with their location in the eukaryotic cells and their function is reported in Table 11.

Utilizing the web server (or the standalone version) of LMSuccSite [21], we successfully identified 18 out of the 25 known succinylated lysine sites (sourced from the "experimental and putative" site collections in the dbPTM database [5]). This resulted in a TPR (sensitivity) of 72% and a FNR (type II error) of 28%. The substantial proportion of TPR over FNR, as illustrated in Fig. 7, underscores the effectiveness of our tool in predicting sites in sequences that actively participate in vital biological pathways like the TCA cycle.

3.6 Limitation and Future Work

Despite the promising results demonstrated by LMSuccSite and pLMSNOSite tools in predicting PTM sites—achieved by merging the global contextual information from pLMs and the local contextual information using word embedding layer—there remain

Table 11
Protein sequences from the tricarboxylic acid (TCA) pathway with their associated information

Accession ID (protein name)	Gene name	Location	Function
O75390 (CISY_HUMAN)	CS	Mitochondrial	Catalyzes the reaction of acetyl-CoA and oxaloacetate to form citrate and CoA in the citric acid cycle
P08559 (ODPA_HUMAN)	PDHA1	Mitochondrial	Part of the pyruvate dehydrogenase complex that catalyzes the overall conversion of pyruvate to acetyl-CoA and CO_2
P40925 (MDHC_HUMAN)	MDH1	Cytoplasmic	Reversibly converts malate into oxaloacetate using NAD+/NADH in the citric acid cycle
Q02218 (ODO1_HUMAN)	OGDH	Mitochondrial	Part of the 2-oxoglutarate dehydrogenase complex which converts 2-oxoglutarate to succinyl-CoA and CO_2 in the TCA cycle
Q99798 (ACON_HUMAN)	ACO2	Mitochondrial	Catalyzes the isomerization of citrate to isocitrate via cis-aconitate in the TCA cycle

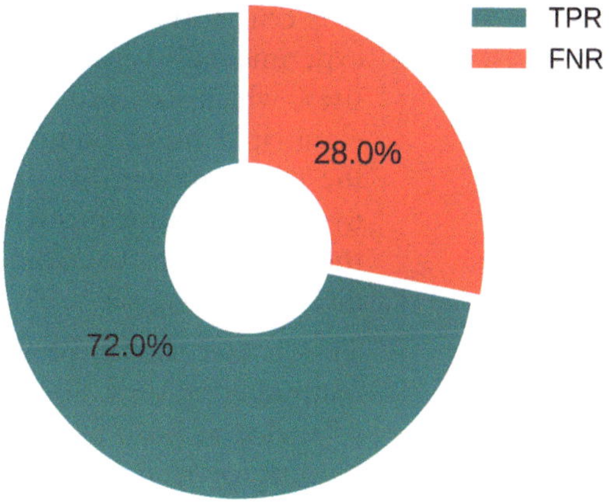

Fig. 7 Donut chart illustrating the distribution of true positive rate (TPR) and false positive rate (FPR) for the prediction of known succinylated sites within proteins involved in the tricarboxylic acid (TCA) cycle pathways, using LMSuccSite

certain limitations and opportunities for enhancement. One limitation is the lack of utilization of structural information. Incorporating structural information can potentially enhance the accuracy and robustness of the predictions, as it provides valuable insights into the spatial arrangement and interactions of amino acids. Furthermore, future work could explore the new pLMs, such as Ankh [38] or ESM-2 [39], which may have improved performance and capture additional contextual information. Additionally, there are alternative methods for obtaining features from pLMs that were not explored in our current approaches. Techniques like FSWE, WSPE, WSWE, and MWVL could be investigated to capture different aspects of protein sequences and potentially improve the predictive performance. Another aspect for future work is the consideration of fine-tuning. While we performed soft fine-tuning (feature-based approach), where we added fully connected layers on top of the pre-trained transformer encoder, hard fine-tuning, which involves jointly training the pLMs and the prediction models, was not explored. Incorporating hard fine-tuning may lead to further improvements in the model's performance. Moreover, an important area of future work includes integrating prediction tools for other types of PTM sites into the LMPTMSite platform.

4 Notes

1. Most existing predictors for S-nitrosylation (SNO) and Succinylation rely on manual or handcrafted feature extraction. Additionally, these predictors operate solely on window sequences, which restricts the computational model to capturing only the local contextual information of the site of interest. In contrast, our approach aims to eliminate these drawbacks while also obtaining the most robust results.

2. It is crucial that positive sites in the dataset are curated from experimentally verified sites to ensure their validity. Redundant sites, which can add noise and cause overfitting, should be eliminated using homology removal algorithms to maintain the uniqueness of the data. Duplicate sites (positive or negative) sharing same protein accession ID must be removed from the training set to avoid potential overfitting and to ensure diverse training set.
3. The dataset should be thoroughly examined for duplicate or contradicting sites. Such inconsistencies can occur when a site is labeled as both positive and negative simultaneously in the training or testing set or if a site's label differs between the training and testing sets. These duplicate or contradictory sites should be identified and removed to maintain the integrity of the dataset.
4. There should be no overlapping sequences between the training and testing sets, i.e., the split between these sets should be based on protein accession IDs. Alternatively, overlapping sequences can be removed from the training set. This is crucial because, especially when using protein language models for feature extraction, similar sequences in the training and testing sets could lead to feature information leakage, thereby inflating performance metrics.
5. Often, the negative set significantly outnumbers the positive set. This imbalance arises because all the sites within the same protein sequences that have not been experimentally verified as positive sites are consequently treated as negative sites. In such cases, data balancing methods should be applied to prevent the model from being biased toward the majority class. However, while addressing imbalance issues, care should be taken to ensure the distributions in the training and testing sets remain consistent and are not distorted, as such changes could inadvertently introduce biases. Furthermore, some sites labeled as negative may be false negatives, meaning they are yet to be experimentally identified as positive sites. This misclassification can result in inaccurate predictions, particularly in situations involving imbalanced learning, such as cost-sensitive learning.
6. In the context of learning from imbalanced datasets, we employ a cost-sensitive (CS) learning approach called class weighting. The strategy counteracts the bias toward the majority class by assigning different weights to each class. These weights serve as a "penalty factor" and aim to influence the learner's decisions. Misclassification of samples of a class with a higher weight will incur a greater penalty, compelling the model to pay more attention to such instances during the learning process. Weight of class i, W_i, is calculated according to the following formula:

$$W_i = \frac{n_s}{n_c \times n_{sc}}$$

where:

n_s refers to the total number of samples

n_c is the total number of unique classes (=2 in binary classification)

n_{sc} denotes the number of samples that belong to a specific class

Please note that the above formula is a common practice and there are variations to it that can also be employed based on the specific requirements of the problem.

7. In the context of post-translational modifications (PTMs), we designate "negative sites" as all potential sites for a given PTM which have not been experimentally annotated as positive. This may appear as a crude approximation, given that it might introduce a degree of false negatives—sites yet to be experimentally confirmed as positive. Although random undersampling could inadvertently eliminate some of these false negatives, it remains an area that requires careful consideration.
8. ESM-1 can accept input sequence length no longer than 1024. In contrast, ProtT5 and ProtBERT do not have this limitation and can handle protein sequences of arbitrary length. This flexibility allows for the analysis of longer protein sequences without the need for truncation or other preprocessing steps.
9. It is worth noting that the learned embeddings from Keras embedding layer are problem-specific. They capture the relationships between amino acids that are relevant to the specific task the model is trained on and may not necessarily be generalizable to other tasks or datasets.
10. We used only the encoder part of ProtT5 in our work, as our experiments showed that the embeddings from the encoder surpassed those from the decoder in performance. To expedite the generation of embeddings, we operated in half-precision mode. Notably, the developers observed no performance degradation in any of their experiments when running in this mode.
11. Within ML/deep learning architectures, three common fusion strategies are employed: early fusion, intermediate fusion, and late fusion. These strategies differ in how they combine input features (or modalities) before inputting them into the final inference model.
12. Early fusion, also known as feature-level fusion, merges input features (or features from the input layer of deep learning architectures) obtained from various representations before inputting them into the final inference model. This method

allows the model to learn joint representations at the input feature level, potentially capturing correlations between different modalities. However, this strategy might be less effective when input features from modalities have differing scales or difficult-to-align characteristics.

13. Intermediate fusion: This strategy involves input features from representations being initially learned by specific layers within their deep learning architectures independently. The resulting learned marginal representations at the intermediate level are combined through concatenation or addition. This method enables the model to learn individual representations for each input source or modality before merging them, which can be advantageous when the inputs possess distinct characteristics that necessitate separate processing.

14. Late fusion, also referred to as score-level fusion, integrates the outputs of models trained on their particular input representations. Each model generates its prediction, and these predictions are combined using methods such as weighted average, voting, or other aggregation techniques to produce the final inference. The limitation of this strategy is that it may not effectively capture correlations between different sources or modalities as well as early or intermediate fusion.

15. In the process of cross-validation, it is crucial to ensure that each fold is constructed in such a way that the training and validation sets do not contain common proteins. This precaution helps prevent any data leakage from the training set into the validation set, thus maintaining the integrity of the evaluation process.

16. In stacked generalization, or stacking, information leakage from base models to the meta-classifier can occur when the same dataset is used for training both. This can lead to overfitting, overly optimistic performance estimates, and potential biases in the final model prediction. To mitigate these issues, strategies like implementing Wolpert's stacking algorithm, using nested cross-validation, utilizing diverse base models, or separating training data for base models and meta-classifiers can be employed.

17. Both pLMSNOSite and LMSuccSite use a standard probability threshold cut-off of 0.5. The user can easily tailor the cut-off value according to their specific preference by following the steps outlined in the *Readme.md* file of the tool's respective repository.

18. The version of the dependencies includes Bio 1.5.2, Keras 2.9.0, Matplotlib 3.5.1, NumPy 1.23.5, Pandas 1.5.0, Requests 2.27.1, scikit-learn 1.2.0, Seaborn 0.11.2, TensorFlow 2.9.1, Torch 1.11.0, tqdm 4.63.0, transformers 4.18.0, and XGBoost 1.5.0.

19. We conducted the runtime and memory usage analysis on a system equipped with an Intel Core i9-10900X processor featuring ten cores, each running at 3.70 GHz (x86_64 architecture), and 128 GB of RAM. This system was operating on Ubuntu 22.04.2 LTS 64-bit. It is crucial to note that runtime and memory usage can significantly vary across different systems. Therefore, when interpreting our results, the specific hardware and software specifications we used should be taken into consideration.

Acknowledgments

This work was partly supported by the National Science Foundation grants number 2210356, 1901793 and MI-SAPPHIRE grant.

References

1. Khoury GA, Baliban RC, Floudas CA (2011) Proteome-wide post-translational modification statistics: frequency analysis and curation of the swiss-prot database. Sci Rep 1:90. https://doi.org/10.1038/srep00090
2. Boeckmann B, Bairoch A, Apweiler R et al (2003) The SWISS-PROT protein knowledgebase and its supplement TrEMBL in 2003. Nucleic Acids Res 31:365–370. https://doi.org/10.1093/nar/gkg095
3. Dinkel H, Chica C, Via A et al (2011) Phospho.ELM: a database of phosphorylation sites—update 2011. Nucleic Acids Res 39: D261–D267. https://doi.org/10.1093/nar/gkq1104
4. O-GLYCBASE version 4.0: a revised database of O-glycosylated proteins | Nucleic Acids Res | Oxford Academic. https://academic.oup.com/nar/article/27/1/370/1241788. Accessed 1 July 2023
5. dbPTM in 2022: an updated database for exploring regulatory networks and functional associations of protein post-translational modifications | Nucleic Acids Research | Oxford Academic. https://academic.oup.com/nar/article/50/D1/D471/6426061. Accessed 1 July 2023
6. Minguez P, Letunic I, Parca L, Bork P (2013) PTMcode: a database of known and predicted functional associations between post-translational modifications in proteins. Nucleic Acids Res 41:D306–D311. https://doi.org/10.1093/nar/gks1230
7. Hornbeck PV, Kornhauser JM, Tkachev S et al (2012) PhosphoSitePlus: a comprehensive resource for investigating the structure and function of experimentally determined post-translational modifications in man and mouse. Nucleic Acids Res 40:D261–D270. https://doi.org/10.1093/nar/gkr1122
8. Pakhrin SC, Pokharel S, Saigo H, KC DB (2022) Deep learning–based advances in protein posttranslational modification site and protein cleavage prediction. In: KC DB (ed) Computational methods for predicting post-translational modification sites. Springer US, New York, pp 285–322
9. Ismail HD, Jones A, Kim JH et al (2016) RF-Phos: a novel general phosphorylation site prediction tool based on random forest. BioMed Res Int 2016:3281590. https://doi.org/10.1155/2016/3281590
10. Larry RM, Jain LC (2001) Recurrent neural networks. Des Appl
11. Hochreiter S, Jürgen S (1997) Long short-term memory. Neural Comput 9:1735–1780
12. Yamashita R, Nishio M, Do RKG, Togashi K (2018) Convolutional neural networks: an overview and application in radiology. Insights Imaging 9:611–629. https://doi.org/10.1007/s13244-018-0639-9
13. Goldberg Y, Levy O (2014) word2vec explained: deriving Mikolov et al.'s negative-sampling word-embedding method
14. Pennington J, Socher R, Manning CD (2014) Glove: global vectors for word representation. Accessed 1 July 2023
15. Devlin J, Chang M W, Lee K, Toutanova K (2019) BERT: pre-training of deep bidirectional transformers for language understanding

16. Pokharel S, Sidorov E, Caragea D, B Kc D (2022) NLP-based encoding techniques for prediction of post-translational modification sites and protein functions. In: Machine learning in bioinformatics of protein sequences. World Scientific, pp 81–127
17. Vaswani A, Shazeer N, Parmar N et al (2017) Attention is all you need. In: Advances in neural information processing systems. Curran Associates, Inc.
18. Radford A, Narasimhan K, Salimans T, Sutskever I (2018) Improving language understanding by generative pre-training
19. Exploring the limits of transfer learning with a unified text-to-text transformer | J Machine Learn Res. https://dl.acm.org/doi/abs/10.5555/3455716.3455856. Accessed 1 July 2023
20. Pratyush P, Pokharel S, Saigo H, Kc DB (2023) pLMSNOSite: an ensemble-based approach for predicting protein S-nitrosylation sites by integrating supervised word embedding and embedding from pre-trained protein language model. BMC Bioinformatics 24:41. https://doi.org/10.1186/s12859-023-05164-9
21. Pokharel S, Pratyush P, Heinzinger M et al (2022) Improving protein succinylation sites prediction using embeddings from protein language model. Sci Rep 12:16933. https://doi.org/10.1038/s41598-022-21366-2
22. Pakhrin SC, Pokharel S, Aoki-Kinoshita KF et al (2023) LMNglyPred: prediction of human N-linked glycosylation sites using embeddings from a pre-trained protein language model. Glycobiology 33:411–422. https://doi.org/10.1093/glycob/cwad033
23. Pakhrin S, Pokharel S, Pratyush P et al (2023) LMPhosSite: A deep learning-based approach for general protein phosphorylation site prediction using embeddings from local window sequence and pre-trained Protein Language Model. J Proteome Res
24. Stomberski CT, Hess DT, Stamler JS (2019) Protein S-nitrosylation: determinants of specificity and enzymatic regulation of S-nitrosothiol-based signaling. Antioxid Redox Signal 30:1331–1351. https://doi.org/10.1089/ars.2017.7403
25. Dai X, Zhou Y, Han F, Li J (2022) Succinylation and redox status in cancer cells. Front Oncol 12
26. Cd-hit: a fast program for clustering and comparing large sets of protein or nucleotide sequences | Bioinformatics | Oxford Academic. https://academic.oup.com/bioinformatics/article/22/13/1658/194225. Accessed 28 June 2023
27. Hasan MM, Manavalan B, Khatun MS, Kurata H (2019) Prediction of S-nitrosylation sites by integrating support vector machines and random forest. Mol Omics 15:451–458. https://doi.org/10.1039/C9MO00098D
28. DeepNitro: prediction of protein nitration and nitrosylation sites by deep learning – ScienceDirect. https://www.sciencedirect.com/science/article/pii/S1672022918303474. Accessed 28 June 2023
29. Hasan MM, Khatun MS, Mollah MNH et al (2017) A systematic identification of species-specific protein succinylation sites using joint element features information. Int J Nanomedicine 12:6303–6315. https://doi.org/10.2147/IJN.S140875
30. The UniProt Consortium (2015) UniProt: a hub for protein information. Nucleic Acids Res 43:D204–D212. https://doi.org/10.1093/nar/gku989
31. Elnaggar A, Heinzinger M, Dallago C et al (2021) ProtTrans: towards cracking the language of Lifes code through self-supervised deep learning and high performance computing. IEEE Trans Pattern Anal Mach Intell PP. https://doi.org/10.1109/TPAMI.2021.3095381
32. Rao R, Meier J, Sercu T, et al (2020) Transformer protein language models are unsupervised structure learners. 2020.12.15.422761
33. Villegas-Morcillo A, Makrodimitris S, van Ham RCHJ et al (2021) Unsupervised protein embeddings outperform hand-crafted sequence and structure features at predicting molecular function. Bioinformatics 37:162–170. https://doi.org/10.1093/bioinformatics/btaa701
34. Wolpert DH (1992) Stacked generalization. Neural Netw 5:241–259. https://doi.org/10.1016/S0893-6080(05)80023-1
35. Hasan MM, Kurata H (2018) GPSuc: global prediction of generic and species-specific succinylation sites by aggregating multiple sequence features. PLOS One 13:e0200283. https://doi.org/10.1371/journal.pone.0200283
36. Yang Y, Gibson GE (2019) Succinylation links metabolism to protein functions. Neurochem Res 44:2346–2359. https://doi.org/10.1007/s11064-019-02780-x
37. Kanehisa M (2002) The KEGG database. In: 'In Silico' simulation of biological processes. Wiley, pp 91–103
38. Elnaggar A, Essam H, Salah-Eldin W, et al (2023) Ankh : optimized protein language model unlocks general-purpose modelling. 2023.01.16.524265

39. Lin Z, Akin H, Rao R, et al (2022) Language models of protein sequences at the scale of evolution enable accurate structure prediction. 2022.07.20.500902
40. Xue Y, Liu Z, Gao X et al (2010) GPS-SNO: Computational Prediction of Protein S-Nitrosylation Sites with a Modified GPS Algorithm. PLOS One 5:e11290. https://doi.org/10.1371/journal.pone.0011290
41. Xu Y, Ding J, Wu L-Y, Chou K-C (2013) iSNO-PseAAC: predict cysteine S-nitrosylation sites in proteins by incorporating position specific amino acid propensity into pseudo amino acid composition. PloS One 8: e55844. https://doi.org/10.1371/journal.pone.0055844
42. SNOSite: exploiting maximal dependence decomposition to identify cysteine S-nitrosylation with substrate site specificity | PLOS One. https://journals.plos.org/plosone/article?id=10.1371/journal.pone.0021849. Accessed 28 June 2023
43. Thapa N, Chaudhari M, McManus S et al (2020) DeepSuccinylSite: a deep learning based approach for protein succinylation site prediction. BMC Bioinformatics 21:63. https://doi.org/10.1186/s12859-020-3342-z
44. pSuc-FFSEA: predicting lysine succinylation sites in proteins based on feature fusion and stacking ensemble algorithm – PubMed. https://pubmed.ncbi.nlm.nih.gov/35686053/. Accessed 28 Jun 2023
45. Ning Q, Zhao X, Bao L et al (2018) Detecting succinylation sites from protein sequences using ensemble support vector machine. BMC Bioinformatics 19:237. https://doi.org/10.1186/s12859-018-2249-4

Chapter 17

Prediction and Evaluation of Protein Aggregation with Computational Methods

Mubashir Hassan, Saba Shahzadi, Mai Suan Li, and Andrzej Kloczkowski

Abstract

Protein and peptide aggregation has recently become one of the most studied biomedical problems due to its central role in several neurodegenerative disorders and of biotechnological importance. Multiple in silico methods, databases, tools, and algorithms have been developed to predict aggregation of proteins and peptides to better understand fundamental mechanisms of various aggregation diseases. Here, we attempt to provide a brief overview of bioinformatic methods and tools to better understand molecular mechanisms of aggregation disorders. Furthermore, through a better understanding of protein aggregation mechanisms, it might be possible to design novel therapeutic agents to treat and hopefully prevent protein aggregation diseases.

Key words Protein aggregation, Protein aggregation diseases, Protein aggregation databases, Computational methods, Bioinformatics

1 Introduction

The human proteome is comprised of 20,000–25,000 proteins having various sizes, amino acid compositions, structures, and functions [1, 2]. Every cellular activity depends upon conformational structures, dynamics, functions, and concentrations of proteins [3, 4]. Proteostasis is a process that regulates proteins within the cell [5] and might be disturbed through internal or external perturbations, which could lead to the loss of protein functions, changes in concentration, and deposition of protein aggregates in the affected tissues [6, 7]. Protein aggregation is a biological process which involves misfolded proteins that assemble into insoluble aggregates [8, 9] which have been implicated in a wide variety of diseases known as amyloidoses, including amyotrophic lateral sclerosis (ALS), Alzheimer's disease, Parkinson's, and prion disease, among others [7, 10, 11]. Despite diverse risk factors such as aging, environmental factors, and genetic mutations, the accumulation of

Andrzej Kloczkowski et al. (eds.), *Prediction of Protein Secondary Structure*, Methods in Molecular Biology, vol. 2867, https://doi.org/10.1007/978-1-0716-4196-5_17,

intracellular and extracellular proteinaceous deposits is considered the key factor in these disorders [12, 13].

Aggregation of proteins is frequently caused by chemical or physical degradation, depending on the protein's thermodynamic stability. It has been observed that reduction in free surface energy by the removal of hydrophobic residues from contact with the solvent is the major driving force in protein aggregation. During this process, there can be a lag phase where the loss of the native structure of a protein is undetectable, and the protein solution remains clear [14]. During protein aggregation, nucleation/ growth occurs; this lag phase arises because of the free energy requirement to build a solid–liquid interface before the nucleation occurs. The energy barrier is the highest when a critical size for the new phase is reached, and the nucleus growth can proceed. Insoluble aggregates are forming when the aggregates are large enough to exceed their solubility limit. The growth of these aggregates occurs in the direction with the lowest free energy to assemble, which can result in an ordered morphology, e.g., fibrils [15].

The manufacturing of protein/peptide-based biotherapeutics is slow and complicated due to protein instability and aggregation. Therefore, the development of biotherapeutic drugs has become increasingly dependent on developing capability assessments and optimizing proteins to increase their stability and solubility while decreasing their viscosity and aggregation [16, 17]. It has been observed that in liquid formulations, the solubility and viscosity of a biopharmaceutical can be affected by their aggregation propensity. However, multiple computational techniques have been designed for developing and optimizing the therapeutic proteins and peptides [18]. The purpose of this review is to provide a collection of computational methods, algorithms, tools, and databases for predicting the aggregation propensity of proteins and peptides and their possible involvement in various aggregation disorders in humans.

2 Protein Aggregation Databases

Databases are online platforms where data can be stored and retrieved for further computational analysis [19]. Recent experiments on protein aggregation have led to an exponential increase in experimental data, so information about protein aggregation must be stored and curated at numerous online platforms. Various online databases are available to assist the community of researchers working on protein aggregation and their therapeutic applications (see Table 1).

Fibril_one was the first amyloidogenic protein database, which contained 250 mutations of 22 proteins accompanied by 50 experimental conditions [20]. This database holds information about

Table 1
Protein aggregation databases

Databases	Online links/URLs	Ref
Fibril_one	http://www.bioinformatics.leeds.ac.uk/group/online/fibril_one	[20]
ZipperDB	https://services.mbi.ucla.edu/zipperdb/	[24]
WALTZ-DB 2.0	http://waltzdb.switchlab.org/	[26]
ProADD*	http://bicmku.in/ProADD	[27]
AmyLoad	http://comprec-lin.iiar.pwr.edu.pl/amyload/	[28]
AmyPro	https://amypro.net/#/	[29]
CPAD 2.0	https://web.iitm.ac.in/bioinfo2/cpad2/	[31]
AmyloBase	http://150.217.63.173/biochimica/bioinfo/amylobase/pages/view.html	[32]
AMYPdb	http://amypdb.genouest.org/e107_plugins/amypdb_project/project.php	[33]
PDB_Amyloid	https://pitgroup.org/amyloid/	[34]
AL-Base	http://albase.bumc.bu.edu/aldb	[35]
A3D	http://biocomp.chem.uw.edu.pl/A3D2/hproteome	[38]
CARs-DB	http://carsdb.ppmclab.com/	[39]
PASTA 2.0	http://protein.bio.unipd.it/pasta2/	[40]

fibril formation and has been populated by extensive literature searches. Fibril_one information is composed of protein sequences linked with various other databases such as GenBank [21], SWISS-PROT [22], and PDB [23], respectively. Moreover, it also explores the single and double pathogenic mutations and provides disease names, the secondary structure location of the mutated residue, and their effect on native structure.

ZipperDB database is another database that predicts the aggregation profiles of 76 genomes [24]. This database contains fibril-forming segment predictions within proteins identified by the 3D profile method. In this approach, over 20,000 putative protein sequences have been analyzed for segments with high fibrillation propensity that could form a "steric zipper" composed of two self-complementary β-sheets, giving rise to the spine of an amyloid fibril [25]. WALTZ-DB contains an assembly of experimentally known amyloid-forming hexapeptides, explored through electron microscopy, dye binding, and Fourier-transform infrared spectroscopy (FTIR) [26]. The updated version of WALTZ-DB is WALTZ-DB 2.0, obtained by expanding the hexapeptide sequence dataset and adding new structural information [26]. ProADD is a database on protein aggregation diseases developed to organize the data under a single platform to facilitate easy access for researchers [27]. This database allows protein aggregation diseases to be classified based on sequence analysis and structural analysis that enables mining patterns of aggregating protein in the dataset [27].

AmyLoad database collects amyloidogenic and non-amyloidogenic sequence fragments from all possible major resources and provides detailed information about each fragment [28]. Similarly, AmyPro is another comprehensive database designed for precursor proteins and their aggregation-prone regions [29]. Besides providing information about these proteins, AmyPro also provides links to other databases and literature references, along with information about phylogenetic annotations of proteins and their role in the amyloid state. This enables the instant visualization of the amyloidogenic sequence fragments within the corresponding protein structures.

Curated Protein Aggregation Database (CPAD) is a comprehensive database that collects results from experimental studies performed by the scientific community aimed at understanding protein/peptide aggregation [30]. The data present in CPAD has been aggregated with additional information such as amyloid fibril-forming peptides of different lengths, amyloid fibril-forming hexapeptides whose crystal structures are available in PDB, and experimentally validated aggregation-prone regions found in amyloidogenic proteins [30]. CPAD has been updated recently to CPAD 2.0 and includes a new category of aggregation-related structures [31].

AmyloBase is a database that collects data on protein aggregation from kinetics experiments [32]. AMYPdb is another database that collects structural information on amyloidogenic proteins [33]. PDB_Amyloid database contains entries on both amyloid proteins and globular proteins having amyloid-like substructures [34]. AL-Base database collects information about amyloidogenic immunoglobulin light chain sequences derived from patients with AL amyloidosis [35].

Aggrescan3D (A3D) database contains AlphaFold-predicted human protein structures particularly focused on their aggregation properties. The A3D algorithm uses 3D atomic models to compute the structurally corrected aggregation values (A3D score) for each amino acid [36, 37]. A3D database provides a detailed analysis of the structure-based aggregation propensity of protein structure. The A3D database provides simple, graphical, user-friendly tools to visualize protein structures. Additionally, it allows users to test the influence of mutations on solubility and stability of proteins [38]. Cryptic amyloidogenic regions database (CARs-DB) focuses on intrinsically disordered proteins (IDPs) with a milder aggregation potential than conventional amyloid regions protected inside globular proteins. However, their presence also appears to be associated with pathologies like cancer or AD. CARs-DB is a database containing precomputed predictions for all CARs present in the

IDPs deposited in the DisProt database. It contains over 8900 unique CARs identified in a total of 1711 IDRs [39].

PASTA is a web server that predicts protein aggregation from sequences. The first version of the PASTA server was developed in 2007. The updated version of PASTA 2.0 provides information on intrinsic disorder and secondary structure predictions that complement the aggregation data. PASTA 2.0 energy function evaluates the stability of putative cross-beta pairings between different sequence stretches [40].

3 Computational Approaches to Protein Aggregation

Protein aggregation has been computationally extensively studied for the last few decades [41]. There are three basic computational techniques used in protein aggregation studies: (i) prediction of aggregation propensity, (ii) prediction of aggregation kinetics, and (iii) molecular dynamic simulations [42]. There are multiple in silico techniques designed to identify the protein aggregation propensity [43, 44]. Generally, these approaches can be divided into sequence-based and structure-based methods, depending on the input data required for the prediction (Table 2).

Table 2
Protein aggregation approaches

Methods	Features	Ref
Amyloidogenic pattern	Pattern derived from positional scanning mutagenesis experiments on amyloidogenic peptide STVIIE	[47]
AGGRESCAN	Aggregation propensity scale for amino acids derived from in vivo experiments on amyloidogenic proteins	[51]
Zyggregator	Amino acid scales for α-helix and β-sheet formation, hydrophobicity and charge, hydrophobic pattern, and presence of gatekeeper residues	[60]
Pafig	41 physicochemical properties of amino acid	[86]
PAGE	Aromaticity, β-sheet propensity, charge, polar–nonpolar surfaces, and solubility	[87, 88]
WALTZ	PSSM, physicochemical properties, position-specific pseudoenergy terms	[61]
AbAmyloid	Amino acid composition, dipeptide composition, and physicochemical properties	[89]
FoldAmyloid	Packing density and hydrogen bond probabilities obtained from protein structures	[90]

(continued)

Table 2
(continued)

Methods	Features	Ref
SALSA β-Strand Contiguity (β-SC)	β-strand propensity	[91]
APPNN	7 amino acid physicochemical and biochemical properties	[54, 57]
Amylogram	17 amino acid properties such as size of residues, hydrophobicity, solvent accessible surface area, frequency of β-sheets, contactivity, and contact site propensities	[92]
ANuPP	Atom compositions of peptides and protein segments	[62]
TANGO	Segmental β-sheet probability derived from empirical and statistically derived energy functions	[63]
SecStr	Secondary structure preferences	[64]
NetCSSP	Residue interactions and solvation energies obtained using AMBER forcefield	[55]
Archcandy	Scoring function derived for steric tension, electrostatic interactions, packing, and hydrogen bond formation	[93]
BetaSerpentine	β-arches (β-strand-loop-β-strand motif from Archcandy), compatibility of β-arches, compactness	[94]
BETASCAN	Pairwise probability tables to identify hydrogen bond forming residues in strand pairs	[95]
AmyloidMutants	Potential energy scoring function derived from observed residue/residue interactions in PDB	[96]
STITCHER	Scoring function addressing enthalpic and entropic changes in protofibril formation and BETASCAN strand pair predictions	[97]
PASTA 2	Hydrogen-bonding energy functions for residue pairs derived from beta-strand structures	[40]
GAP	Residue pair potentials derived from hexapeptide sequences	[66]
FISH Amyloid	Residue co-occurrence matrix derived from amyloidogenic and non-amyloidogenic peptides of length (4–10)	[98]
AgMata	Statistical potentials derived for residue position, secondary structure probabilities, and interaction energies	[99]
3D PROFILE (ZipperDB)	Microcrystal structure of the NNQQNY peptide and atomic-level potential ROSETTADESIGN	[24]
Pre-Amyl	Template ensemble obtained from microcrystal structures of the NNQQNY peptide and KBP, atomic distance-dependent knowledge-based pairwise residue potentials	[100]
CORDAX	Thermodynamic stability calculated by threading over 140 amyloid fibril cores	[101]
PATH	Modeller Dope score and Rosetta (REF15) energy values from homology models of 7 template structures	[102]

(continued)

Table 2 (continued)

Methods	Features	Ref
AMYLPRED2	Consensus predictor includes outputs from AGGRESCAN, NetCSSP, AmyloidMutants, Pafig, Amyloidogenic Pattern, SecStr, Average Packing Density, TANGO, Beta-strand contiguity, WALTZ, Hexapeptide Conformational Energy	[57]
MetAmyl	Consensus predictor that includes PAFIG, SALSA, WALTZ, and FoldAmyloid	[58]
SAP	Residue hydrophobicity, solvent accessible area over time obtained from MD	[58]
Developability Index	SAP and PROPKA values	[103]
AggScore	Hydrophobic and hydrophilic patches obtained by using atom partial charges and logP values	[104]
AGGRESCAN3D 2.0	AGGRESCAN residue score, exposed surface area, FoldX energy-minimized protein structure or ensemble from CABS-flex simulations	[36]

3.1 Sequence-Based Methodology for Protein Aggregation Prediction

Protein aggregation can be predicted using sequence-based approaches by analyzing amino acid physiochemical properties, sequence patterns, statistically determined propensity values, scoring functions based on knowledge, secondary structure propensities, residue–residue contact potentials, and threading [45, 46]. Pattern matching, including amyloidogenic pattern matching, is the sequence-based approach usually employed to analyze linear sequences. One of such study has been focused on the STVIIE peptide sequence through positional scanning mutagenesis and identifying sequence patterns of hexapeptides, which form amyloid-like fibrils [47].

The first generation of algorithms designed to predict protein aggregation was based on the identification of linear polypeptide sequences, and more than 20 sequence-based algorithms have been developed [48]. Their design assumed that protein aggregation is triggered by short and well-defined sequential segments, which tend to be hydrophobic and low in net charge. Based on underlying principles, several different subclasses of algorithms can be defined. The first class of algorithms belongs to the phenomenological class, related to experimental data which define the determinants of aggregation [49]. AGGRESCAN and Zyggregator are considered the best examples of phenomenological algorithms that are based on the rationalization of experimentally determined factors influencing protein aggregation [50, 51]. The second class of algorithms is based on theoretical assessments of sequence properties that have been linked with aggregation. This class includes the programs

TANGO, PASTA 2.0, FoldAmyloid, Waltz, and Amyloid mutants, which assess a sequence propensity to adopt a specific aggregation-prone conformation, the density of protein packing, the makeup and pattern of residues, or the suitability to adopt the topologically restricted conformations that characterize amyloid-like states [52].

Furthermore, a growing number of machine learning (ML)-based methods have been developed for the prediction of protein aggregation. These ML-based methods utilize neural networks to identify the sequential features that are highly correlated with aggregation [53]. ML-based approaches attain performances comparable to or better than the best traditional predictors. APPNN [54], netCSSP [55], and FISH Amyloid are examples of ML-based algorithms [56]. There are also consensus algorithms based on combining and weighting the outputs of other predictors into a single output. These consensus predictors assemble different concepts behind each predictor to increase robustness while reducing method-specific biases. AMYLPRED 2 and MetAmyl exemplify this kind of algorithms [57, 58].

3.2 Amino Acid Characteristics as Lead Parameters for Protein Aggregation

Amino acids are molecules considered as building blocks of life that combine together with peptide bonds to form protein chains [59]. Amino acid characteristics such as hydrophobicity, size, surface area, charge, aromaticity, contact frequency, β-sheet propensity, and several other physiochemical characteristics are used to identify aggregation-prone regions. The best example of algorithms that utilize the amino acid aggregation-propensity scale derived from in vivo experiments on amyloidogenic proteins is AGGRESCAN [51].

Similarly, Zyggregator is another algorithm that uses amino acid propensities of α-helix and β-sheet formation, residue charges, hydrophobicities, hydrophobic pattern, and the presence of gatekeeper amino acids [60]. WALTZ algorithm utilizes a hybrid approach by combining a position-specific scoring matrix derived from amyloidogenic peptides and amino acid physicochemical properties and position-specific pseudo-energy values obtained from modeled structures [61]. ANuPP is an ensemble classifier composed of nine independently trained logistic regression models on groups of amyloidogenic peptides, taking into account the diversity of aggregate nucleation, diffusion, and fibrillation processes [62].

3.3 Secondary Structure Propensity in Protein Aggregation

β-sheet is one of the key features of amyloid-fibril-forming peptides that is being used in developing various predicting algorithms. TANGO is the best example of such algorithm that uses empirically and statistically derived potential functions to estimate the probability of a segment forming mediated aggregates [63]. Additionally, TANGO compares the probabilities of a segment to be in various secondary structure states such as α-helix, β-sheet, coil, and turn. Similarly, SecStr and NetCSSP algorithms use conformational switches from other secondary states to β-sheet formation to predict protein aggregation [55, 64]. It was recently shown that the β-content of the monomer determines the tendency to aggregation, and the higher the β content, the faster protein aggregation occurs [65]. Moreover, when molecular dynamics simulations and experimental data obtained for various variants of the Aβ42 peptide were combined, the aggregation rate depended exponentially on the β-content.

3.4 Residue Pairs and Aggregation

In amyloid fibrils cross-β spine, a double β-sheet, with each sheet formed from parallel segments stacked in-register, is a common structural motif. Side chains protruding from the two sheets form a tightly self-complementing steric zipper, bonding the sheets with interdigitated side chains and axial hydrogen bonds. Furthermore, aromatic residues such as Asn, Gln, Thr, and Ser contribute additional stability by forming stacking and ladders of hydrogen bonds. These residue–residue interactions play a significant role in some protein aggregation prediction methods [66], such as PASTA2 and BETASCAN, that use residue–residue probabilities and scoring functions for β-sheet hydrogen bond formation and contacts derived from protein structure databases [41].

3.5 Structure-Based Approaches

SAP, Developability Index, AGGRESCAN3D, and AggScore are examples of structure-based prediction methods that depend upon 3D protein structures [67]. These methods predominantly consider the solvent accessibility of protein residues and atoms to estimate surface hydrophobicity. In addition to the static structure, short molecular dynamics (MD) simulations are performed to calculate the ensemble statistics over time. However, compared to sequence-based methods, structure-based methods take into account both the folding and the native state of a protein [68].

4 Systematic Coarse-Graining Approaches

Systematic coarse-graining provides a bottom-up methodology for obtaining optimized coarse-grained potentials that fully match atomistic simulations or experimental data. The simplest coarse-grained models are the lattice models developed by Li et al. [69] and Vacha and Frenkel [70]. Although these models are simple, they can predict key factors that control protein aggregation [71] and the important role of nonspecific interactions in amyloid nucleation [72]. More systematic coarse-graining techniques include the relative entropy method [73], multiscale coarse-graining method [74], and the iterative Boltzmann inversion method, respectively [75]. Many systematic coarse-graining methods have been applied to protein aggregation [76–78]. The primary drawback of these methods is that they require reliable all-atom statistics to parameterize the coarse-grained models [79].

4.1 Thermodynamic Methods

Both atomistic and coarse-grained simulations with improved sampling techniques are needed to effectively compute free energy profiles. The biased sampling methods of prespecified collective variables include metadynamics [80] and umbrella sampling [81]. Metadynamics and umbrella sampling are computer simulation methods to estimate a system's free energy and other state functions, where ergodicity is hindered by the form of the system's energy landscape. Parallel tempering methods create multiple copies of a system, called replicas, that are randomly initialized and run at different temperatures. This does not require the specification of collective variables; instead, it increases sampling by exchanging trapped systems with replicas run at higher temperature [82].

4.2 Kinetic Methods

The kinetics of many molecular processes are dominated by rare events on the molecular time scales. Many simulation methods have been developed to enhance the sampling of the barrier between states. Some sampling methods address kinetics by generating an ensemble of dynamical short, parallel trajectories that enhance sampling. This includes the Markov state model (MSM) formalism that has been applied to the study of biomolecular processes [83, 84]. This approach increases sampling by launching parallel trajectories. MSM provides a complementary approach to free energy methods such as replica exchange or metadynamics, disregarding kinetics in favor of sampling the energy landscape (Table 3).

Table 3
Selection of computational methods applied to studying protein aggregation [85]

Computational technique	Method	Description
Thermodynamic enhanced sampling methods	Metadynamics	Enhanced sampling of free energy landscape using collective variables
	Umbrella sampling	Enhanced sampling of free energy landscape using collective variables
	Parallel tempering	Enhanced sampling of free energy landscape using runs at multiple temperatures
	Replica exchange molecular dynamics (REMD)	Enhanced sampling of free energy landscape using multiple temperatures and/or varying Hamiltonian parameters
	Replica exchange statistical temperature MD	Replica minimization required for good sampling in REMD
	Replica exchange Monte Carlo	As for REMD, but for Monte Carlo sampling

(continued)

Table 3
(continued)

Computational technique	Method	Description
Kinetic enhanced sampling methods	Markov state model	Enhanced sampling of kinetics using collective variables; launches short trajectories from configurations requiring the greatest sampling
	Free energy guided sampling	Enhanced sampling of kinetics using approximated free energy surface; launches short trajectories in parallel with Boltzmann distribution
	WExplore	Enhanced sampling of kinetics using multiple replicas in dynamically assigned tessellations of configuration space; clones/merges replicas as needed
	Transition path sampling	Rare-event sampling method, well suited to study processes, such as protein folding and aggregation, that have complex underlying energy landscapes; involves generating a set of reactive trajectories between an initial and final state, typically followed by likelihood maximization to obtain optimized reaction coordinates
	String method	Free energy landscape method to find lowest barrier path ("mountain pass") between two states; parallel runs are tied together in a string in state space to prevent them from merging/separating; the string converges to the transition path between the ends
Data/trajectory analysis methods	Secondary nucleation kinetic analysis	Analysis of trajectory/experimental fibrillization kinetic data; involves fitting to a master equation, including secondary pathway terms
	Normal mode analysis	Trajectory analysis method using elastic network models; used to study the low-frequency collective motions of biomolecules
Systematic coarse graining	Relative entropy coarse graining	Systematic coarse-graining method matching coarse-graining to all-atom configurational probability distributions for minimal information loss
	Multiscale coarse graining	Systematic coarse-graining method matching coarse-graining to all-atom momenta
	Iterative Boltzmann inversion	Systematic coarse-graining method matching coarse-graining to all-atom Boltzmann distributions
Fast molecular dynamics integrator	Discontinuous molecular dynamics	Molecular dynamics integrator allowing for discontinuous potentials

5 Conclusions and Future Prospects

The widespread nature of protein aggregation influences almost all aspects of protein functions, from disease onset to the production of biotherapeutics. It is crucial to develop novel methods to address and solve protein aggregation problems quickly and reliably to change this almost random and unpredictable phenomenon into a more predictable process. Presently, in silico approaches are being gradually introduced into the daily operations of many laboratories as efficient tools to support and supplement experimental efforts. In the future, we can expect the development of novel, improved, and more robust algorithms based on protein sequence and protein structure that will be commonly implemented in most laboratories to streamline research workflow by providing an easy and fast optimization step for the design of protein-based therapeutics, impacting all biomedical and clinical applications.

Acknowledgments

AK acknowledges financial support from NIH grants R01GM127701 and R01HG012117. MH acknowledges the Ohio State University for providing the "President's Postdoctoral Scholars Program (PPSP)" award and for financial support to complete this computational research.

References

1. Thul PJ, Lindskog C (2018) The human protein atlas: a spatial map of the human proteome. Protein Sci 27(1):233–244
2. Ponomarenko EA, Poverennaya EV, Ilgisonis EV, Pyatnitskiy MA, Kopylov AT, Zgoda VG, Lisitsa AV, Archakov AI (2016) The size of the human proteome: the width and depth. Int J Anal Chem 2016
3. Alberts B, Johnson A, Lewis J, Raff M, Roberts K, Walter P (2002) Protein function. In: Molecular biology of the cell, 4th edn. Garland Science
4. Alberts B, Johnson A, Lewis J, Raff M, Roberts K, Walter P (2002) The shape and structure of proteins. In: Molecular biology of the cell, 4th edn. Garland Science
5. Toyama BH, Hetzer MW (2013) Protein homeostasis: live long, won't prosper. Nat Rev Mol Cell Biol 14(1):55–61
6. Díaz-Villanueva JF, Díaz-Molina R, García-González V (2015) Protein folding and mechanisms of proteostasis. Int J Mol Sci 16(8):17193–17230
7. Tanzi RE, Bertram L (2005) Twenty years of the Alzheimer's disease amyloid hypothesis: a genetic perspective. Cell 120(4):545–555
8. Ye S, Hsiung C-H, Tang Y, Zhang X (2022) Visualizing the multistep process of protein aggregation in live cells. Acc Chem Res 55(3):381–390
9. Shahzadi S, Yasir M, Aftab B, Babar S, Hassan M (2022) Exploration of protein aggregations in Parkinson's disease through computational approaches and big data analytics. In: Computer simulations of aggregation of proteins and peptides. Springer, pp 449–467
10. Felice FGD, Vieira MN, Meirelles MNL, Morozova-Roche LA, Dobson CM, Ferreira ST (2004) Formation of amyloid aggregates from human lysozyme and its disease-associated variants using hydrostatic pressure. FASEB J 18(10):1099–1101
11. Nguyen PH, Ramamoorthy A, Sahoo BR, Zheng J, Faller P, Straub JE, Dominguez L, Shea J-E, Dokholyan NV, De Simone A (2021) Amyloid oligomers: a joint

experimental/computational perspective on Alzheimer's disease, Parkinson's disease, type II diabetes, and amyotrophic lateral sclerosis. Chem Rev 121(4):2545–2647

12. Stefani M (2004) Protein misfolding and aggregation: new examples in medicine and biology of the dark side of the protein world. Biochim Biophys Acta 1739(1):5–25
13. Imbimbo BP, Lombard J, Pomara N (2005) Pathophysiology of Alzheimer's disease. Neuroimaging Clin 15(4):727–753
14. Carpenter JF, Randolph TW, Jiskoot W, Crommelin DJ, Middaugh CR, Winter G, Fan Y-X, Kirshner S, Verthelyi D, Kozlowski S (2009) Overlooking subvisible particles in therapeutic protein products: gaps that may compromise product quality. J Pharm Sci 98(4):1201–1205
15. Berrill A, Biddlecombe J, Bracewell D (2011) Product quality during manufacture and supply. In: Peptide and protein delivery. Elsevier, pp 313–339
16. Pham NB, Meng WS (2020) Protein aggregation and immunogenicity of biotherapeutics. Int J Pharm 585:119523
17. Wang W (2005) Protein aggregation and its inhibition in biopharmaceutics. Int J Pharm 289(1–2):1–30
18. Kalita P, Tripathi T, Padhi AK (2023) Computational protein design for COVID-19 research and emerging therapeutics. ACS Central Science 9(4):602–613
19. Revesz P (2010) Introduction to databases. Springer
20. Siepen JA, Westhead DR (2002) The fibril_one on-line database: mutations, experimental conditions, and trends associated with amyloid fibril formation. Protein Sci 11(7):1862–1866
21. Benson DA, Karsch-Mizrachi I, Lipman DJ, Ostell J, Rapp BA, Wheeler DL (2000) GenBank. Nucleic Acids Res 28(1):15–18
22. Bairoch A, Apweiler R (2000) The SWISS-PROT protein sequence database and its supplement TrEMBL in 2000. Nucleic Acids Res 28(1):45–48
23. Berman HM, Westbrook J, Feng Z, Gilliland G, Bhat TN, Weissig H, Shindyalov IN, Bourne PE (2000) The protein data bank. Nucleic Acids Res 28(1):235–242
24. Thompson MJ, Sievers SA, Karanicolas J, Ivanova MI, Baker D, Eisenberg D (2006) The 3D profile method for identifying fibril-forming segments of proteins. Proc Natl Acad Sci 103(11):4074–4078
25. Sawaya MR, Sambashivan S, Nelson R, Ivanova MI, Sievers SA, Apostol MI, Thompson MJ, Balbirnie M, Wiltzius JJ, McFarlane HT (2007) Atomic structures of amyloid cross-β spines reveal varied steric zippers. Nature 447(7143):453–457
26. Louros N, Konstantoulea K, De Vleeschouwer M, Ramakers M, Schymkowitz J, Rousseau F (2020) WALTZ-DB 2.0: an updated database containing structural information of experimentally determined amyloid-forming peptides. Nucleic Acids Res 48(D1):D389–D393
27. Shobana R, Pandaranayaka EP (2014) ProADD: a database on protein aggregation diseases. Bioinformation 10(6):390
28. Wozniak PP, Kotulska M (2015) AmyLoad: website dedicated to amyloidogenic protein fragments. Bioinformatics 31(20): 3395–3397
29. Varadi M, De Baets G, Vranken WF, Tompa P, Pancsa R (2018) AmyPro: a database of proteins with validated amyloidogenic regions. Nucleic Acids Res 46(D1):D387–D392
30. Thangakani AM, Nagarajan R, Kumar S, Sakthivel R, Velmurugan D, Gromiha MM (2016) CPAD, curated protein aggregation database: a repository of manually curated experimental data on protein and peptide aggregation. PLoS One 11(4):e0152949
31. Rawat P, Prabakaran R, Sakthivel R, Mary Thangakani A, Kumar S, Gromiha MM (2020) CPAD 2.0: a repository of curated experimental data on aggregating proteins and peptides. Amyloid 27(2):128–133
32. Belli M, Ramazzotti M, Chiti F (2011) Prediction of amyloid aggregation in vivo. EMBO Rep 12(7):657–663
33. Pawlicki S, Le Béchec A, Delamarche C (2008) AMYPdb: a database dedicated to amyloid precursor proteins. BMC Bioinformatics 9:1–11
34. Takács K, Varga B, Grolmusz V (2019) PDB _amyloid: an extended live amyloid structure list from the PDB. FEBS Open Bio 9(1): 185–190
35. Bodi K, Prokaeva T, Spencer B, Eberhard M, Connors LH, Seldin DC (2009) AL-Base: a visual platform analysis tool for the study of amyloidogenic immunoglobulin light chain sequences. Amyloid 16(1):1–8
36. Kuriata A, Iglesias V, Pujols J, Kurcinski M, Kmiecik S, Ventura S (2019) Aggrescan3D (A3D) 2.0: prediction and engineering of protein solubility. Nucleic Acids Res 47(W1): W300–W307
37. Pujols J, Peña-Díaz S, Ventura S (2018) AGGRESCAN3D: toward the prediction of the aggregation propensities of protein

structures. In: Computational drug discovery and design. Humana Press, New York, pp 427–443

38. Badaczewska-Dawid AE, Garcia-Pardo J, Kuriata A, Pujols J, Ventura S, Kmiecik S (2022) A3D database: structure-based predictions of protein aggregation for the human proteome. Bioinformatics 38(11): 3121–3123
39. Pintado-Grima C, Bárcenas O, Manglano-Artuñedo Z, Vilaça R, Macedo-Ribeiro S, Pallares I, Santos J, Ventura S (2022) CARs-DB: a database of cryptic amyloidogenic regions in intrinsically disordered proteins. Front Mol Biosci 9
40. Walsh I, Seno F, Tosatto SC, Trovato A (2014) PASTA 2.0: an improved server for protein aggregation prediction. Nucleic Acids Res 42(W1):W301–W307
41. Prabakaran R, Rawat P, Thangakani AM, Kumar S, Gromiha MM (2021) Protein aggregation: in silico algorithms and applications. Biophys Rev 13:71–89
42. Housmans JA, Wu G, Schymkowitz J, Rousseau F (2023) A guide to studying protein aggregation. FEBS J 290(3):554–583
43. Navarro S, Ventura S (2022) Computational methods to predict protein aggregation. Curr Opin Struct Biol 73:102343
44. Meric G, Robinson AS, Roberts CJ (2017) Driving forces for nonnative protein aggregation and approaches to predict aggregation-prone regions. Annu Rev Chem Biomol Eng 8:139–159
45. Ventura S (2005) Sequence determinants of protein aggregation: tools to increase protein solubility. Microb Cell Factories 4(1):1–8
46. Gsponer J, Vendruscolo M (2006) Theoretical approaches to protein aggregation. Protein Pept Lett 13(3):287–293
47. López de la Paz M, Serrano L (2004) Sequence determinants of amyloid fibril formation. Proc Natl Acad Sci 101(1):87–92
48. Santos J, Pujols J, Pallarès I, Iglesias V, Ventura S (2020) Computational prediction of protein aggregation: advances in proteomics, conformation-specific algorithms and biotechnological applications. Comput Struct Biotechnol J 18:1403–1413
49. Pallarès I, Ventura S (2016) Understanding and predicting protein misfolding and aggregation: insights from proteomics. Proteomics 16(19):2570–2581
50. Pallarès I, Ventura S (2019) Advances in the prediction of protein aggregation propensity. Curr Med Chem 26(21):3911–3920
51. Conchillo-Solé O, de Groot NS, Avilés FX, Vendrell J, Daura X, Ventura S (2007) AGGRESCAN: a server for the prediction and evaluation of "hot spots" of aggregation in polypeptides. BMC Bioinformatics 8:1–17
52. Dovidchenko NV, Galzitskaya OV (2015) Computational approaches to identification of aggregation sites and the mechanism of amyloid growth. In: Lipids in protein misfolding. Springer, Cham, pp 213–239
53. Bhattacharya D, Kleeblatt DC, Statt A, Reinhart WF (2022) Predicting aggregate morphology of sequence-defined macromolecules with recurrent neural networks. Soft Matter 18(27):5037–5051
54. Família C, Dennison SR, Quintas A, Phoenix DA (2015) Prediction of peptide and protein propensity for amyloid formation. PLoS One 10(8):e0134679
55. Kim C, Choi J, Lee SJ, Welsh WJ, Yoon S (2009) NetCSSP: web application for predicting chameleon sequences and amyloid fibril formation. Nucleic Acids Res 37(suppl_2): W469–W473
56. Moreira CA, Philot EA, Lima AN, Scott AL (2019) Predicting regions prone to protein aggregation based on SVM algorithm. Appl Math Comput 359:502–511
57. Tsolis AC, Papandreou NC, Iconomidou VA, Hamodrakas SJ (2013) A consensus method for the prediction of 'aggregation-prone' peptides in globular proteins. PLoS One 8(1): e54175
58. Emily M, Talvas A, Delamarche C (2013) MetAmyl: a METa-predictor for AMYLoid proteins. PLoS One 8(11):e79722
59. Wilson RP (2003) Amino acids and proteins. In: Fish nutrition. Elsevier, pp 143–179
60. Tartaglia GG, Vendruscolo M (2008) The Zyggregator method for predicting protein aggregation propensities. Chem Soc Rev 37(7):1395–1401
61. Oliveberg M (2010) Waltz, an exciting new move in amyloid prediction. Nat Methods 7(3):187–188
62. Prabakaran R, Rawat P, Kumar S, Gromiha MM (2021) ANuPP: a versatile tool to predict aggregation nucleating regions in peptides and proteins. J Mol Biol 433(11):166707
63. Fernandez-Escamilla A-M, Rousseau F, Schymkowitz J, Serrano L (2004) Prediction of sequence-dependent and mutational effects on the aggregation of peptides and proteins. Nat Biotechnol 22(10):1302–1306
64. Hamodrakas SJ (1988) A protein secondary structure prediction scheme for the IBM PC

and compatibles. Bioinformatics 4(4): 473–477
65. Thu TTM, Co NT, Tu LA, Li MS (2019) Aggregation rate of amyloid beta peptide is controlled by beta-content in monomeric state. J Chem Phys 150(22):225101
66. Thangakani AM, Kumar S, Nagarajan R, Velmurugan D, Gromiha MM (2014) GAP: towards almost 100 percent prediction for β-strand-mediated aggregating peptides with distinct morphologies. Bioinformatics 30(14):1983–1990
67. Waibl F, Fernández-Quintero ML, Wedl FS, Kettenberger H, Georges G, Liedl KR (2022) Comparison of hydrophobicity scales for predicting biophysical properties of antibodies. Front Mol Biosci 9:960194
68. Wu S, Zhang Y (2008) A comprehensive assessment of sequence-based and template-based methods for protein contact prediction. Bioinformatics 24(7):924–931
69. Li MS, Klimov D, Straub J, Thirumalai D (2008) Probing the mechanisms of fibril formation using lattice models. J Chem Phys 129(17):175101
70. Vácha R, Frenkel D (2011) Relation between molecular shape and the morphology of self-assembling aggregates: a simulation study. Biophys J 101(6):1432–1439
71. Li MS, Reddy G, Hu C-K, Straub J, Thirumalai D (2010) Factors governing fibrillogenesis of polypeptide chains revealed by lattice models. Phys Rev Lett 105(21):218101
72. Šarić A, Chebaro YC, Knowles TP, Frenkel D (2014) Crucial role of nonspecific interactions in amyloid nucleation. Proc Natl Acad Sci 111(50):17869–17874
73. Shell MS (2008) The relative entropy is fundamental to multiscale and inverse thermodynamic problems. J Chem Phys 129(14): 144108
74. Izvekov S, Voth GA (2005) A multiscale coarse-graining method for biomolecular systems. J Phys Chem B 109(7):2469–2473
75. Reith D, Pütz M, Müller-Plathe F (2003) Deriving effective mesoscale potentials from atomistic simulations. J Comput Chem 24(13):1624–1636
76. Bezkorovaynaya O, Lukyanov A, Kremer K, Peter C (2012) Multiscale simulation of small peptides: consistent conformational sampling in atomistic and coarse-grained models. J Comput Chem 33(9):937–949
77. Wang Y, Voth GA (2010) Molecular dynamics simulations of polyglutamine aggregation using solvent-free multiscale coarse-grained models. J Phys Chem B 114(26):8735–8743
78. Simunovic M, Mim C, Marlovits TC, Resch G, Unger VM, Voth GA (2013) Protein-mediated transformation of lipid vesicles into tubular networks. Biophys J 105(3): 711–719
79. Larini L, Shea J-E (2012) Coarse-grained modeling of simple molecules at different resolutions in the absence of good sampling. J Phys Chem B 116(29):8337–8349
80. Laio A, Parrinello M (2002) Escaping free-energy minima. Proc Natl Acad Sci 99(20): 12562–12566
81. Torrie GM, Valleau JP (1977) Nonphysical sampling distributions in Monte Carlo free-energy estimation: umbrella sampling. J Comput Phys 23(2):187–199
82. Sugita Y, Okamoto Y (1999) Replica-exchange molecular dynamics method for protein folding. Chem Phys Lett 314(1–2): 141–151
83. Swope WC, Pitera JW, Suits F (2004) Describing protein folding kinetics by molecular dynamics simulations. 1. Theory. J Phys Chem B 108(21):6571–6581
84. Singhal N, Snow CD, Pande VS (2004) Using path sampling to build better Markovian state models: predicting the folding rate and mechanism of a tryptophan zipper beta hairpin. J Chem Phys 121(1):415–425
85. Morriss-Andrews A, Shea J-E (2015) Computational studies of protein aggregation: methods and applications. Annu Rev Phys Chem 66:643–666
86. Kotulska M, Unold O (2013) On the amyloid datasets used for training PAFIG how (not) to extend the experimental dataset of hexapeptides. BMC Bioinformatics 14:1–8
87. Tartaglia GG, Cavalli A, Pellarin R, Caflisch A (2004) The role of aromaticity, exposed surface, and dipole moment in determining protein aggregation rates. Protein Sci 13(7): 1939–1941
88. Tartaglia GG, Cavalli A, Pellarin R, Caflisch A (2005) Prediction of aggregation rate and aggregation-prone segments in polypeptide sequences. Protein Sci 14(10):2723–2734
89. Liaw C, Tung C-W, Ho S-Y (2013) Prediction and analysis of antibody amyloidogenesis from sequences. PLoS One 8(1):e53235
90. Garbuzynskiy SO, Lobanov MY, Galzitskaya OV (2010) FoldAmyloid: a method of prediction of amyloidogenic regions from protein sequence. Bioinformatics 26(3):326–332
91. Zibaee S, Makin OS, Goedert M, Serpell LC (2007) A simple algorithm locates β-strands in the amyloid fibril core of α-synuclein, Aβ,

and tau using the amino acid sequence alone. Protein Sci 16(5):906–918

92. Burdukiewicz M, Sobczyk P, Rödiger S, Duda-Madej A, Mackiewicz P, Kotulska M (2017) Amyloidogenic motifs revealed by n-gram analysis. Sci Rep 7(1):12961
93. Ahmed AB, Znassi N, Château M-T, Kajava AV (2015) A structure-based approach to predict predisposition to amyloidosis. Alzheimers Dement 11(6):681–690
94. Bondarev SA, Bondareva OV, Zhouravleva GA, Kajava AV (2018) BetaSerpentine: a bioinformatics tool for reconstruction of amyloid structures. Bioinformatics 34(4):599–608
95. Bryan AW Jr, Menke M, Cowen LJ, Lindquist SL, Berger B (2009) BETASCAN: probable β-amyloids identified by pairwise probabilistic analysis. PLoS Comput Biol 5(3):e1000333
96. O'Donnell CW, Waldispühl J, Lis M, Halfmann R, Devadas S, Lindquist S, Berger B (2011) A method for probing the mutational landscape of amyloid structure. Bioinformatics 27(13):i34–i42
97. Bryan AW Jr, O'Donnell CW, Menke M, Cowen LJ, Lindquist S, Berger B (2012) STITCHER: dynamic assembly of likely amyloid and prion β-structures from secondary structure predictions. Proteins 80(2): 410–420
98. Gasior P, Kotulska M (2014) FISH amyloid–a new method for finding amyloidogenic segments in proteins based on site specific co-occurrence of amino acids. BMC Bioinformatics 15(1):1–8
99. Orlando G, Silva A, Macedo-Ribeiro S, Raimondi D, Vranken W (2020) Accurate prediction of protein beta-aggregation with generalized statistical potentials. Bioinformatics 36(7):2076–2081
100. Zhang Z, Chen H, Lai L (2007) Identification of amyloid fibril-forming segments based on structure and residue-based statistical potential. Bioinformatics 23(17):2218–2225
101. Louros N, Orlando G, De Vleeschouwer M, Rousseau F, Schymkowitz J (2020) Structure-based machine-guided mapping of amyloid sequence space reveals uncharted sequence clusters with higher solubilities. Nat Commun 11(1):3314
102. Wojciechowski JW, Kotulska M (2020) Path-prediction of amyloidogenicity by threading and machine learning. Sci Rep 10(1):1–9
103. Lauer TM, Agrawal NJ, Chennamsetty N, Egodage K, Helk B, Trout BL (2012) Developability index: a rapid in silico tool for the screening of antibody aggregation propensity. J Pharm Sci 101(1):102–115
104. Sankar K, Krystek SR Jr, Carl SM, Day T, Maier JK (2018) AggScore: prediction of aggregation-prone regions in proteins based on the distribution of surface patches. Proteins 86(11):1147–1156

Chapter 18

Bioinformatics Approaches for Understanding the Binding Affinity of Protein–Nucleic Acid Complexes

K. Harini, Masakazu Sekijima, and M. Michael Gromiha

Abstract

Protein–nucleic acid interactions are involved in various biological processes such as gene expression, replication, transcription, translation, and packaging. Understanding the recognition mechanism of the protein–nucleic acid complexes has been investigated from different perspectives, including the binding affinities of protein–DNA and protein–RNA complexes. Experimentally, protein–nucleic acid interactions are analyzed using X-ray crystallography, Isothermal Titration Calorimetry (ITC), DNA/RNA pull-down assays, DNA/RNA footprinting, and systematic evolution of ligands by exponential enrichment (SELEX). On the other hand, numerous databases and computational tools have been developed to study protein–nucleic acid complexes based on their binding sites, specific interactions between them, and binding affinity. In this chapter, we discuss various databases for protein–nucleic acid complex structures and the tools available to extract features from them. Further, we provide details on databases and prediction methods reported for exploring the binding affinity of protein–nucleic acid complexes along with important structure-based parameters, which govern the binding affinity.

Key words Protein-DNA complexes, Binding affinity, Prediction methods, Computational tools, Machine learning, Protein-RNA complexes

1 Introduction

Protein–nucleic acid interactions are important in various biological processes, including DNA replication, transcription, translation, RNA pre-processing, and translocation. Quantitatively, the strength of these interactions is defined by the term, binding affinity as free energy of binding (ΔG) or dissociation constant (K_d). Understanding the affinity of protein–DNA interactions is important to study the effects on gene expressions and delineate the recognition mechanism based on interactions between protein and DNA [1, 2]. On the other hand, binding sites in protein–nucleic acid complexes are vital for influencing their specificity and affinity. Alteration in binding sites leads to the loss of binding and/or change in affinity, which may impact the function of the complex.

Andrzej Kloczkowski et al. (eds.), *Prediction of Protein Secondary Structure*, Methods in Molecular Biology, vol. 2867, https://doi.org/10.1007/978-1-0716-4196-5_18,

It is reported that a single amino acid mutation in a protein–nucleic acid binding site can cause deadly diseases, including cancer [3]. Hence, investigating the affinity of the protein–nucleic acid complexes has a broad spectrum of applications, such as designing complexes with desired affinities, predictive methods for the target sites, and the quantitative simulation of gene regulation networks.

Experimentally, the binding affinity of protein–nucleic acid complexes is studied with electrophoretic mobility shift assay, filter binding assay, fluorescence spectroscopy, isothermal titration calorimetry, surface plasmon resonance, and bio-layer interferometry. Although these methods are highly accurate, they are labor-intensive, expensive, and time-consuming. Hence, computational methods have been proposed for predicting the affinity of protein-DNA [4] and RNA complexes [5, 6], as well as changes in the binding affinity upon mutations [7–14].

In addition, various investigations have been carried out to study protein–nucleic acid complexes, such as identifying the DNA/RNA binding proteins [15–18]; predicting their binding sites; locating key residues, which are involved in both binding and stabilizing [19]; predicting hotspot residues [20, 21]; and understanding the recognition mechanism of protein–nucleic acid complexes [22–25].

In this review, we survey various databases available for the structure of protein–DNA and protein–RNA complexes, along with tools for extracting features from protein–nucleic acid complex structures. In addition, we discuss the databases specific for the binding affinity of protein–nucleic acid complexes and insights obtained for relating structure-based parameters with protein–nucleic acid binding affinity. Further, computational methods available for predicting the binding affinity of protein–nucleic acid complexes will be discussed. Figure 1 illustrates the overall workflow of the review.

2 Databases for Structures of Protein–Nucleic Acid Complexes and Their Interactions

Several databases are available in the literature for protein–nucleic acid complex structures and exploring the interactions between protein and DNA/RNA. They are grouped into two categories: (i) structures of protein–DNA and protein–RNA complexes and (ii) interactions between protein and nucleic acid in a complex. The databases available for protein–nucleic acid complex structures are listed in Table 1.

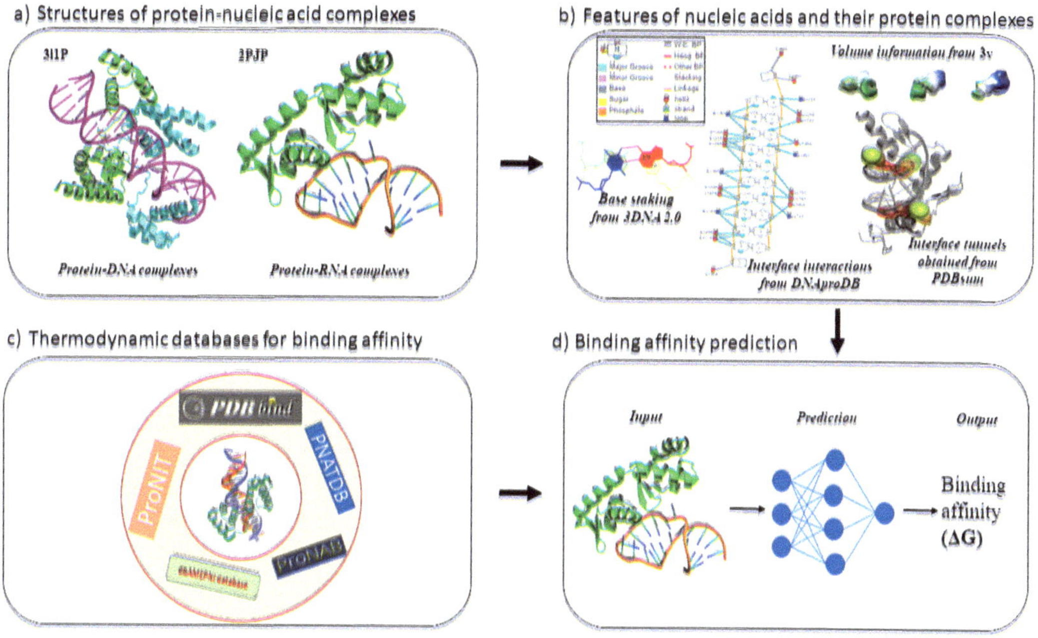

Fig. 1 Overall workflow of the review: (**a**) collection of experimentally determined protein–nucleic acid complex structures, (**b**) extracting structural features of nucleic acids and interface of protein–nucleic acid complexes, (**c**) databases for binding affinity of protein–DNA and protein–RNA complexes, and (**d**) prediction of binding affinity

2.1 Protein–Nucleic Acid Complex Structures

Nucleic Acid Database (NDB) [26] contains structural information specific to nucleic acids and protein–nucleic acid complexes. It has geometric data, motifs, structural and functional classifications of nucleic acid containing complexes, as well as tools and software for the analysis of nucleic acids. NDB is linked to external tools such as RNA 3D motif atlas, WebFR3D for the geometry of RNA 3D structures, and DNAproDB for the analysis of DNA. It has options to search data based on annotations relating to the sequence, structure, and function of nucleic acids and nucleic acids containing complexes. NDB is available at http://ndbserver.rutgers.edu/.

PDBsum [28] has a pictorial representation of 3D structures of proteins and their complexes deposited in the Protein Data Bank, and it contains details of protein domains, protein–nucleic acid interactions, and secondary structure analysis of proteins. Additionally, separate options are available for specifically analyzing SARS-CoV-2 virus protein structures, along with AlphaFold predicted structures for proteins. PDBsum is available at http://www.ebi.ac.uk/thornton-srv/databases/pdbsum/.

ZifBASE database has various natural and engineered zinc finger proteins obtained from UniProtKB, GenBank, Protein Data Bank, ModBase, Protein Model Portal, and the literature [29]. It contains specific features such as finger sequences and positions, number of fingers, physicochemical properties,

Table 1
Databases for protein–nucleic acid complex structures and protein–nucleic acid interactions

Databases	Description	Link	References
Protein–DNA and protein–RNA complex structures			
Nucleic Acid Database	3D structures of nucleic acids and protein–nucleic acid complexes from PDB	http://ndbserver.rutgers.edu/	[26]
Nucleic Acid-Protein Interaction DataBase[a]	Database of experimentally determined nucleic acids and complex assemblies	http://npidb.belozersky.msu.ru/	[27]
PDBsum	Pictorial database of 3D structures in Protein Data Bank with interaction details	http://www.ebi.ac.uk/thornton-srv/databases/pdbsum/	[28]
ZifBase	Database of zinc finger proteins and associated resources	https://web.iitd.ac.in/~sundar/zifbase/	[29]
PDIdb	The Protein-DNA Interface database	http://melolab.org/pdidb/web/content/home	[30]
3D-footprint	Database of DNA-binding protein structures, structure-based binding specificities, and sequence logos	https://3dfootprint.eead.csic.es/	[31]
PRIDB	Protein–RNA Interface DataBase	http://bindr.gdcb.iastate.edu/PRIDB.	[32]
RBPDB	RNA-Binding Protein Database	http://rbpdb.ccbr.utoronto.ca/	[33]
Database with interactions between protein and nucleic acids			
ProtCID	Database with structures of interactions of proteins with proteins, peptides, nucleic acids, and ligands	http://dunbrack2.fccc.edu/ProtCiD/	[34]
DBBP	Binding Pairs in protein–nucleic acid interactions	http://bclab.inha.ac.kr/dbbp	[35]
JASPAR 2022	Database of transcription factor binding profiles	https://jaspar.genereg.net/	[36]
DNAproDB	Protein–DNA database with structural analysis and visualization for features of protein–DNA interactions	https://dnaprodb.usc.edu/	[37]
RsiteDB[a]	Database of protein binding pockets that interact with RNA nucleotide bases	http://bioinfo3d.cs.tau.ac.il/RsiteDB	[38]
RPpocket	RNA-Protein Intuitive Database with RNA Pocket Topology Resources	http://zhaoserver.com.cn/RPpocket/RPpocket.html	[39]

Last accessed on 15th Feb 2023; [a]currently not available

framework, classes, and PubMed. The database is freely accessible at https://web.iitd.ac.in/~sundar/zifbase/.

The Protein–DNA Interface database (PdIdb) contains structural information of protein–DNA complexes in PDB with three levels of functional classification such as class, type, and sub-type. It includes the features for DNA (number of strands, gapped/nicked DNA), protein (number of protein chains, classification of proteins), and interaction features at the interface [30]. It is available at http://melolab.org/pdidb/web/content/home.

3D footprint database is the database of DNA-binding protein structures with structure-based binding specificities, sequence logos, structural classification, and clustering of protein–DNA interfaces [31]. It contains interface graphs and footprint diagrams, a novel graphical representation that summarizes contacts in the protein–DNA interface. It is available at https://3dfootprint.eead.csic.es/.

PRIDB is a secondary database for protein–RNA complexes with options to display interfacial amino acids and ribonucleotides in protein and RNA chains, respectively, and to visualize the interacting atoms in the 3D structure of the complex [32]. It is linked to different motif databases, such as ProSite motifs in protein chains and FR3D motifs in RNA chains, along with visualization options for viewing the 3D structure of the complex. PRIDB is available at http://bindr.gdcb.iastate.edu/PRIDB.

RBPDB, RNA-Binding Protein Database, contains information on RNA-binding sites, RNA binding domains, experimental methods, and sequences [33]. It has RNA binding domains in different organisms and options to search and download the data based on organism or domain and scan the sequence for binding sites in the protein. It is available at http://rbpdb.ccbr.utoronto.ca/.

2.2 Databases with Interactions Between Proteins and Nucleic Acids

The interactions between protein and nucleic acid in a complex provide information on the preference of residues and nucleic acids at the interface and their interaction patterns. Protein Common Interface Database (ProtCID) provides structural analyses of proteins and biomolecular interactions, and it contains clustered structures of interactions between proteins and nucleic acids [34]. The clusters are annotated with information such as PDB code, protein name, species, UniProt ID, protein domain family identifiers (Pfam), surface areas, crystal forms, and biological assembly annotations. It is available at http://dunbrack2.fccc.edu/ProtCiD/.

Database of binding pairs in protein–nucleic acid interactions (DBBP) contains details of contacts between protein and nucleic acid and specifically hydrogen bonding interactions at the interface of the complexes [35]. It has interaction details at both atomic and residue levels. The link to the web interface of the database is http://bclab.inha.ac.kr/dbbp.

JASPAR 2022 is a recently updated database with transcription factor binding profiles, names, structural class, and family of the DNA binding domain and links to external databases such as UniProt, PDB, DRV, and ReMAP [36]. In addition, it has the sequence logo of the binding site. JASPAR is available at https://jaspar.genereg.net/.

DNAproDB includes structural features and annotations of protein–DNA complexes such as single-stranded, double-stranded, and G-quadruplex DNA, as well as protein secondary structure elements [37]. It is supplemented with function and sequence-based search and visualization options. DNAProDB database is available at https://dnaprodb.usc.edu/index.html.

Rppocket is a database with RNA–protein interface residues (binding sites) with options to analyze the secondary structure and topological features such as volume, area, centroid, and effective radius [39]. It is freely available at http://zhaoserver.com.cn/RPpocket/RPpocket.html.

3 Tools for Extracting Features from Protein–Nucleic Acid Complex Structures

Several computational algorithms have been reported in the literature for extracting features from DNA, RNA, protein–DNA, and protein–RNA complexes. The list of the tools available for analyzing protein–nucleic acid complexes is provided in Table 2.

3.1 Tools for Analyzing Protein–DNA and Protein–RNA Complexes

The availability of protein–DNA and protein–RNA complex structures has been effectively used to derive several structure-based parameters. NUCPLOT includes a schematic 2D plot of protein–DNA and protein–RNA interactions using the PDB structures of the complexes [40]. It provides a list of hydrogen bonds and van der Waals contacts in the interface using the HBPLUS program. Also, interacting residues and the atoms in the protein are depicted in a 2D diagram. The source code is available upon request, and the program details are available at https://www.ebi.ac.uk/thornton-srv/software/NUCPLOT/. HBPLUS is a program that provides the details of the hydrogen bonds in the given structure, such as geometries and positions of hydrogen bonds, donor and acceptor atoms, and conformations [41]. HBPLUS can be downloaded from https://www.ebi.ac.uk/thornton-srv/software/HBPLUS/. Naccess calculates the solvent-accessible surface area of each atom and residue in a complex [42]. The program is available at http://www.bioinf.manchester.ac.uk/naccess/.

FoldX is an empirical force field developed for calculating the free energy of a macromolecule based on a given 3D structure [43]. It is also helpful for calculating the stability of a protein, repairing PDB files, analyzing the effect of mutations, and calculating the interaction energies of protein–protein, protein–DNA, and

Table 2
Tools for analyzing protein–nucleic acid complexes

Tools	Features	Link	References
Protein–nucleic acid interaction features			
NUCPLOT	Hydrogen bonds and van der Waals contacts	https://www.ebi.ac.uk/thornton-srv/software/NUCPLOT/	[40]
HBPLUS	Hydrogen bonds in a given PDB structure	http://www.ebi.ac.uk/thornton-srv/software/HBPLUS/	[41]
NACCESS	Accessible area of a molecule from a PDB	http://www.bioinf.manchester.ac.uk/naccess/	[42]
FoldX	Interaction energies of the complex, stability of the protein, water bridge, and so on	http://foldx.embl.de/	[43]
RNAmap2D	Contact and distance maps for RNA and protein–RNA complex structures	https://genesilico.pl/software/stand-alone/rnamap2d	[44]
PDBsum1	Secondary structure, interacting residues, hydrogen bonds, motifs, and cavities in the complex	https://www.ebi.ac.uk/thornton-srv/software/PDBsum1/	[45]
DNAproDB	Function, biophysical, and structural features	https://dnaprodb.usc.edu/	[37]
ProDFace	Hydrogen bond donors and conservation of the residues	http://structbioinfo.iitj.ac.in/resources/bioinfo/pd_interface/	[46]
PDA[a]	Restoration of double-stranded DNA, base pair detection, annotations of the DNA binding site	bioinfozen.uncc.edu/webpda	[47]
3v: webserver	Internal volumes from input RNA and protein structures	http://3vee.molmovdb.org/	[48]
PRince	Solvent accessible surface area, hydration, and structural and physicochemical properties of the protein–RNA interface	http://www.facweb.iitkgp.ac.in/~rbahadur/prince/home.html	[49]
Nucleic acid structural features			
Curves+	Conformation of nucleic acid structures	http://gbio-pbil.ibcp.fr/cgi/Curves_plus/	[50]
w3DNA	Analysis, visualization, and modeling of 3D nucleic acid structures	http://web.x3dna.org/	[51]

Last accessed on 15th Feb 2023; [a]currently not available

protein–RNA complexes. FoldX is available for downloading at foldxsuite.crg.eu. RNAmap2D provides information about interactions between proteins and RNA along with contact maps [44]. It is available at http://iimcb.genesilico.pl/rnamap2d.html.

PDBsum1 is a standalone program that provides the pictorial representation of a 3D structure along with possible interactions in the complex [45]. The software can be downloaded and installed from https://www.ebi.ac.uk/thornton-srv/software/PDBsum1. DNAproDB is a visualization tool that maps the interactions between the protein and DNA in a complex and is available at https://dnaprodb.usc.edu/ [37].

ProDFace is a web server to understand the protein–DNA interface structures [46]. It takes the protein–DNA complex in PDB format as input and provides physicochemical parameters of the interface, details of hydrogen bond donors, water-mediated hydrogen bonds, and conservation of residues in the interface core and rim of the protein as output. The web tool is available at http://structbioinfo.iitj.ac.in/resources/bioinfo/pd_interface/. Protein–DNA complex analyzer (PDA) is a structure analyzer that performs a comprehensive analysis of protein–DNA complex structures and provides information on the structural restoration of double-stranded DNA, DNA base-pair detection, annotations of the interactions, and DNA binding units [47]. The web server can be accessed using the link bioinfozen.uncc.edu/webpda.

3V is a cavity, channel, and cleft volume calculator and extractor [48]. It automatically extracts and analyzes all the internal volumes of RNA and protein structures by taking the difference between two rolling probes. The server is available at http://3vee.molmovdb.org/. PRince is a web server for analyzing the interface of protein–RNA complexes, which provides solvent-accessible surface area and structural, physicochemical, and hydration properties of the interface [49]. It is available at http://www.facweb.iitkgp.ernet.in/~rbahadur/prince/home.html.

3.2 Extraction of Features from Nucleic Acid Structures

The features based on DNA/RNA structures are also reported to be important for understanding the interactions between protein and nucleic acids as well as the binding affinity. Several tools are currently available in the literature to extract these features. Curves + is a software for analyzing the conformation of nucleic acid structures, which takes the nucleic acid structure files as input and provides helical, backbone, and groove parameters, inter-base pair and intra-base pair parameters, and sugar puckers of nucleic acids [50]. The software is freely available at http://gbio-pbil.ibcp.fr/cgi/Curves_plus/.

w3DNA 2.0 has six main modules for analysis, visualization, rebuilding, composite, fiber model, and mutation [51]. The "analysis module" provides information about the conformational features of DNA/RNA, such as base step parameters (shift, slide, rise, tilt, roll, twist), local base-pair parameters (shear, stretch, stagger, buckle, propeller, opening), groove width, and torsional angles, whereas "visualization module" is used to create high-resolution base-block schematic images. The last four modules focus on modelling atomic-level simulations of proteins or DNA/RNA structures. The w3DNA is available at http://web.x3dna.org/.

3.3 Prediction of Protein–Nucleic Acid Complex Structures

In this section, we outline the methods available for modelling protein–nucleic acid complex structures (Table 3). Nucleic Acid-Protein Dock (NPDock) [23] performs a series of tasks, such as modelling protein–nucleic acid complexes, scoring poses with statistical potentials, clustering the best models, and refinement of the selected best models. The web server is available at https://genesilico.pl/NPDock/.

HDOCK utilizes homology-based searching and template-based modelling for structure prediction with an option to incorporate biological information [52]. The server is available at http://hdock.phys.hust.edu.cn/. Tfmodeller is also developed based on the information about protein–DNA interfaces in the PDB [53]. The software is available at http://www.ccg.unam.mx/tfmodeller. pyDockDNA takes the apo-protein and DNA molecules as input, utilizes FTDOCK to generate protein–DNA orientations and DNA-based scoring function, and outputs the best 10 or 100 docking models along with options for visualizations [54]. It is available at https://model3dbio.csic.es/pydockdna. MELD-DNA has additional options to identify preferred binding modes from multiple poses and provide qualitative binding

Table 3
Docking tools for protein–nucleic acid complexes

Tools	Description	Link	References
HDOCK	Protein–protein and protein–DNA/RNA docking based on a hybrid algorithm of template-based modeling and ab initio free docking	http://hdock.phys.hust.edu.cn/	[52]
NPDock	Modeling of RNA–protein and DNA–protein complex structures	https://genesilico.pl/NPDock/	[23]
Tfmodeller	Provides evolutionary contact matrix, a schematic representation of the putative binding interface and atomic coordinates of the modelled complex	http://www.ccg.unam.mx/tfmodeller	[53]
pyDockDNA	Energy-based protein–DNA docking and scoring	https://model3dbio.csic.es/pydockdna	[54]
MELD-DNA	Predict the structures of protein–DNA complexes	github.com/maccallumlab/meld	[55]

preferences between DNA sequences [55]. The software can be downloaded from the GitHub repository, github.com/maccallumlab/meld.

4 Databases for Protein–Nucleic Acid Binding Affinity

Binding affinity (free energy upon binding, ΔG, or dissociation constant, K_d) is a parameter that can quantitatively provide information on the strength of interactions between protein and nucleic acids. The experimental data on ΔG and/or K_d available in the literature are stored in the form of databases with options to search and download the data. Table 4 provides a comprehensive list of the databases available for protein–nucleic acid binding affinities.

ProNIT [56] is the first thermodynamic database developed for protein–nucleic acid complexes. It has a user-friendly web interface with search, display, and sorting options. PDBbind [57] contains the experimentally determined binding affinity of biomolecules, including protein–ligand, protein–DNA/RNA, and protein–protein complexes. The web server is available at http://www.pdbbind.org.cn/. dbAMEPNI [58] is a database with binding affinity data specific for alanine mutations in protein–nucleic acid complexes. Each entry is provided with a dissociation constant (K_d), Gibbs free energy change ($\Delta\Delta G$), experimental conditions, and structural parameters. The database is available at http://zhulab.ahu.edu.cn/dbAMEPNI.

Table 4
Thermodynamic databases for protein–nucleic acid complexes

Database	Description	Number of data	Links	References
ProNIT	Experimentally determined thermodynamic interaction data	12,174	http://dna00.bio.kyutech.ac.jp/pronit/	[56]
PDBbind	Protein–ligand complexes with known three-dimensional structures	1052	http://www.pdbbind.org.cn/	[57]
dbAMEPNI	Alanine mutagenic effects on protein–nucleic acid interactions	578	http://zhulab.ahu.edu.cn/dbAMEPNI	[58]
PNATDB	Experimentally determined binding affinities of protein–nucleic acid interactions	12,635	http://chemyang.ccnu.edu.cn/ccb/database/PNAT/	[59]
ProNAB	Experimentally determined binding affinities of protein–nucleic acid complexes and their mutants	20,219	https://web.iitm.ac.in/bioinfo2/pronab/	[60]

Protein–Nucleic Acid Thermodynamics Database (PNATDB) has about 12,000 binding affinity values, along with details of molecular interactions in the complexes [59]. It is available at http://chemyang.ccnu.edu.cn/ccb/database/PNAT/.

ProNAB [60] is the largest protein–nucleic acid binding affinity database available in the literature, with more than 20,000 data. It has both wild-type and mutant data with sequence and structural features of the protein, nucleic acid and its complex, experimental conditions, thermodynamic parameters such as dissociation constant, the free energy of binding, and change in free energy upon mutation. Each entry in the database is also cross-linked with other databases such as GenBank, UniProt, PDB, ProthermDB, PROSITE, DisProt, and PubMed. The database has options for uploading and downloading the data. The database is available at https://web.iitm.ac.in/bioinfo2/pronab/.

5 Analysis on Binding Affinity of Protein–Nucleic Acid Complex Structures

Several analyses have been carried out to understand the relationship between structure-based features and the binding affinity of protein–nucleic acid complexes, and specific important features are discussed below.

Chen et al. [61] reported that binding affinity is directly related to the buried surface area at the interface. Further, it is shown to be influenced by the number of contacts at the interface [62] and electrostatic interactions between proteins and DNA [63]. In addition, secondary structural information and several local structural motifs are shown to be important for DNA recognition, specifically in the major groove of the DNA [64].

Intrinsic disorder regions are also considered to be important in modulating the affinity of the DNA-binding proteins. The disordered regions help in the formation of transient and dynamic interactions by supplementing the folded globular protein interactions with DNA, improving the affinity of the complex [65]. Similarly, in protein–RNA interactions, RNA binding proteins are often extended with intrinsically disordered regions that mediate additional interactions with proteins or RNA, increasing the affinity and specificity of interactions [66].

Pant et al. [67] revealed that the bicyclo-nucleotide modifications in DNA increase the affinity of proteins to DNA. Substitution of oxygen with sulfur (dithioation) in a protein increased the affinity to the DNA. Further, each phosphate that formed an ion pair with a protein side chain increased the binding affinity of the complex by ~0.6–0.8 kcal/mol [68].

Table 5
Methods for predicting the binding affinity of protein–nucleic acid complexes

Methods	Type of the complex	Data size	Links	References
PredPRBA	Protein–RNA	103	http://predprba.denglab.org/	[6]
PRA-Pred	Protein–RNA	217	https://web.iitm.ac.in/bioinfo2/prapred/	[69]
PreDBA	Protein–DNA	100	http://predba.denglab.org/	[4]
DNAffinity	Transcription factor–DNA	600	https://github.com/Jalbiti/DNAffinity.	[70]
PDA-Pred	Protein–DNA	398	https://web.iitm.ac.in/bioinfo2/pdapred/	[71]
PNAB	Protein–DNA/RNA	203	http://pnab.denglab.org/	[72]

6 Prediction of Protein–Nucleic Acid Binding Affinity

Several methods have been reported in the literature for predicting the binding affinity of protein–DNA/RNA complexes. These methods are based on molecular dynamics simulations, statistical methods, and machine learning techniques [67], and a list of prediction methods is presented in Table 5.

Nithin et al. [5] reported a regression model with 40 protein–RNA complexes to predict the binding affinity using structural and physicochemical features of the complexes. Deng et al. [6] developed a web server, PredPRBA, to predict the binding affinity of protein–RNA complexes using gradient-boosted regression techniques. The model was trained with 103 protein–RNA complexes and 37 sequence and structure-based features. PredPRBA web server is freely available at http://PredPRBA.denglab.org/. Recently, a structure-based regression model has been developed using 217 protein–RNA complexes using RNA base step parameters, protein–RNA contacts, and energy-based features [69]. The server, PRA-Pred, can be accessed through https://web.iitm.ac.in/bioinfo2/prapred/.

Yang et al. [4] developed a web server, PredDBA, for predicting the binding affinity of protein–DNA complexes using 52 sequence and structure-based features. Barissi et al. [70] developed a physics-based machine learning method, DNAffinity, for predicting the binding affinity of transcription factors using the features derived from molecular dynamics simulations. It is available at https://github.com/Jalbiti/DNAffinity. Recently, Harini et al. [71] developed regression-based models based on the structural and functional class of proteins and binding site residues. The web server, PDA-Pred, is available at https://web.iitm.ac.in/bioinfo2/pdapred/.

Yang et al. [72] developed a combined model, PNAB, for protein–DNA and protein–RNA complexes. They classified the complexes into 11 classes based on the type of DNA or RNA and utilized the heterogeneous stacking ensemble for prediction. The server is available at http://pnab.denglab.org/.

7 Conclusions

The development of computational tools for analyzing protein–nucleic acid interactions is helpful for understanding the recognition mechanism, thereby developing therapeutic strategies. We have reviewed various databases available for protein–nucleic acid complex structures as well as interactions between proteins and DNA/RNA. Further, computational tools available for analyzing protein–nucleic acid complexes have been discussed with specific parameters for nucleic acid and the interface of protein–DNA or protein–RNA complexes. Moreover, various algorithms available for predicting the binding affinity of protein–nucleic acid complexes have been outlined. We suggest that the performance of prediction methods will be improved upon the availability of experimental data for a large number of complexes.

Acknowledgments

The authors acknowledge the Indian Institute of Technology Madras and High-Performance Computing Environment (HPCE) for computational facilities. We thank the members of Protein Bioinformatics Lab for providing valuable suggestions. The work is partially supported by the Science and Engineering Research Board (SERB), Ministry of Science and Technology, Government of India, to MMG (No. CRG/2020/000314).

References

1. Ladbury JE (1995) Counting the calories to stay in the groove. Structure 3(7):635–639
2. Rastogi C, Rube HT, Kribelbauer JF, Crocker J, Loker RE, Martini GD, Laptenko O, Freed-Pastor WA, Prives C, Stern DL, Mann RS, Bussemaker HJ (2018) Accurate and sensitive quantification of protein-DNA binding affinity. Proc Natl Acad Sci USA 115(16):E3692–E3701
3. Taniguchi H, Fujimoto A, Kono H, Furuta M, Fujita M, Nakagawa H (2018) Loss-of-function mutations in Zn-finger DNA-binding domain of HNF4A cause aberrant transcriptional regulation in liver cancer. Oncotarget 9(40):26144–26156
4. Yang W, Deng L (2020) PreDBA: a heterogeneous ensemble approach for predicting protein-DNA binding affinity. Sci Rep 10(1):1278
5. Nithin C, Mukherjee S, Bahadur RP (2019) A structure-based model for the prediction of

protein-RNA binding affinity. RNA 25(12): 1628–1645

6. Deng L, Yang W, Liu H (2019) PredPRBA: prediction of protein-RNA binding affinity using gradient boosted regression trees. Front Genet 10:637
7. Pires DEV, Ascher DB (2017) mCSM-NA: predicting the effects of mutations on protein-nucleic acids interactions. Nucleic Acids Res 45(W1):W241–W246
8. Zhang N, Chen Y, Zhao F, Yang Q, Simonetti FL, Li M (2018) PremPDI estimates and interprets the effects of missense mutations on protein-DNA interactions. PLoS Comput Biol 14(12):e1006615
9. Zhang N, Lu H, Chen Y, Zhu Z, Yang Q, Wang S, Li M (2020) PremPRI: predicting the effects of missense mutations on protein-RNA interactions. Int J Mol Sci 21(15):5560
10. Jiang Y, Liu HF, Liu R (2021) Systematic comparison and prediction of the effects of missense mutations on protein-DNA and protein-RNA interactions. PLoS Comput Biol 17(4): e1008951
11. Li G, Panday SK, Peng Y, Alexov E (2021) SAMPDI-3D: predicting the effects of protein and DNA mutations on protein-DNA interactions. Bioinformatics 37(21):3760–3765
12. Mei LC, Wang YL, Wu FX, Wang F, Hao GF, Yang GF (2021) HISNAPI: a bioinformatic tool for dynamic hot spot analysis in nucleic acid-protein interface with a case study. Brief Bioinform 22(5):bbaa373
13. Harini K, Phogat A, Gromiha MM (2024) Bioinformatics approaches for understanding the consequences of mutations on the binding affinity of protein-DNA complexes. In: Gromiha MM (ed) Protein mutations: consequences on structure, function and diseases. World Scientific, Singapore, pp 123–150
14. Harini K, Krishnan SR, Sekijima M, Gromiha MM (2024) Computational resources for understanding the effect of mutations in binding affinities of protein–RNA complexes. In: Gromiha MM (ed) Protein mutations: consequences on structure, function and diseases. World Scientific, Singapore, pp 151–170
15. Kumar M, Gromiha MM, Raghava GP (2007) Identification of DNA-binding proteins using support vector machines and evolutionary profiles. BMC Bioinf 8:463
16. Mishra A, Pokhrel P, Hoque MT (2019) StackDPPred: a stacking based prediction of DNA-binding protein from sequence. Bioinformatics 35(3):433–441
17. Paz I, Kosti I, Ares M Jr, Cline M, Mandel-Gutfreund Y (2014) RBPmap: a web server for mapping binding sites of RNA-binding proteins. Nucleic Acids Res 42(Web Server issue): W361–W367
18. Bressin A, Schulte-Sasse R, Figini D, Urdaneta EC, Beckmann BM, Marsico A (2019) TriPepSVM: de novo prediction of RNA-binding proteins based on short amino acid motifs. Nucleic Acids Res 47(9):4406–4417
19. Kulandaisamy A, Srivastava A, Nagarajan R, Gromiha MM (2018) Dissecting and analyzing key residues in protein-DNA complexes. J Mol Recognit 31(4):e2692
20. Gronning AGB, Doktor TK, Larsen SJ, Petersen USS, Holm LL, Bruun GH, Hansen MB, Hartung AM, Baumbach J, Andresen BS (2020) DeepCLIP: predicting the effect of mutations on protein-RNA binding with deep learning. Nucleic Acids Res 48(13): 7099–7118
21. Zhang S, Wang L, Zhao L, Li M, Liu M, Li K, Bin Y, Xia J (2021) An improved DNA-binding hot spot residues prediction method by exploring interfacial neighbor properties. BMC Bioinf 22(Suppl 3):253
22. Gromiha MM, Siebers JG, Selvaraj S, Kono H, Sarai A (2004) Intermolecular and intramolecular readout mechanisms in protein-DNA recognition. J Mol Biol 337(2):285–294
23. Tuszynska I, Magnus M, Jonak K, Dawson W, Bujnicki JM (2015) NPDock: a web server for protein-nucleic acid docking. Nucleic Acids Res 43(W1):W425–W430
24. Yan Y, Zhang D, Zhou P, Li B, Huang SY (2017) HDOCK: a web server for protein-protein and protein-DNA/RNA docking based on a hybrid strategy. Nucleic Acids Res 45(W1):W365–W373
25. van Zundert GCP, Rodrigues J, Trellet M, Schmitz C, Kastritis PL, Karaca E, Melquiond ASJ, van Dijk M, de Vries SJ, Bonvin A (2016) The HADDOCK2.2 Web Server: user-friendly integrative modeling of biomolecular complexes. J Mol Biol 428(4):720–725
26. Narayanan CB, Westbrook J, Ghosh S, Petrov AI, Sweeney B, Zirbel CL, Leontis NB, Berman HM (2014) The nucleic acid database: new features and capabilities. Nucleic Acids Res 42(Database issue):D114–D122
27. Kirsanov DD, Zanegina ON, Aksianov EA, Spirin SA, Karyagina AS, Alexeevski AV (2013) NPIDB: nucleic acid-protein interaction DataBase. Nucleic Acids Res 41(Database issue):D517–D523
28. Laskowski RA, Jablonska J, Pravda L, Varekova RS, Thornton JM (2018) PDBsum: structural summaries of PDB entries. Protein Sci 27(1): 129–134

29. Jayakanthan M, Muthukumaran J, Chandrasekar S, Chawla K, Punetha A, Sundar D (2009) ZifBASE: a database of zinc finger proteins and associated resources. BMC Genomics 10:421
30. Norambuena T, Melo F (2010) The protein-DNA Interface database. BMC Bioinf 11:262
31. Contreras-Moreira B (2010) 3D-footprint: a database for the structural analysis of protein-DNA complexes. Nucleic Acids Res 38(Database issue):D91–D97
32. Lewis BA, Walia RR, Terribilini M, Ferguson J, Zheng C, Honavar V, Dobbs D (2011) PRIDB: a protein-RNA interface database. Nucleic Acids Res 39(Database issue):D277–D282
33. Cook KB, Kazan H, Zuberi K, Morris Q, Hughes TR (2011) RBPDB: a database of RNA-binding specificities. Nucleic Acids Res 39(Database issue):D301–D308
34. Xu Q, Dunbrack RL Jr (2020) ProtCID: a data resource for structural information on protein interactions. Nat Commun 11(1):711
35. Park B, Kim H, Han K (2014) DBBP: database of binding pairs in protein-nucleic acid interactions. BMC Bioinf 15(Suppl 15):S5
36. Castro-Mondragon JA, Riudavets-Puig R, Rauluseviciute I, Lemma RB, Turchi L, Blanc-Mathieu R, Lucas J, Boddie P, Khan A, Manosalva Perez N, Fornes O, Leung TY, Aguirre A, Hammal F, Schmelter D, Baranasic D, Ballester B, Sandelin A, Lenhard B, Vandepoele K, Wasserman WW, Parcy F, Mathelier A (2022) JASPAR 2022: the 9th release of the open-access database of transcription factor binding profiles. Nucleic Acids Res 50(D1):D165–D173
37. Sagendorf JM, Markarian N, Berman HM, Rohs R (2020) DNAproDB: an expanded database and web-based tool for structural analysis of DNA-protein complexes. Nucleic Acids Res 48(D1):D277–D287
38. Shulman-Peleg A, Nussinov R, Wolfson HJ (2009) RsiteDB: a database of protein binding pockets that interact with RNA nucleotide bases. Nucleic Acids Res 37(Database issue): D369–D373
39. Yang R, Liu H, Yang L, Zhou T, Li X, Zhao Y (2022) RPpocket: an RNA-protein intuitive database with RNA pocket topology resources. Int J Mol Sci 23(13):6903
40. Luscombe NM, Laskowski RA, Thornton JM (1997) NUCPLOT: a program to generate schematic diagrams of protein-nucleic acid interactions. Nucleic Acids Res 25(24): 4940–4945
41. McDonald IK, Thornton JM (1994) Satisfying hydrogen bonding potential in proteins. J Mol Biol 238(5):777–793
42. Hubbard SJ, Thornton JM (1993) NACCESS. Department of Biochemistry and Molecular Biology, University College London
43. Schymkowitz J, Borg J, Stricher F, Nys R, Rousseau F, Serrano L (2005) The FoldX web server: an online force field. Nucleic Acids Res 33(Web Server issue):W382–W388
44. Pietal MJ, Szostak N, Rother KM, Bujnicki JM (2012) RNAmap2D – calculation, visualization and analysis of contact and distance maps for RNA and protein-RNA complex structures. BMC Bioinf 13:333
45. Laskowski RA (2022) PDBsum1: a standalone program for generating PDBsum analyses. Protein Sci 31(12):e4473
46. Pal A, Chakrabarti P, Dey S (2022) ProDFace: a web-tool for the dissection of protein-DNA interfaces. Front Mol Biosci 9:978310
47. Kim R, Guo JT (2009) PDA: an automatic and comprehensive analysis program for protein-DNA complex structures. BMC Genomics 10 (Suppl 1):S13
48. Voss NR, Gerstein M (2010) 3V: cavity, channel and cleft volume calculator and extractor. Nucleic Acids Res 38(Web Server issue): W555–W562
49. Barik A, Mishra A, Bahadur RP (2012) PRince: a web server for structural and physicochemical analysis of protein-RNA interface. Nucleic Acids Res 40(Web Server issue):W440–W444
50. Blanchet C, Pasi M, Zakrzewska K, Lavery R (2011) CURVES+ web server for analyzing and visualizing the helical, backbone and groove parameters of nucleic acid structures. Nucleic Acids Res 39(Web Server issue):W68–W73
51. Li S, Olson WK, Lu XJ (2019) Web 3DNA 2.0 for the analysis, visualization, and modeling of 3D nucleic acid structures. Nucleic Acids Res 47(W1):W26–W34
52. Li H, Huang E, Zhang Y, Huang SY, Xiao Y (2022) HDOCK update for modeling protein-RNA/DNA complex structures. Protein Sci 31(11):e4441
53. Contreras-Moreira B, Branger PA, Collado-Vides J (2007) TFmodeller: comparative modelling of protein-DNA complexes. Bioinformatics 23(13):1694–1696
54. Rodriguez-Lumbreras LA, Jimenez-Garcia B, Gimenez-Santamarina S, Fernandez-Recio J (2022) pyDockDNA: a new web server for energy-based protein-DNA docking and scoring. Front Mol Biosci 9:988996

55. Esmaeeli R, Bauza A, Perez A (2023) Structural predictions of protein-DNA binding: MELD-DNA. Nucleic Acids Res 51(4): 1625–1636
56. Kumar MD, Bava KA, Gromiha MM, Prabakaran P, Kitajima K, Uedaira H, Sarai A (2006) ProTherm and ProNIT: thermodynamic databases for proteins and protein-nucleic acid interactions. Nucleic Acids Res 34 (Database issue):D204–D206
57. Liu Z, Li Y, Han L, Li J, Liu J, Zhao Z, Nie W, Liu Y, Wang R (2015) PDB-wide collection of binding data: current status of the PDBbind database. Bioinformatics 31(3):405–412
58. Liu L, Xiong Y, Gao H, Wei DQ, Mitchell JC, Zhu X (2018) dbAMEPNI: a database of alanine mutagenic effects for protein-nucleic acid interactions. Database (Oxford) 2018:bay034
59. Mei LC, Hao GF, Yang GF (2023) Thermodynamic database supports deciphering protein-nucleic acid interactions. Trends Biotechnol 41(2):140–143
60. Harini K, Srivastava A, Kulandaisamy A, Gromiha MM (2022) ProNAB: database for binding affinities of protein-nucleic acid complexes and their mutants. Nucleic Acids Res 50(D1): D1528–D1534
61. Chen J, Sawyer N, Regan L (2013) Protein-protein interactions: general trends in the relationship between binding affinity and interfacial buried surface area. Protein Sci 22(4): 510–515
62. Vangone A, Bonvin AM (2015) Contacts-based prediction of binding affinity in protein-protein complexes. elife 4:e07454
63. Pal A, Levy Y (2019) Structure, stability and specificity of the binding of ssDNA and ssRNA with proteins. PLoS Comput Biol 15(4): e1006768
64. Rohs R, Jin X, West SM, Joshi R, Honig B, Mann RS (2010) Origins of specificity in protein-DNA recognition. Annu Rev Biochem 79:233–269
65. Vuzman D, Levy Y (2012) Intrinsically disordered regions as affinity tuners in protein-DNA interactions. Mol BioSyst 8(1):47–57
66. Ottoz DSM, Berchowitz LE (2020) The role of disorder in RNA binding affinity and specificity. Open Biol 10(12):200328
67. Pant P, Pathak A, Jayaram B (2023) Bicyclo-DNA mimics with enhanced protein binding affinities: insights from molecular dynamics simulations. J Biomol Struct Dyn 41(9): 4040–4047
68. Nguyen D, Zandarashvili L, White MA, Iwahara J (2016) Stereospecific effects of oxygen-to-sulfur substitution in DNA phosphate on ion pair dynamics and protein-DNA affinity. Chembiochem 17(17):1636–1642
69. Harini K, Sekijima M, Gromiha MM (2024) PRA-Pred: structure-based prediction of protein-RNA binding affinity. Int J Biol Macromol 259:129490
70. Barissi S, Sala A, Wieczor M, Battistini F, Orozco M (2022) DNAffinity: a machine-learning approach to predict DNA binding affinities of transcription factors. Nucleic Acids Res 50(16):9105–9114
71. Harini K, Kihara D, Gromiha MM (2023) PDA-Pred: predicting the binding affinity of protein-DNA complexes using machine learning techniques and structural features. Methods 213:10–17
72. Yang W, Deng L (2019) PNAB: Prediction of protein-nucleic acid binding affinity using heterogeneous ensemble models. In: 2019 IEEE international conference on bioinformatics and biomedicine (BIBM), San Diego, CA, USA, pp 58–63

Chapter 19

A Glimpse of Noncoding RNAs: Secondary Structure, Emerging Trends, and Potential Applications in Human Diseases

Mubashir Hassan, Amal Malik, Zainab Yaseen, Saba Shahzadi, Muhammad Yasir, and Andrzej Kloczkowski

Abstract

An appealing strategy for the treatment of several diseases is the therapeutic targeting of noncoding RNAs (ncRNAs), such as microRNAs (miRNAs) and long noncoding RNAs (lncRNAs). Many antisense oligonucleotides and small interfering RNAs have been tested in clinical studies over the past 10 years, and several of these have received FDA approval. However, trial results have thus far been mixed, with some studies reporting strong effects and others showing low effectiveness or side effects, including toxicity. Clinical trials for alternative entities like antimiRNAs are underway, and interest in lncRNA-based therapies is constantly growing. From this perspective, we discuss the basic overview of ncRNAs, their significant role as therapeutic biomarkers against different diseases, and the role of secondary structure in noncoding RNAs.

Key words Noncoding RNA, Long noncoding RNA, Small nuclear RNA, Small nuclear ribonucleic acid RNA, Circular RNA, Transfer RNA, Secondary structure

1 Introduction

Noncoding RNA (ncRNA) refers to RNA sequences that are not translated into proteins, excluding them from participating in the translation step of protein synthesis [1]. Small nuclear RNAs (snRNAs) and ribosomal ribonucleic acids (rRNAs), which are involved in protein translation and transcription as well as RNA modification, are thought to be involved in generic cell processes that are mediated by noncoding RNAs (ncRNAs) [2]. It has been shown that most genes within mammals and other complex organisms are located within the sequences of ncRNAs [3]. These sequences may be spliced and processed to form smaller sequences to produce specific products within the organism's system [3]. Other functions of ncRNAs within the human genome include

Andrzej Kloczkowski et al. (eds.), *Prediction of Protein Secondary Structure*, Methods in Molecular Biology, vol. 2867, https://doi.org/10.1007/978-1-0716-4196-5_19,

epigenetic regulation, protein modification, chromatin remodeling, RNA degeneration, and the promotion or repression of gene expression dependent on genome interaction [4]. These ncRNAs may also interact with various regulatory complexes and enzymatic activities, which in turn can change the chromatin landscapes through nucleosome modifications, DNA methylation, and histone modification [4, 5].

1.1 Types of ncRNAs

ncRNAs are a diverse group of molecules with different regulatory branch classes [6]. These fragments of ncRNAs can be classified into three possible categories by the length of RNA: (i) ncRNAs longer than 200 nucleotides, these may include ribosomal RNA (rRNA) and long noncoding RNA (lncRNA) [7]. (ii) ncRNAs longer than 40 nucleotides but shorter than 200, such as transfer RNA (tRNA), small nucleolar RNA (snoRNA), and small nuclear ribonucleic acid RNA (snRNA) [8]. (iii) ncRNAs shorter than 40 nucleotides in length, these include microRNA (miRNA), piwi-interacting RNA (piRNA), and transfer ribonucleic acid (tRNA)-derived small RNA (tsRNA), respectively [9]. (iv) The fourth, newly defined category of ncRNAs established, recently consists of circular RNAs (cirRNA) dissimilar to the linear RNA sequences. This RNA forms a covalently closed loop and does not have 5′ and 3′ polar ends. Its size may vary anywhere between 100 nucleotides and over 10,000 nucleotides in length [10].

Small RNA molecules known as miRNAs, usually 21–23 nucleotides long, are essential for post-transcriptional gene control. They function by attaching themselves to the 3′ untranslated regions (UTRs) of the target messenger RNA (mRNA), which inhibits translation or causes mRNA destruction [11]. Moreover, small interfering RNAs (siRNAs) are essential parts of the RNA interference (RNAi) process and range in length from 20 to 25 nucleotides. Their facilitation of target mRNA cleavage and degradation contributes to gene silencing and protection against viral infections [12]. The piwi-interacting RNAs (usually 24–30 nucleotides long) are recognized for their transposon silencing and germ cell development functions. Comparatively, the long noncoding RNAs (lncRNAs) are longer and frequently exceed 200 nucleotides. They play a crucial part in several biological processes, with some having clearly established roles and others having unclear character, requiring more research [13, 14].

1.2 ncRNAs Secondary Structure and Functionality

RNA secondary structure is formed by the interactions between the nucleotide bases within the same RNA molecule [15, 16]. The secondary structures of ncRNAs are auxiliary design, which alludes to the particular shapes that RNA atoms overlay into because of intramolecular base matching and design various structural motifs, such as hairpins, bulges, internal loops, multibranch loops, and junctions, respectively [17–19]. Hairpins: A hairpin loop is a

U-shaped or loop-like structure formed when a segment of an RNA molecule folds back on itself and base pairs with another complementary segment within the same strand [20, 21]. Bulges: Regions where extra nucleotides are inserted into the stem, disrupting the base pairing [22]. The non-base-paired areas in between two stems are known as internal loops. Because it affects the RNA's stability, capacity to interact with proteins and other RNAs, and its participation in a variety of cellular activities, the secondary structure of RNA is essential to its function [23]. For RNAs, a few secondary structural functions are significant. Part of the RNA molecules' lifetime and functional stability within the cell is due to the secondary structure that shields them from nuclease destruction [24]. The secondary structures in ncRNAs can regulate gene expression by influencing mRNA stability, translation, and splicing [25]. Moreover, specific secondary structures enable ncRNAs to bind to proteins, forming ribonucleoprotein complexes essential for various cellular processes [26]. The secondary structures facilitate interactions between different RNA molecules, enabling ncRNAs to act as guides, scaffolds, or decoys [27].

2 Databases and Tools to Analyze ncRNAs

A long list of databases and tools is currently being used for ncRNA sequence retrieval and evaluation. A few of the most popular ones are listed below, and they are frequently used to understand better the structure and functions of ncRNAs in associations with different signaling pathways and their medicated diseases.

2.1 miRBase

miRBase is a comprehensive database focusing on microRNAs, providing sequences and annotations for these small regulatory RNAs. It provides a wealth of information about miRNAs, making it an invaluable tool for researchers studying gene expression and disease. miRBase contains entries for over 58,000 mature miRNA sequences from 271 organisms, including humans, animals, and plants, providing information like genomic location, predicted targets, expression levels, functional annotations, and links to relevant literature [28].

2.2 RNAcentral

RNAcentral is a database of all known noncoding RNA sequences of all types and across all organisms. RNAcentral is a comprehensive, open-access platform dedicated to ncRNA research. It is a central hub for scientists, offering a wealth of resources and tools to explore, analyze, and utilize ncRNA data. RNAcentral houses diverse databases covering various ncRNA types, including miRNAs, lncRNAs, rRNAs, and tRNAs. Most importantly, it provides sequenced ncRNA entries with detailed annotations (e.g., genomic location, function, expression levels). Furthermore, RNAcentral offers powerful tools for ncRNA analysis, including sequence search, target prediction, expression analysis, and network analysis [29, 30].

2.3 Rfam

Rfam is a database that catalogs and annotates various families of noncoding RNAs, facilitating the studies of miRNAs, siRNAs, and other small RNAs. Rfam is a treasure trove for anyone interested in ncRNA families. It's a comprehensive database that houses a wealth of information about these fascinating molecules, serving as a crucial resource for researchers studying RNA structure, function, and evolution. Rfam categorizes ncRNAs into families based on shared structural features and evolutionary relationships based on multiple sequence alignments of conserved and divergent regions. Rfam compiles information about known or predicted functions of each ncRNA family, linking them to specific biological processes [31, 32].

2.4 TargetScan

TargetScan is a widely used tool for miRNA target prediction, aiding in identifying potential target sites within mRNA 3′ UTRs. TargetScan is a leading online resource for predicting the targets of miRNAs, those tiny molecules that regulate gene expression. It's a powerful tool for researchers studying gene regulation, disease mechanisms, and the complex interplay between miRNAs and their targets. The TargetScan provides potential target genes in various organisms, including humans, mice, and zebrafish, against a specific miRNA sequence. TargetScan utilizes several algorithms with different strengths and focuses, comprehensively analyzing potential targets [33].

2.5 DAVID

DAVID stands for "Database for Annotation, Visualization, and Integrated Discovery." It is a versatile tool for functional annotation and pathway analysis, helping to identify the biological processes and pathways associated with noncoding RNA targets. It is a one-stop shop for researchers exploring functional enrichment, gene ontology (GO) analysis, pathway mapping, and network visualization. The core functionality includes finding experimental data for a query gene list. DAVID identifies GO terms, KEGG pathways, and other functional annotations that are statistically overrepresented in the query gene list compared to a background set of all genes [34].

2.6 miRDeep

miRDeep is a tool for analyzing deep-sequencing data to discover novel miRNAs, which is crucial for detecting miRNAs from small RNA sequencing datasets [35].

Bioinformatic tools have become indispensable for the study of ncRNA fragments. They play a critical role in the identification, annotation, target prediction, and functional analysis of these small but influential RNA molecules. ncRNA databases are invaluable resources for researchers. They house curated data on various

Table 1
Databases and tools for ncRNA analysis

Database	Function	Web link/URLs
GENCODE	Gene annotations	https://www.gencodegenes.org/
NRED	ncRNA expression	http://jsm-research.imb.uq.edu.au/nred/cgi-bin/ncrnadb.pl
Ensemble ncRNA	ncRNA annotation	https://useast.ensembl.org/info/genome/genebuild/ncrna.html
miRBAse	miRNAs database	https://mirbase.org/
IncRNAdb	For functional IncRNA	http://www.lncrnadb.org/
Rfam	RNAs families	https://rfam.org/
RMDB	Mapping of RNAs	https://rmdb.stanford.edu/
NONCODE	ncRNAs except tRNA and tRNAs	http://www.noncode.org/

classes of ncRNA fragments, facilitating the identification and characterization of these molecules. A detailed list of databases and tools is shown in Table 1.

3 ncRNA Target Prediction

Understanding the regulatory role of ncRNA fragments often involves predicting their target mRNAs. Tools for target prediction are instrumental in this regard. TargetScan is a widely used tool for miRNA target prediction, employing sophisticated algorithms to predict potential miRNA target sites within mRNA 3′ UTRs. miRanda is another tool for miRNA target prediction that considers both sequence complementarity and thermodynamic stability, aiding in identifying putative miRNA–mRNA interactions. One can use various packages available in R for lncRNA analysis. incRScan is a specific tool within the R package designed for automated analysis of incubation behavior. It primarily helps researchers evaluate nest temperature data to quantify and understand incubation patterns. iSeeRNA is a computational tool developed by researchers at The Chinese University of Hong Kong. It uses a support vector machine (SVM) algorithm to classify transcripts from RNA-seq data as either protein-coding RNAs (pcRNAs) or long intergenic noncoding RNAs (lincRNAs). iSeeRNA demonstrates high prediction accuracy compared to similar tools, making it a reliable option for lincRNA identification. iSeeRNA runs significantly faster than other programs, saving valuable time during analysis. iSeeRNA can be integrated into larger data analysis pipelines or run as a web server for smaller datasets [36] (Table 2).

Table 2
Computational tools for lncRNA investigation

Tools	Applications	Ref
lncRScan	Detect lncRNA from the complex assemblies; distinguish lncRNA from mRNAs	[60]
iSeeRNA	Detection of lincRNA from large datasets	[36]
Annocript	Detect lncRNA by leveraging public databases	[61]
LncRNA2Function	Annotate lncRNA based on similar function and biological pathways	[62]
DARIO	Quantify and annotate ncRNAs	[63]
CPSS	Annotate ncRNAs	[64]
ncPRO-seq	Detect known small ncRNAs	[65]
CoRAL	Divide small ncRNA into functional categories based on biologically interpretations	[66]
RNA-CODE	Combine secondary structure with de novo assembly	[67]

4 Application of ncRNAs in the Treatment of Different Diseases

ncRNA fragments, including miRNAs, siRNAs, and other small noncoding RNAs, have shown immense promise in various medical applications. Their ability to regulate gene expression and influence cellular processes has led to significant medical advancements [37].

4.1 Diagnostic Biomarkers

ncRNA fragments have been identified as potential diagnostic biomarkers for various diseases [38]. ncRNAs can be detected in body fluids such as blood, urine, and saliva, making them accessible for non-invasive diagnostics [39].

4.2 Cancer Diagnosis and Prognosis

Cancer diagnosis and prognosis are undergoing a fascinating transformation, where noncoding ncRNAs are taking center stage. These once-mysterious molecules are emerging as powerful tools for understanding, detecting, and potentially even predicting the course of the disease. Specific ncRNAs can act as unique fingerprint-like signatures for different cancer types, allowing for earlier and more accurate diagnosis. The levels and types of ncRNAs present in a tumor enable us to gain insights into its potential behavior and response to treatment. Researchers are actively exploring ways to target ncRNAs involved in cancer progression. They aim to disrupt cancer pathways and halt tumor growth by silencing or activating specific ncRNAs. Imagine a future where a single injection could control cancer by targeting its hidden RNA regulators [40] (Table 3).

Table 3
Dysregulated ncRNAs in cancer [41]

	Biomarker	Sample	Clinical information
Disease/ Diagnostic	Seven paired miRNA panel	Plasma	Distinguishes early-stage LUAD + benign disease from control
	Five paired miRNA panel	Plasma	Distinguishes early-stage LUAD from benign disease
	Ten paired miRNA panel	Plasma	Distinguishes NSCLC (LUAD + LUSC) from controls (healthy/endobronchitis patients)
	SNHG1 + RMRP	Plasma	Distinguishes NSCLC from cancer-free controls
	Let-7b-5p, let-7e-5p, miR-24-5p, and miR-21-5p	Exosome	Distinguishes stage-I NSCLC patients from healthy controls
	miR-181-5p + miR-361-5p	Exosome	Can discern LUAD from other NSCLC histologies
	miR-320b + miR-10b-5p	Exosome	Can discern LUSC from other NSCLC histologies
	circRNA-0001073 + circRNA-0001495	Tissue	Differentiate LUAD and LUSC
	miR-126	Exosome	Distinguishes early-stage NSCLC patients from healthy controls
	RNA panels including ncRNAs	Tumor-educated platelets	Early- and late-stage NSCLC detection
	Edited miR-411-5p	Tissue + exosomes	Distinguishes late-stage NSCLC from controls
	Edited miR-99a-5p	Tissue	Distinguishes LUAD from controls
Prognostic/ Treatment	Edited miR-99a-5p	Tissue	Informs of shorter overall survival and recurrence-free survival after resection of LUAD
	miR-17-3pmiR-1268b + miR-6075	Plasma	Predicts resectable lung cancers regardless of histology and staging
	Let-7	Tissue	Informs of poor prognosis following tumor resection
	Let-7a + miR-155	Tissue	Informs of overall survival
	MALAT1	Tissue	Informs of poor prognosis for LUSC patients

4.3 Therapeutic Targets

ncRNA fragments are promising targets for therapeutic interventions. Small interfering RNAs (siRNAs) can be designed to target and silence specific disease-related genes. This approach, known as RNA interference (RNAi), is being explored to treat various diseases, including viral infections and genetic disorders [42, 43]. In clinical trials, siRNAs have shown potential for conditions like amyloidosis and hypercholesterolemia [44].

4.4 Cancer Therapy

miRNAs have been investigated as potential agents for cancer therapy and reported as oncomiRs that promote cancer development by inhibiting tumor-suppressor genes. Researchers develop miRNA-based therapies to restore the balance between oncomiRs and tumor-suppressor miRNAs [45, 46]. Colon cancer-associated transcript 2 (CCAT2) is a lncRNA that has been studied and determined to be overexpressed in colorectal cancer, thus causing resistance toward most common cancer treatments [47]. Targeting this lncRNA can be counteracted by another lncRNA, metastasis-associated in colon cancer 1 Antisense RNA 1 (MACC1-AS1), which in turn mediates chemoresistance by inhibiting and depressing cancer-causing RNAs [48]. ncRNAs play a significant role in cancer pathogenesis, and lncRNAs can be targeted by multiple approaches: (i) Post-transcriptional RNA degradation pathways can knock down pathogenic RNAs. Alternatively, antisense oligonucleotides (ASOs) with chemical modifications can target the RNA of interest for degradation via a ribonuclease (Rnase) H-dependent mechanism. (ii) Modulation of lncRNA genes by steric blockade of the promoter or genome-editing techniques. (iii) Finally, one can also achieve loss of function by creating steric inhibition of RNA–protein interactions or preventing secondary structure formation. RNA binding small molecules, or ASOs, can be used in this case. Here, we describe examples of current advances in each of the strategies mentioned above to target lncRNAs and their potential therapeutic applications (Fig. 1).

4.5 Clinical Implementations

Antisense oligonucleotides (ASOs), RNAi-based technologies (e.g., miRNAs and siRNAs), mRNA therapies, and single-guide RNAs (sgRNAs), the latter one used to direct gene targeting of CRISPR-Cas9-based systems, are among the RNA-based medications that have become very popular in recent years [50]. RNA medicine has used two main approaches to cure diseases using noncoding RNAs. Restoring tumor-suppressive ncRNAs is one strategy. This can be achieved by delivering artificial analogs of miRNAs known as mimetics. An analog of the tumor suppressor miR-34, which is frequently decreased in human malignancies, MRX34 was the first mimic to enter clinical trials for cancer treatment [51]. The second strategy seeks to mute genes related to the etiology of disease. To do this, exogenous siRNAs and artificial miRNAs (amiRNAs) that momentarily suppress target gene expression at the mRNA level can be engineered using our understanding of intrinsic RNAi pathways [52] (Table 4).

4.6 Drug Delivery

Drug delivery techniques depend heavily on ncRNA fragments, especially siRNAs. To target particular tissues or cells, they can be encapsulated in nanoparticles or other carriers. This method reduces adverse effects and enables accurate drug distribution. It has great promise in treating chronic illnesses and hereditary problems [53].

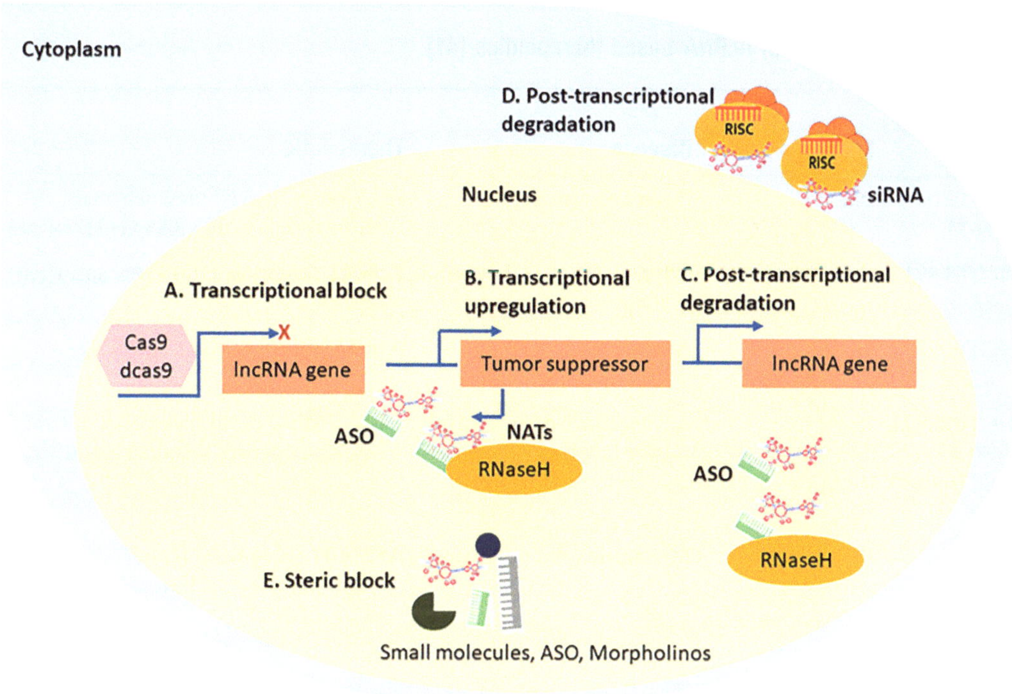

Fig. 1 Therapeutic targeting of long noncoding RNAs in cancer [49]. The figure summarizes different approaches to target lncRNAs in the nucleus and cytoplasm. (**a**) Transcriptional inhibition can be achieved through classical CRISPR/Cas9 to delete regions of interest in lncRNA loci. Alternatively, dead-Cas9 fused to a repressor complex can inhibit transcription of lncRNA genes. (**b**) Transcriptional upregulation of tumor suppressors can be achieved through knockdown of the corresponding natural antisense transcripts (NATs). NATs can be targeted post-transcriptionally using antisense oligonucleotides (ASOs). (**c**) ASOs can also be used to post-transcriptionally knock down lncRNAs that are overexpressed in cancers. (**d**) Post-transcriptional silencing can also be achieved by siRNAs targeting lncRNAs. siRNAs stimulate dicer activity in the cytoplasm and recruit the RISC complex (RNA-induced silencing complex) to post-transcriptionally degrade target RNAs. (**e**) Steric inhibition of lncRNA–protein interactions can be achieved using small molecules, morpholinos, or uniformly modified ASOs that cannot stimulate an RNA degradation pathway

4.7 Neurodegenerative Diseases

Neurodegenerative illnesses like Alzheimer's and Parkinson's disease (AD and PD) have been related to miRNAs. miRNA-based therapies aim to slow disease progression and restore regular miRNA expression. miRNA profiles can also be used as early diagnostic tools [54]. BACE1-AS lncRNA is a potential biomarker for Parkinson's and Alzheimer's diseases. Amyotrophic lateral sclerosis (ALS), PD, AD, and Huntington's disease (HD) can all be detected, and their progressions can be determined using lncRNA NEAT1 [55].

Table 4
Current clinical trials with ncRNA-based therapeutics [41]

Clinical trial ID	Stage	Disease	Therapeutic
NCT03608631	Phase I	Pancreatic cancer	**iExosomes** derived mesenchymal stromal cells with KRAS G12D siRNA
NCT01591356	Phase I	Advanced or recurrent solid tumors	**EphA2-targeting** DOPC-encapsulated siRNA
NCT00938574	Phase I	Advanced solid tumors	**Atu027** (siRNA targeting PKN3)
NCT00882180 + NCT01158079	Phase I	Advanced solid tumors with liver involvement	**ALN-VSP02** (lipid nanoparticle with siRNA targeting VEGF-A + KSP)
NCT03087591 + NCT02166255	Phase I	Metastatic solid neoplasms	**APN401** (peripheral blood mononuclear cells transfected with cbl-b siRNA)
NCT02369198	Phase I	Pleural mesothelioma, NSCLC	**TargomiRs** (miR-16 mimic)
NCT00689065	Phase I (terminated)	Solid tumors	**CALAA-01** (siRNA targeting RRM2)
NCT01829971 + NCT02862145	Phase I (terminated)	Solid tumors, liver cancer	**MRX34** (miR-34 mimic)
NCT01188785 + NCT01676259	Phase I + II	Pancreatic cancer	**siG12D-LODER** (siRNA targeting KRAS G12D)
NCT01437007	Phase I/II	Primary/secondary liver cancer	**TKM-080301** (siRNA targeting PLK1)
NCT02110563 + NCT02314052	Phase I/II (terminated)	Solid tumor, multiple myeloma, lymphoma	**DCR-MYC** (siRNA targeting MYC)
NCT03713320 + NCT03837457	Phase II (terminated)	Cutaneous T cell lymphoma/mycosis fungoides	**Cobomarsen/MRG-106** (oligonucleotide inhibitor of miR-155)

4.8 Cardiovascular Diseases

ncRNA fragments play a significant role in the control of cardiovascular functions. Myocardial infarction and heart failure are two heart disorders that may be exacerbated by dysregulation of miRNAs. miRNA-based diagnostics can help with risk assessment, and identifying disease-specific miRNAs provides prospective treatment targets. Several miRNAs, such as miR25 and miR22, have been established to control the force of cardiac contractions. Even more miRNAs have been associated with heart regeneration, including the miR-302-367 family, the miR99 family, and the miR100 family. miR155 and miR221/222 are implicated in inflammation, and miR21, miR125b, and miR208b are implicated in fibrosis. These miRNAs are crucial targets for therapy because it has been shown that they are also involved in other vascular functions [56–58].

4.9 Infectious Diseases

siRNAs and other small noncoding RNAs have been explored for their antiviral properties. By targeting viral RNA, siRNAs can inhibit viral replication and limit infection. This approach holds promise in developing therapies against diseases caused by RNA viruses [59].

5 Conclusion and Future Prospectus

Even though they are tiny, noncoding RNA fragments significantly impact how genes are regulated and how cells function. With the use of bioinformatic databases and methods, their research has advanced dramatically, allowing scientists to discover, annotate, forecast targets, and comprehend the roles of these vital molecules. Our understanding of noncoding RNA fragments and their significance in influencing the biology of life is growing along with our awareness of these fragments. In summary, noncoding RNA fragments have transformed medicine by presenting novel treatment, diagnosis, and illness prevention approaches. In conclusion, noncoding RNA fragments have revolutionized medicine by offering new diagnostics, therapeutics, and disease management opportunities. Their ability to fine-tune gene expression and influence cellular processes makes them powerful tools in the fight against various diseases, from cancer to neurodegenerative disorders and infectious diseases. Valuable treatment plans and therapies can be developed by determining which ncRNAs are expressed and how they behave across various diseases and disorders. It can also help to determine the survival rates and disease progression rates with how much a ncRNA is regulated or dysregulated, which can buy value treatment time for patients. ncRNAs are a powerful tool, with endless possibilities of applications and further conduction of studies.

Acknowledgments

AK acknowledges the financial support from NIH grants R01GM127701 and R01HG012117. MH acknowledges the Ohio State University for providing the "President's Postdoctoral Scholars Program (PPSP)" award for financial support. Amal Malik and Zainab Yaseen have equally contributed in this book chapter.

References

1. Mattick JS, Makunin IV (2006) Non-coding RNA. Hum Mol Genet 15(suppl_1):R17–R29
2. Zhang P, Wu W, Chen Q, Chen M (2019) Noncoding RNAs and their integrated networks. J Integr Bioinform 16(3):20190027
3. Mattick JS (2003) Challenging the dogma: the hidden layer of non-protein-coding RNAs in complex organisms. BioEssays 25(10): 930–939
4. Bhan A, Mandal SS (2014) Long noncoding RNAs: emerging stars in gene regulation, epigenetics and human disease. ChemMedChem 9(9):1932–1956
5. Keller C, Bühler M (2013) Chromatin-associated ncRNA activities. Chromosom Res 21:627–641
6. Hombach S, Kretz M (2016) Noncoding RNAs: classification, biology and functioning. In: Noncoding RNAs in colorectal cancer. Springer, Cham, pp 3–17
7. Wu H, Yang L, Chen L-L (2017) The diversity of long noncoding RNAs and their generation. Trends Genet 33(8):540–552
8. Storz G (2002) An expanding universe of noncoding RNAs. Science 296(5571):1260–1263
9. Romano G, Veneziano D, Acunzo M, Croce CM (2017) Small noncoding RNA and cancer. Carcinogenesis 38(5):485–491
10. Belousova E, Filipenko M, Kushlinskii N (2018) Circular RNA: new regulatory molecules. Bull Exp Biol Med 164:803–815
11. O'Brien J, Hayder H, Zayed Y, Peng C (2018) Overview of microRNA biogenesis, mechanisms of actions, and circulation. Front Endocrinol 9:402
12. de Fougerolles A, Vornlocher H-P, Maraganore J, Lieberman J (2007) Interfering with disease: a progress report on siRNA-based therapeutics. Nat Rev Drug Discov 6(6): 443–453
13. Ulitsky I (2016) Evolution to the rescue: using comparative genomics to understand long noncoding RNAs. Nat Rev Genet 17(10): 601–614
14. Kazimierczyk M, Wrzesinski J (2021) Long non-coding RNA epigenetics. Int J Mol Sci 22(11):6166
15. Andronescu M, Zhang ZC, Condon A (2005) Secondary structure prediction of interacting RNA molecules. J Mol Biol 345(5):987–1001
16. Mathews DH, Moss WN, Turner DH (2010) Folding and finding RNA secondary structure. Cold Spring Harb Perspect Biol 2(12): a003665
17. O'Leary CA (2022) Uncovering the structure and function of RNAs using computational and experimental approaches. Iowa State University
18. Laing C, Schlick T (2009) Analysis of four-way junctions in RNA structures. J Mol Biol 390(3):547–559
19. Butcher SE, Pyle AM (2011) The molecular interactions that stabilize RNA tertiary structure: RNA motifs, patterns, and networks. Acc Chem Res 44(12):1302–1311
20. Svoboda P, Cara AD (2006) Hairpin RNA: a secondary structure of primary importance. Cell Mol Life Sci 63:901–908
21. Bugnon LA, Edera AA, Prochetto S, Gerard M, Raad J, Fenoy E, Rubiolo M, Chorostecki U, Gabaldón T, Ariel F (2022) Secondary structure prediction of long noncoding RNA: review and experimental comparison of existing approaches. Brief Bioinform 23(4): bbac205
22. Georgakopoulos-Soares I, Parada GE, Hemberg M (2022) Secondary structures in RNA synthesis, splicing and translation. Comput Struct Biotechnol J 20:2871–2884
23. Vandivier LE, Anderson SJ, Foley SW, Gregory BD (2016) The conservation and function of RNA secondary structure in plants. Annu Rev Plant Biol 67:463–488
24. Bochman ML, Paeschke K, Zakian VA (2012) DNA secondary structures: stability and function of G-quadruplex structures. Nat Rev Genet 13(11):770–780

25. Karapetyan AR, Buiting C, Kuiper RA, Coolen MW (2013) Regulatory roles for long ncRNA and mRNA. Cancers 5(2):462–490
26. Mercer TR, Mattick JS (2013) Structure and function of long noncoding RNAs in epigenetic regulation. Nat Struct Mol Biol 20(3): 300–307
27. Zampetaki A, Albrecht A, Steinhofel K (2018) Long noncoding RNA structure and function: is there a link? Front Physiol 9:405446
28. Griffiths-Jones S, Grocock RJ, Van Dongen S, Bateman A, Enright AJ (2006) miRBase: microRNA sequences, targets and gene nomenclature. Nucleic Acids Res 34(suppl_1): D140–D144
29. RNAcentral (2015) An international database of ncRNA sequences. Nucleic Acids Res 43 (D1):D123–D129
30. RNAcentral (2019) A hub of information for noncoding RNA sequences. Nucleic Acids Res 47(D1):D221–D229
31. Kalvari I, Nawrocki EP, Argasinska J, Quinones-Olvera N, Finn RD, Bateman A, Petrov AI (2018) Non-coding RNA analysis using the Rfam database. Curr Protoc Bioinformatics 62(1):e51
32. Griffiths-Jones S, Bateman A, Marshall M, Khanna A, Eddy SR (2003) Rfam: an RNA family database. Nucleic Acids Res 31(1): 439–441
33. Singh NK (2017) miRNAs target databases: developmental methods and target identification techniques with functional annotations. Cell Mol Life Sci 74:2239–2261
34. Dennis G Jr, Sherman BT, Hosack DA, Yang J, Gao W, Lane HC, Lempicki RAD (2003) Database for annotation, visualization, and integrated discovery. Genome Biol 4(5)
35. An J, Lai J, Lehman ML, Nelson CC (2013) miRDeep*: an integrated application tool for miRNA identification from RNA sequencing data. Nucleic Acids Res 41(2):727–737
36. Sun K, Chen X, Jiang P, Song X, Wang H, Sun H (2013) iSeeRNA: identification of long intergenic noncoding RNA transcripts from transcriptome sequencing data. BMC Genomics 14(2):1–10
37. Bayraktar E, Bayraktar R, Oztatlici H, Lopez-Berestein G, Amero P, Rodriguez-Aguayo C (2023) Targeting miRNAs and other non-coding RNAs as a therapeutic approach: an update. Non-coding RNA 9(2):27
38. Aftabi Y, Ansarin K, Shanehbandi D, Khalili M, Seyedrezazadeh E, Rahbarnia L, Asadi M, Amiri-Sadeghan A, Zafari V, Eyvazi S (2021) Long noncoding RNAs as potential biomarkers in the prognosis and diagnosis of lung cancer: a review and target analysis. IUBMB Life 73(2): 307–327
39. Majem B, Rigau M, Reventós J, Wong DT (2015) Noncoding RNAs in saliva: emerging biomarkers for molecular diagnostics. Int J Mol Sci 16(4):8676–8698
40. Tano K, Akimitsu N (2012) Long noncoding RNAs in cancer progression. Front Genet 3: 219
41. Le P, Romano G, Nana-Sinkam P, Acunzo M (2021) Noncoding RNAs in cancer diagnosis and therapy: focus on lung cancer. Cancers 13(6):1372
42. Toden S, Zumwalt TJ, Goel A (2021) Noncoding RNAs and potential therapeutic targeting in cancer. Biochim Biophys Acta Rev Cancer 1875(1):188491
43. Cheng JC, Moore TB, Sakamoto KM (2003) RNA interference and human disease. Mol Genet Metab 80(1–2):121–128
44. Zhang MM, Bahal R, Rasmussen TP, Manautou JE, Zhong X-b (2021) The growth of siRNA-based therapeutics: updated clinical studies. Biochem Pharmacol 189:114432
45. Nguyen D-D, Chang S (2017) Development of novel therapeutic agents by inhibition of oncogenic microRNAs. Int J Mol Sci 19(1):65
46. Hassan M, Shahzadi S, Malik A, Su D, Yasir M, Chun W, Kloczkowski A (2023, 1849) Oncomeric profiles of microRNAs as new therapeutic targets for treatment of Ewing's sarcoma: a composite review. Genes 14(10)
47. Pirlog R, Drula R, Nutu A, Calin GA, Berindan-Neagoe I (2021) The roles of the colon cancer associated transcript 2 (CCAT2) long noncoding RNA in cancer: a comprehensive characterization of the tumorigenic and molecular functions. Int J Mol Sci 22(22): 12491
48. Chen B, Dragomir MP, Yang C, Li Q, Horst D, Calin GA (2022) Targeting noncoding RNAs to overcome cancer therapy resistance. Signal Transduct Target Ther 7(1):121
49. Arun G, Diermeier SD, Spector DL (2018) Therapeutic targeting of long noncoding RNAs in cancer. Trends Mol Med 24(3): 257–277
50. Ray RM, Lazar AD, Balahura LR, Mocanu-Dobranici AE, Costache M, Dinescu S (2023) Therapeutic targeting noncoding RNAs. In: Navigating noncoding RNA. Elsevier, pp 349–417
51. Hong DS, Kang Y-K, Borad M, Sachdev J, Ejadi S, Lim HY, Brenner AJ, Park K, Lee J-L, Kim T-Y (2020) Phase 1 study of MRX34, a liposomal miR-34a mimic, in

patients with advanced solid tumours. Br J Cancer 122(11):1630–1637

52. Voloudakis AE, Kaldis A, Patil BL (2022) RNA-based vaccination of plants for control of viruses. Annu Rev Virol 9:521–548
53. Subhan MA, Torchilin V (2020) siRNA based drug design, quality, delivery and clinical translation. Nanomedicine 29:102239
54. Mushtaq G, Greig NH, Anwar F, Zamzami MA, Choudhry H, Shaik MM, Tamargo IA, Kamal MA (2016) miRNAs as circulating biomarkers for Alzheimer's disease and Parkinson's disease. Med Chem 12(3):217–225
55. Zhang M, He P, Bian Z (2021) Long noncoding RNAs in neurodegenerative diseases: pathogenesis and potential implications as clinical biomarkers. Front Mol Neurosci 14:685143
56. Uchida S, Dimmeler S (2015) Long noncoding RNAs in cardiovascular diseases. Circ Res 116(4):737–750
57. Gurha P, Abreu-Goodger C, Wang T, Ramirez MO, Drumond AL, Van Dongen S, Chen Y, Bartonicek N, Enright AJ, Lee B (2012) Targeted deletion of microRNA-22 promotes stress-induced cardiac dilation and contractile dysfunction. Circulation 125(22):2751–2761
58. Abkhooie L, Sarabi MM, Kahroba H, Eyvazi S, Montazersaheb S, Tarhriz V, Hejazi MS (2021) Potential roles of MyomiRs in cardiac development and related diseases. Curr Cardiol Rev 17(4):e010621188335
59. Kang H, Ga YJ, Kim SH, Cho YH, Kim JW, Kim C, Yeh J-Y (2023) Small interfering RNA (siRNA)-based therapeutic applications against viruses: principles, potential, and challenges. J Biomed Sci 30(1):88
60. Capilla-Lasheras P (2018) incR: a new R package to analyse incubation behaviour. J Avian Biol 49(8):e01710
61. Musacchia F, Basu S, Petrosino G, Salvemini M, Sanges R (2015) Annocript: a flexible pipeline for the annotation of transcriptomes able to identify putative long noncoding RNAs. Bioinformatics 31(13):2199–2201
62. Jiang Q, Ma R, Wang J, Wu X, Jin S, Peng J, Tan R, Zhang T, Li Y, Wang Y (2015) LncRNA2Function: a comprehensive resource for functional investigation of human lncRNAs based on RNA-seq data. BMC Genomics 3:1–11
63. Fasold M, Langenberger D, Binder H, Stadler PF, Hoffmann S (2011) DARIO: a ncRNA detection and analysis tool for next-generation sequencing experiments. Nucleic Acids Res 39 (suppl_2):W112–W117
64. Wan C, Gao J, Zhang H, Jiang X, Zang Q, Ban R, Zhang Y, Shi Q (2017) CPSS 2.0: a computational platform update for the analysis of small RNA sequencing data. Bioinformatics 33(20):3289–3291
65. Chen C-J, Servant N, Toedling J, Sarazin A, Marchais A, Duvernois-Berthet E, Cognat V, Colot V, Voinnet O, Heard E (2012) ncPRO-seq: a tool for annotation and profiling of ncRNAs in sRNA-seq data. Bioinformatics 28(23):3147–3149
66. Leung YY, Ryvkin P, Ungar LH, Gregory BD, Wang L-S (2013) CoRAL: predicting noncoding RNAs from small RNA-sequencing data. Nucleic Acids Res 41(14):e137–e137
67. Yuan C, Sun Y (2013) RNA-CODE: a noncoding RNA classification tool for short reads in NGS data lacking reference genomes. PLoS One 8(10):e77596

Index

A

AAindex 250
Accessible surface area (ASA) 142, 144–146, 148, 149, 320, 321
Accession number 172–175
Accuracy score 108, 113, 115
Accurate Predictor Of DLs (APOD) 219–229
Acetate transport 163
Acetylation 262
Acquired immunodeficiency syndrome (AIDS) 185–188, 195
Activation function 82, 91–94
Active site 189–195
ADRM1 .... 227, 228
Aggregates 299, 300, 306
Aggregation 300, 302, 303, 305–307, 309
Aggregation kinetics 302, 303
Aggregation of proteins 299–310
Aggregation prediction 299–310
Aggregation-prone regions 302, 306
Aggregation propensity 300, 302, 303, 306
Aggregation rate 307
Aggregation state 307
AGGRESCAN 303, 305, 306
Aggrescan3D (A3D) 301, 302, 305, 307
AggScore 305, 307
AgMata 304
Alanine mutations 324
Algorithms 320, 323, 327
All-atom molecular dynamics 58
All-atom simulations 36
Allosteric regulation 220
AlphaFold 84–86, 123, 131, 132, 235, 236, 238, 317
AlphaFold DB 169–170
AlphaFold2 (AF2) 12–14, 121–135, 141–150, 155, 160, 163, 201
Alpha helix (α-helix) 1–14, 43, 44, 81, 106, 107, 113, 121, 144
α keratin 2
α_L motif 4, 5
Alzheimer's disease 299
Amber forcefield 304
Amino acid 1, 6–9, 55–57, 59–61, 71
Amino acid composition (AAC) 221, 263, 268
AmyLoad 301, 302
AmyloGram 304
Amyloid 301, 302, 304–307
Amyloid fibrils 301, 302, 304, 306, 307
Amyloid mutants 306
Amyloidogenic 300, 302–306
Amyloidogenic peptides 306
Amyloidose 299
AmylPred 306
AMYPdb 301, 302
AmyPro 301, 302
Analytical 22, 24, 26, 34, 287
Analytics 195
ANCHOR2 .... 236, 241
Anfinsen, C.B. 122
Annocript 336
Annotation 302
Annotations of the unknown genes 154
Apocytochrome 5
Apo form 143
Apoptosis 227
APPNN 304, 306
ArchCandy 304
Area under the cross-prediction curve (AUCPC) 248, 249
Area under the ROC curve (AUC) 220, 222, 236–241, 248
Aromaticity 303, 306
Artificial intelligence 12–13
Artificial intelligence (AI) methods 59
Artificial neural networks (ANN) 268, 277–279, 282, 283
ASAquick 250
Asparagine 4, 5
Assembly of proteins 7, 65, 142, 262
Assisted modeling 19–36
Association 8, 333
Atomic force field 124
Atomic-level 169
Attention mechanism 49, 50
AWSEM coarse-grained force field 26
Axial symmetry 22, 23

Andrzej Kloczkowski et al. (eds.), *Prediction of Protein Secondary Structure*, Methods in Molecular Biology, vol. 2867, https://doi.org/10.1007/978-1-0716-4196-5,

B

Backbone dihedral angles 20
Backbone geometry 1
Backbone-hydrogen-bonding 20
Backbone-local interactions 23
Backbone refinement 126
Backbone torsion angles 143
Backbone-virtual-bond-dihedral angles 22, 26, 28, 34
Backbone virtual-bond-torsional potentials 28
Back-propagation 145
Barnase 5
Base step 322, 326
Bend 144
BetaSerpentine 304
Beta sheet (β-sheet) 81, 121, 301, 303, 304, 306, 307
Beta strand (β-strand) 43–45, 106, 107, 113, 301, 303, 304, 306, 307
Beta-turn 306
BFD database 46
Biasing potentials 20
Bidirectional Encoder Representations from Transformers (BERT) 264, 272, 273
Big data 142
Big protein systems 57
Binding affintiy 315–327
BioGRID 248
Bioinformatics data 57
BioLip 177, 248
Biomine 251, 256, 257
Biotherapeutics 300, 310
Bipolar activation function 144
BLAST 155–159, 161–164, 167, 174, 179
Boltzmann factors 21
Bond-deformation potentials 22
Bovine pancreatic ribonuclease 7
Bridge 142
Bulges 332, 333

C

CABS-flex 305
CABS model 57–60, 73, 74, 76
CABS statistical force field 24
Cancer 316
Cantilever 189, 191, 192
Capping motif 3, 4, 6
CASP13 .... 123, 132
CASP14 .... 85, 86, 123, 124, 132, 135
CASP15 .... 132, 134, 135
Catalysis 84, 87
Catalytic residues 170
Catalytic sites 220
Catalytic triad 189
Cell cycle regulation 202
Cell metabolism 247
Characterization 82, 178
Charge 303, 305, 306
Chemical shifts 33, 34
Circular dichroism (CD) 5, 8
Circular RNAs (cirRNA) 332
Class imbalance 279–280
CLIP 236, 238
Closed conformation 190
Clustering 68–71, 74
Coarse graining 307, 309
Coarse-grained modeling 57, 59, 60, 73, 76
Coarse-grained models 20, 26, 34–36, 307
CodeOcean capsule 108, 113, 115
Co-evolutionary 123, 125, 129, 132, 134
Coevolution information 128, 135
Coil 43–45, 51, 106, 142, 147, 149
Coiled-coil equilibrium 9
Computational 57–59, 61, 62, 64–66
Computational biology 90, 234
Computational modeling 291
Conformation 306
Conformational changes 57
Conformational energy 305
Conformational flexibility 191
Conformational sampling 124, 125
Conformational space 56, 62, 65, 66, 68, 74, 76
Conserved hydrogen bonds 190, 191
Contact distance 31
Contact frequency 306
Contact maps 83, 123, 142, 145, 322, 332
Contact potentials 62, 305
Contacts 300, 304, 305, 307
Contour level 108, 111, 112, 115, 117
Convolutional filter 106
Convolutional neural networks (CNNs) 44, 80, 82, 83, 89–95, 97, 99, 106, 108, 123, 125, 131, 133, 263, 268, 277
CoRAL 336
CPAD 2.0 301
C-peptide 8
CPSS 336
CRISPR-Cpf1 89
Critical Assessment of Intrinsic Disorder (CAID) 202, 234–242
Critical Assessment of Protein Structural Prediction (CASP) 141–150
Critical Assessment of Techniques for Protein Structure Prediction (CASP3) 24
Critical Assessment of Techniques for Protein Structure Prediction (CASP13) 45–48
Cross-attention component 50
Cross-prediction 248, 250, 255, 256

Cross validation 50
Cryo-electron microscopy (Cryo-EM) 105–119, 126, 127, 130, 131, 135
Cryo-electron tomogram (Cryo-ET) 105–119
Cryo-EM maps 105, 106, 108, 111, 114, 117–119
C-terminus 3, 210, 211, 255
Curated Protein Aggregation Database (CPAD) 302
Curled flap 191
Curled states 191
Cutoff distance 145
CYANA software 34
Cysteine 64, 266, 268, 275, 285

D

D-amino acids 12
DARIO 336
Dark proteomes 219
Database 169–180, 300–303, 316–320, 324–325, 327
Data-driven 127–132
Deep convolutional neural networks 106
DeepDISOBind 236, 238, 241, 248
Deep language models 43–51
Deep learning 13, 49, 51, 87, 88, 99, 119, 123, 125–130, 132, 135, 141–143, 154, 236, 242, 263–266, 269, 275–277, 281, 283, 293, 294
DeepMSA 46
DeepPRObind 248
Designing complexes 316
Diagnostic biomarkers 336
Dihedral angles 2, 10, 20–23, 26, 28, 30–32, 34, 85, 142, 144, 145
Dimer 189, 255, 256
Dimer interface 189
Direct coupling analysis 129
Disorder content 221
Disordered linker 204, 206, 208, 210, 211, 214, 219–229
Disordered protein 121, 122
Disordered residue 250
Disorder predictors 303
Disorder regions 325
DisoRDPBind 238, 240, 241
DisProt 177, 220, 222, 227, 228, 248
Dissociation 315, 324, 325
Dissociation constant 315, 324
Distance biases 57
Disulfide bonds 23
DNA 105–119
DNA-binding 173, 177, 180, 202, 204, 208, 210, 213, 234, 236, 250
DNA binding domain 320
DNA damage repair 227
DNA recombination protein 165, 167
Docker 205, 209, 212
Docking 143, 196, 247, 323
Dropout 94, 97
Drug designing 43, 81, 84, 122, 135, 195, 247
Drug resistance 185–197
DSSP 144, 146
DSSP algorithm 44, 48
Dynamic properties 59

E

ECEPP/3 force field 24, 28
Eight state DSSP assignments 144
Elastic network model 309
Electron microscopy (EM) 301
Electrostatic interactions 304
Electrostatics 304
ELMs 233
Emap2sec 105–119
Emap2sec+ 105–119
EM-GAN 119
Encoder maps 49
Energetic barrier 63, 65–68
Energetic minima 63
Ensemble ncRNA 335
ENSHROUD 236, 238, 241, 242
Entropic chain 220
Environmental factors 299
Epimerase 163
ESCASA algorithm 34, 36
ESM-1b 271–273
ESM1b large protein language model 154
Eukaryotic linear motif 233
Evoformer 85
Evolutionary conservation 80, 250
Evolutionary information 43–45
Excluded volume 57, 61
Experimental 300, 305, 307, 309, 310
Experimental data 61, 71
Expert-curated statistical profiles 44
Extended-strand 144
Extended time frame 62

F

FASTA 205, 209, 212, 223, 226, 229, 252, 253, 285, 287, 288, 290
FASTA format 148
FATCAT tool 154–156, 158, 162, 164–166
Feed-forward neural network 145
Fibril 300–302, 304, 305, 307
Fibril formation 301
Fibrillation 301, 306

Fictitious dipole moment 26, 28
FISH amyloid 304, 306
Flanking residues 3
Flap 189, 191–193, 197
FIDPnn 201–214
Flexibility 57, 74
Flexible structural alignment 155, 158
Fluorescence 316
Folate transporter 164, 167
FoldAmyloid 303, 305, 306
Folded states 66
Folding transition 57
Foldseek 155–159, 161–164, 167
FoldX 320–322
Force fields 14, 19–36, 57, 59–63, 71, 73, 74, 85, 124, 135, 320
Fourier transform infrared spectroscopy (FTIR) 301
Fraction of alanines 6
Fraction of native contacts 69
Free energy 300, 308, 309
Free energy landscape (FEL) 308, 309
Free energy surfaces 300, 309
Free modeling 33, 45, 142
Fulcrum 189, 191
Fully connected deep neural network (DNN) 204
Functionally annotated homologs 155

G

Gag and gag-pol polyproteins 188–190
Gated recurrent neural networks (GRU) 44, 95
GENCODE 335
Gene ontology (GO) 154, 157, 159–162
GEneral Neural Network (GENN) 143
Generative modeling 127, 129, 135
Generative Pre-trained Transformer (GPT) 264
GitHub 108, 115
Global Distance Test Total Score (GDT_TS) 31–35
Globular proteins 302
Glycine 3–5
Glycosylation 262
GO annotation 154, 159, 161, 162
Google Colab Notebook website 108, 109, 115–117
G-quadruplex 320
Gradient descent 123, 126
Graph neural networks (GNNs) 81

H

Hairpins 332
Half-sphere exposure 170
Handedness of a helix 10
Helical parameters 322
Helical turn 2–4
Helicity 5, 8
Helix 142
Helix capping 3–6
Helix-coil theory 7, 9
Hemoglobin 2, 19
HHblits 46, 250
Hidden layers 144
Hidden Markov models 44
Hidden nodes 144
Hierarchical modeling 58
High-resolution 60, 61, 105, 131, 323
Hinge 189, 191–193, 195
HIV infection 185–188, 193, 197
HIV protease 185–197
HMGB1 210, 211
HMG domain 210
Holo form 143
Homodimeric structure 189
Homolog-based predictions 155, 163
Homolog detection 154
Homologous sequences 45
Homology search 323
Hotspot 316
Human Immunodeficiency virus 185–197
Hydrogen bonded-turn 144
Hydrogen bonds 1–5, 7, 9, 56–58, 61, 62, 68, 190–192, 194, 196, 303, 304, 307, 319–322
Hydrophobic interactions 5, 191
Hydrophobicity 250, 303–307
Hydrophobic patches 305

I

Implicit solvent 57
In vitro 55
Inception block 97, 98
IncRNAdb 335
IncRScan 335, 336
Insulin 122
Integrative modeling 58
Interaction energy 304
Interaction with water molecules 191
Intermediate-resolution 105, 106
Internal loops 332, 333
Intra-chain hydrogen-bonds 2
Intrinsically disordered linker 219–229
Intrinsically disordered proteins (IDPs) 201, 202, 233, 302, 303
Intrinsically disordered regions (IDRs) 36, 121, 143, 201, 202, 204, 208–211, 233–235, 241
Intrinsic disorder 170–173, 177, 178, 201–214, 227, 234
Ions 125, 163, 325
Ion transport 163

ISeeRNA 335, 336
Isolation 8
I-TASSER 87

J

JCVI-Syn3a organism 154, 167
JSON 175, 178, 180
JSONWP tool 157
Junctions 332

K

KAKSI 144
KEGG 290
Knowledge-based 57, 59–62, 71, 73, 80
Known proteins 59, 62
Kubo cluster-cumulant expansion 22, 23, 26
Kullback-Leibler divergence 50

L

Lactate transporter 163, 167
L amino acids 11
Language translation problem 49
Large protein systems 57
Large-scale transition 57, 73
Lattice model 307
Levenberg-Marquardt algorithm 145
Levinthal paradox 56, 58
Ligand-binding 248–250, 254
Ligands 82, 87, 88, 170, 177, 191, 196, 247–250, 254–256, 318
Linear interacting peptides (LIPs) 233–242
Lipid membrane 187
Liquid 300
Liquid-liquid phase separation 202
LMPTMSite 261–295
LMSuccSite 265, 266, 268, 270, 275–288, 290, 291, 294
LncRNA2Function 336
Local density pattern 106
Logistic regression 250
Long noncoding RNAs (lncRNAs) 332, 333, 335, 336, 338, 339
Long-range contacts 83, 84
Long simulation 62
Long-term memory networks 44
Long-time dynamics 57
Loop 143
Loss function 50
Low-resolution 55–76, 105, 106, 118
Low resolution coarse-grained modeling 58

M

Machine learning (ML) 306, 326
Major groove parameter 322
Major mutations 193, 195
Markov random field 123
Markov state models (MSM) 306, 308, 309
Matthew's correlation coefficient (MCC) 212, 220, 222, 236, 238, 240, 241, 248, 249
Maximum-likelihood calibration 23, 28
MC simulations 65
MD simulations 307
Mean-field electrostatic interaction potentials 22
Medium-resolution 57, 58
Medium resolution coarse-grained modeling 57–59, 73
Membraneless organelles 202
Mentha 248
Metagenomic database 129
MetAmyl 305, 306
Methylation 262
Metropolis criterion 63, 66
Microscopy 106
Minimal genome 154, 155
Minimal organism 154, 156, 162, 163, 167
Minor mutations 193
MiRBAse 333, 335
MiRNAs 332–335, 337–339
Misfolding 266
MMSeq2 algorithm 155
MMseqs2 48
Modeling tool 57
Modeling workflow 59
Modeller 304
Models of protein 57, 59, 71, 73
Molecular assembly 202
Molecular dynamics (MD) 124, 125, 130, 303, 305, 307, 309
Molecular recognition 202
Monomer 307
Monte Carlo (MC) 308
Monte Carlo Dynamics (MCD) 57, 58, 62, 63, 65, 68, 73, 76
Monte Carlo sampling 62, 75, 76
MoRF 241
MoRFchibi 236, 238, 240–242
Morphology 300
Multi-body interactions 61, 62
Multibody potentials 23
Multibranch loops 332
Multimer 127, 134
Multiple functional states 105
Multiple sequence alignments (MSAs) 12, 13, 44–49, 51, 123, 125, 126, 128–130, 134, 135
Multiplexed replica exchange molecular dynamics (MREMD) 32
Multiscale modeling 58–60
Mutations 299–302, 316, 320, 322, 324, 325
Mycoplasma mycoides capri 153, 161, 162

N

NADPH....255
NAD(P)H pathway....166
Na+ /H+ antiporter....163, 167
Native conformation....70
Native-like topology....68, 69
Native structure....56, 68, 72, 73
Natural language processing (NLP)....44, 80–82, 90, 91, 94–96, 99, 141, 263, 264, 270
Natural proteins....55–58
Ncap mutations....5
NCBI....157–159
NcPRO-seq....336
NDH-1 complex....166
Nerve cells....12
Net charge....305
NetCSSP....304–306
Network model....203, 309
Neural networks....44
Neurodegenerative diseases....266, 339
Neurons....12, 13, 94, 97
Nonbonded interaction....22
Nonbonded interaction potentials....22
NONCODE....335
Noncoding RNAs (ncRNAs)....331–341
NRED....335
N-terminus....4, 5
Nuclear magnetic resonance (NMR)....13, 20, 23, 26, 33–36, 80
Nuclear Overhauser Effect (NOE)....34
Nucleation....300, 306
Nucleic acids....106
Nucleus....300
Number of effective sequences (Neff)....46–48

O

Oligomer....56
Oligonucleotides....340
One-dimensional structural features....142
Open conformation....190–192

P

P53....174, 175
Packing....303–306
Pafig....303, 305
Parkinson's disease (PD)....299
PASTA....303
PASTA 2.0....301, 303, 306
PATH....304
Pathogenic mutations....301
Pauling-Corey-Brandson α-helix model....1
Pearson's correlation coefficient....144, 145
Peptide aggregation....302
Peptide-binding....172, 173, 179, 180
Peptide bond....2
Peptides....1, 4–9, 300, 302–307
Phase separation....219
pH-dependent switch....5
PHD program....6
Phosphatidate phosphatase....164, 167
Phosphorylation....262, 263
Physicochemical features....326
Physics-based energy....35
Physiological conditions....79, 233
Pi-helix....144
Platelets....337
pLMSNOSite....266, 268, 270, 277–290, 294
Polarity....250
Poly-γ-benzyl-L-glutamate....2
Polypeptide chain....20–23, 34
Pooling layer....88, 91, 93
Position-specific scoring matrix (PSSM)....44, 81–83, 144, 148, 173, 179, 203, 212, 213, 263, 268
Post-translational modifications (PTMs)....234, 261–295
Potential applications....331–341
Potential of mean force (PMF)....21–23, 61, 62
Precision-recall curve....237, 239
Predicted secondary structure....57, 58, 72
Pressure....192
Pre-trained CNN....106
Primary mutations....193, 195
Principal component analysis (PCA)....125
ProADD....301
Probabilistic criterion....62
Profile-based prediction....44
ProNAB....324, 325
PROST....154–164, 166, 167
ProtBERT....271, 273, 275, 283, 293
Protein aggregation....299–310
Protein assembly....299
Protein-binding....173, 177, 179, 180, 202, 204, 208, 210, 211, 213
Protein-coding genes....155
Protein Data Bank (PDB)....24, 45, 48, 123, 124, 133, 169, 170, 177, 247, 255
Protein degradation....227
Protein design....302, 303, 305, 310
Protein-DNA complexes....319, 320, 322, 326
Protein docking....323
Protein dynamics....57, 65, 66, 71, 73
Protein flexibility....105, 122
Protein folding....61, 62, 67–69, 71–73, 307, 309
Protein language embeddings....155
Protein language models (pLMs)....141, 261–295
Protein-like folding....58
Protein misfolding....299

Protein modeling 55–76
Protein-nucleic acid interactions 315, 318, 319, 321, 324
Protein-protein interaction (PPIs) 81, 82, 84, 90, 95, 96, 127, 247, 248, 271
Protein representation 73
Protein-RNA complexes 316–322, 326, 327
Proteins 299–310
Protein stability 300, 302
Protein structure 58–62, 71, 75
Protein structure initiative 143
Protein therapeutics 300, 310
Protein unfolding 57, 130
Proteolytic cleavage 7
Protofibrils 304
ProtSub matrix 158, 160
ProtT5 ....271–273, 275, 277, 278, 283, 285, 288, 293
P-SEA 144
Pseudo atoms 58–60, 63
Pseudo-trajectory 62, 65, 68
PSI-BLAST 203, 212, 214
PSIPRED 30–33, 45–48, 50, 72, 73, 81, 107, 170, 171, 173, 178, 203, 214, 250, 257
Purine ribonucleoside efflux pump 163
p-value 155, 158, 162, 164–166
PyMOL 125, 192, 194
Python 209, 214, 285

Q

Q3 accuracy score 45–48, 50
Quantitative structure-activity relationship (QSAR) 195–197
Quaternary structure 142

R

Ramachandran map 24, 26, 27
Random forest (RF) 203, 204, 263, 268, 282, 283
Real-SPINE 3.0 144
Reconstruction of atomistic details 60
Recurrent neural networks (RNNs) 81–83, 89, 263, 277
Redox 210
Reduced temperature factor 63, 65, 66, 68, 75, 76
Regression models 326
Regulation 219, 220
Relative ASA (RASA) 142–147, 149
Remote homologs 154, 155, 163
Replica exchange 308
Replica exchange molecular dynamics 308
Replica Exchange Monte Carlo (REMC) 75, 76
Replica exchange rate 66
Residue propensities 6
Residue-residue (RR) contact 82, 83, 99
ResNet 83, 97, 98
Rfam 334, 335
Riboflavin transporter 164
RMDB 335
RmuC 164, 165
RNA 105–119
RNA-binding 173, 177–180, 202, 204, 214, 234, 236, 250
RNA-CODE 336
Root mean square deviation 144, 146
Rosetta 304
RoseTTAFold 85, 86

S

Sanger, F. 122
Scaling distance 144
Schellman motif 4, 6
SCRIBER 247–257
Secondary mutations 193
Secondary structure (SS) 105–119, 142–145, 170, 172, 173, 177, 179, 221, 225, 226, 250, 301, 303–307, 317, 320, 321, 331–341
Secondary structure assignment 56–58, 72
Secondary structure assignment method 144
Secondary structure prediction 56, 81–82
Secondary structure preferences 56, 58, 61, 62
Secondary-structure scaffold 19, 35
SECSTR 144
Seder 141–150
Self-assembly 7
Self-attention component 49, 50
Self-replication 153
Semi-open state 191
Sensitivity 239
Sequence conservation 203, 221
Sequence profile 221
Shape complementarity 191
Shear 323
Sheet 142, 145, 147, 149
Short linear motifs 233
Short-range interactions 61
Short-term memory networks 44
Sidechain 307
Sidechain-rotamer potentials 23
Sidechain-sidechain interaction potentials 22
Signaling 202, 219
Signaling pathway 234
Signal peptides 170, 172, 173, 177, 179
Simulations 57, 58, 60, 62–66, 68, 71, 74–76, 316, 323, 326
Single-sequence-based method 45, 46
Single-sequence based prediction 44
SKSP+ 144
SLiMs 233

Small nucleolar RNA (snoRNA) 332
S-nitrosylation (SNO) 262, 265–270, 275, 282, 284, 291
Solubility 300, 302, 303
Solvent 300, 305, 307
Solvent accessibility 81–83, 143, 145, 170, 173, 177, 180, 221, 249, 263
Solvent accessible surface area 142, 191, 304
Specificity 10, 89, 163, 190–191, 193, 239, 280, 283, 318, 319, 325
S-peptide 8
Spherical symmetry 22, 35
SPINE-X (SPX) 141–150
SPOT-1D 45, 46, 50
Squeeze-and-excitation network (SE-network) 97, 98
Stability 300, 302–304, 307
Statistical force field 59, 73, 74
Statistical model 57
Statistical potential 59–61, 73
Stiffness 56
STRIDE 144
Structural accuracy 59, 64
Structural alignments 154–156, 158, 159, 162, 163, 165, 166
Structural biology 56, 59
Structural clustering 68–71
Structural data 59
Structural disorder 81
Structural flexibility 105
Structural model 143
Structural motifs 325
Structural parameters 316, 320, 324
Structural regularities 57, 61
Structural transition 68
Structure prediction 57, 59
Structure refinement 124, 126–134
Structure representation 60
Subcellular localization 89, 99
Sub-nanometer resolution 106
Substrate specificity 190
Subtomogram 106
Succinylation 262, 265–270, 275, 284
Sugar puckers 322
SUMOylation 262
Supervised learning 195
Supervised machine learning technique 185
Support vector machine (SVM) 221, 222, 226, 236, 263, 268, 282, 283
SURPASS model 58, 60–64, 71–74
Swiss-Prot 170
SwissProt database 155, 159, 161
Synthesis 164, 331
System's energy function 62
System's energy landscape 62, 68
System's energy minimum 68

T

TANGO 304–306
Template-based modeling (TBM) 122
Template-free modeling 122
Tertiary structure 142, 143, 146, 147, 149
Therapeutic targets 337, 339
Thermodynamics 324, 325
3D Profiles 301, 304
Three-dimensional structure 142, 143
3-helix 144
3_{10}-helices 5
Threonine 4, 5
TIGRfam classification 158, 159, 167
T4 lysozyme 6
TM align method 160
TM-score 143–149, 155, 162
Torsion angles 81, 83, 170
Transaldolase (TALA) 163
Transcription 202, 219
Transcription factors 318, 326
Transfer ribonucleic acid (tRNA)-derived small RNA (tsRNA) 332
Transformer 261–295
Transformer networks 43–51
Translation 202, 219
Transmembrane proteins 154
TransPross predictor 45, 48, 51
TrEMBL 169, 170
Tricarboxylic acid (TCA) cycle 265, 290, 291
23S rRNA 164
Two-dimensional structural property 142
Two-phase network 107

U

Ubiquitin 227
Ubiquitination 262
Umbrella sampling 308
Uniprot 158–162
UniProtKB 169
United atom 59, 61, 73
UNited RESidue (UNRES) 19–36, 57, 58, 60, 73
UNRES force field 19–36
UNRES model of polypeptide chains 24–26

V

Van der Waals contacts 321
VESPER 119
Viral proteins 188

Virtual-bond 21–23, 25, 26, 28, 30–32, 34
Virtual-bond length 21, 22
Visualization 302

W

Waltz 305–306
Waltz DB 301
WALTZ-DB 2.0 301
Web server 202, 204–210, 220, 222–227, 229, 249–257, 321–324, 326, 327
Word embedding 268, 270, 271, 275–277, 284, 290

X

X-ray 1, 2, 19, 34, 35, 80, 127
X-ray crystallography 1, 2
XGBoost 242

Y

YceD 164

Z

Zimm-Bragg parameter 7, 8
Zinc finger 317, 318
ZipperDB 301, 304

MIX
Papier aus verantwortungsvollen Quellen
Paper from responsible sources
FSC® C105338

If you have any concerns about our products,
you can contact us on
ProductSafety@springernature.com

In case Publisher is established outside the EU,
the EU authorized representative is:
Springer Nature Customer Service Center GmbH
Europaplatz 3, 69115 Heidelberg, Germany

Printed by Libri Plureos GmbH
in Hamburg, Germany